Mercury Stories

Mercury Stories

Understanding Sustainability through a Volatile Element

Henrik Selin and Noelle Eckley Selin

The MIT Press
Cambridge, Massachusetts
London, England

This book was set in Stone Serif and Stone Sans by Westchester Publishing Services. Printed and bound in the United States of America.

Library of Congress Cataloging-in-Publication Data

Names: Selin, Henrik, 1971– author. | Eckley, Noelle, author.
Title: Mercury stories : understanding sustainability through a volatile element / Henrik Selin and Noelle Eckley Selin.
Description: Cambridge, Massachusetts : The MIT Press, [2020] | Includes bibliographical references and index.
Identifiers: LCCN 2019049225 | ISBN 9780262539203 (paperback)
Subjects: LCSH: Mercury—Environmental aspects. | Mercury industry and trade—Environmental aspects. | Sustainable development.
Classification: LCC TD196.M38 S45 2020 | DDC 363.17/91—dc23
LC record available at https://lccn.loc.gov/2019049225

10 9 8 7 6 5 4 3 2 1

Contents

Acknowledgments

Mercury is a slippery substance. It is difficult to control in its physical form; it is also an analytical and practical challenge to examine more than 8,000 years of human interactions with this elusive compound. In writing this book, we were fortunate to be able to draw on the support of many colleagues whose suggestions and comments greatly helped us in our efforts to trace the often surprising flows of quicksilver and its many impacts on the environment and human well-being through time and across all regions of the world.

Both of us, individually and together, have worked on mercury for more than 15 years, but we wrote much of this book while we were on sabbatical from Boston University and MIT, respectively, in 2018. We visited the Department of Thematic Studies, Environmental Change, at Linköping University in Sweden in spring 2018, fostering engaging and useful discussions on our analytical framework and issues of sustainability transitions and transformations, often over fika. We especially thank Björn-Ola Linnér for graciously hosting our stay at Linköping University, and for his insightful comments on early drafts, along with those of Olof Hjelm, Mattias Hjerpe, Julie Wilk, Henrik Kylin, and Joy Routh. Noelle also spent time at Stockholm University's Department of Analytical Chemistry and Environmental Sciences as a visiting professor while we were in Sweden, and thanks Matt MacLeod and colleagues for helpful discussions of chemical pollution and mercury science.

Our work on the book manuscript was greatly facilitated by our stay at the Bavarian School of Public Policy at the Technical University of Munich (TUM) as Hans Fischer Senior Fellows for 2018–2021, including during fall 2018 and summer 2019. As part of our appointment as Hans Fischer Senior

Fellows, we acknowledge the support of the TUM Institute for Advanced Study. The fellowship program is funded by the German Excellence Initiative and the European Union Seventh Framework Programme under grant agreement #291763. We particularly thank our TUM faculty host Miranda Schreurs for her friendly welcome to Munich, for generously including us in her large and dynamic research group during our stays in Bavaria, for her thoughtful comments, and for organizing helpful discussions and seminars. Visits to local biergartens (mostly) helped advance our thinking and analysis.

We thank MIT's Policy Lab at the Center for International Studies for supporting our efforts to connect science and policy in the context of mercury governance, which we built on while researching and writing this book. We thank Dan Pomeroy for his expert policy knowledge and assistance through the Policy Lab. The Pardee Center for the Study of the Longer-Range Future at Boston University provided critical resources for Henrik's research and outreach at negotiations of the Minamata Convention on Mercury. Center Director Tony Janetos offered much valuable support and encouragement for this project, including reading and commenting on an early draft of our analytical framework. Sadly, Tony passed away in summer 2019, and his caring and sharp insights are sorely missed at Boston University and beyond.

During the later stages of the book project, we received grant support from the US National Science Foundation, Dynamics of Integrated Socio-Environmental Systems (CNH2) program, under award #1924148. This grant assisted us in finalizing our analytical framework as well as the chapter on artisanal and small-scale gold mining (ASGM). It has also allowed us to expand our collaborations with Ruth Goldstein at University of Wisconsin–Madison, and to further analyze issues of ASGM and sustainability, with a focus on Peru. Particular thanks are due to Steven Barrett at MIT, the MIT Center for Global Change Science, and the Boston University Global Development Policy Center for helping with the grant application to make this continuing support possible.

This book on mercury combines perspectives from a range of different disciplines and analytical approaches. We were highly fortunate to have been able to draw on many of our colleagues who are experts on different aspects of the mercury issue, who read and commented on drafts of individual chapters. Special thanks go to Amanda Giang, Kathleen Mulvaney, Hélène Angot, Ruth Goldstein, Ken Davis, Susan Keane, and Celia Chen. Ken Davis is further responsible for (guilty of?) alerting us to the use of mercury in "Thunderclappers," which we discuss in chapter 4. We express our apologies to Warner Brothers for borrowing from films in the Matrix franchise for our chapter title and section headings in chapter 8.

We presented earlier versions of individual chapters at the International Studies Association (ISA) annual conference in San Francisco, California, in April 2018, and at the Earth System Governance meeting in Utrecht, the Netherlands, in October 2018. We particularly thank Pam Chasek for her extensive comments as panel discussant at ISA. We also thank all those who provided helpful comments during seminars at Linköping University; the Swedish University of Agricultural Sciences in Uppsala; the Technical University of Munich; the Rachel Carson Center for Environment and Society at the Ludwig-Maximilians-Universität in Munich; and the Institute for Advanced Sustainability Studies in Potsdam, Germany.

We thank three anonymous peer reviewers for their comments and suggestions on both the initial book proposal and the first full draft of the book. We give special thanks to Pia Kohler, Leah Stokes, and Stacy VanDeveer for taking time to read through our text and for providing critical comments on the main arguments and the structure of the book, as we were revising the full manuscript. We further acknowledge Beth Clevenger at MIT Press for her support through the writing and publication process, and Virginia Crossman and Mary Bagg for their detailed and valuable assistance with copyediting.

A particular thank you goes to Bill Clark, whose extensive and incisive comments on multiple iterations of the text helped us to sharpen our arguments and better address a broader academic community interested in sustainability science. We benefited from and enjoyed wide-ranging and thought-provoking discussions on mercury and sustainability with Bill in Cambridge as well as in Munich, and we look forward to many more discussions in the future (preferably over more good food and wine, and with cheese served at the appropriate temperature, of course).

At this point in the acknowledgments section, it is typical to thank one's spouse for their support and encouragement. Many such acknowledgments reveal contributions to the research and writing process as well as manuscript preparation that ought to have been credited with coauthorship. We are, however, married to each other. We thus each thank our spouse for her/his contributions to typing, editing, and analysis during this book project. Accordingly, for any remaining errors, we blame each other.

1 Mercury Elementary

People have developed increasingly modern societies in their efforts to improve standards of living. At the same time, they have modified the environment in ways that pose challenges for maintaining and advancing human well-being, today and in the future. In this book, we focus on how people attempt to provide better lives for themselves and their descendants on a finite planet. We draw insights from human interactions with mercury to illustrate how people have made beneficial use of this unique element, how they have been harmed by its toxic properties, and how they have taken actions to protect human health and the environment from its impacts. Analyzing the mercury issue offers an opportunity to trace a millennial-scale history of human interactions with an element whose use has both benefited and harmed human well-being in complex and interacting ways. We develop and apply an analytical framework throughout this book—using the perspective of a human-technical-environmental system—to learn from the long history of human mercury use and exposure, and to inform strategies by which people can effect change toward greater sustainability.

On a sweltering Wednesday afternoon in early October 2013, hundreds of representatives of countries, intergovernmental organizations, and civil society groups from around the world filed into a crowded auditorium in Minamata, Japan, to hear Rimiko Yoshinaga tell a profoundly personal story about the human toll of mercury pollution. She related how fishers and their family members living in small hamlets around Minamata began to suffer devastating health consequences more than half a century ago after a local chemical company, the Chisso Corporation, discharged methylmercury, a particularly toxic form of mercury, into Minamata Bay. Their illness, discovered in the 1950s, resulted from methylmercury exposure, which causes permanent and sometimes fatal damage to the brain and central nervous system, and is now known the world over as Minamata disease.

Rimiko Yoshinaga is one of the storytellers affiliated with the Minamata Disease Municipal Museum, which is dedicated to informing visitors about the disease and its enduring legacy. A quiet and respectful hush quickly descended on the auditorium on that October day as she began to describe her childhood in Minamata and the mysterious illness that began to affect her father, and two years later led to his death, as a result of eating fish from Minamata Bay contaminated with methylmercury. Her mother and two of her grandparents also suffered from Minamata disease. Although many people in Minamata attempted to hide that they were victims of the disease to avoid social stigma and discrimination, a growing number of survivors have spoken out publicly to raise awareness of mercury poisoning, using the Japanese oral tradition of storytelling to share their experiences. Public awareness about Minamata disease also spread around the world through photographs of victims taken by the American photojournalist W. Eugene Smith for *Life* magazine in the early 1970s.

The participants who heard Rimiko Yoshinaga's story were there to attend the ceremonial opening of a conference to adopt the Minamata Convention on Mercury, a new global treaty (H. Selin 2014). They also visited the eco-park that was created on top of contaminated sediments dredged from Minamata Bay during the cleanup project, and ate a lunch of now-safe local fish and vegetables prepared by community members. The next day, October 10, 2013, country representatives formally adopted the final text of the Minamata Convention in the prefectural capital Kumamoto, 90 kilometers north of Minamata. The objective of the Minamata Convention is to protect human health and the environment from anthropogenic emissions and releases of mercury. Its adoption concluded three years of treaty negotiations under the auspices of the United Nations Environment Programme (UNEP). The Preamble of the Minamata Convention recognizes not only "the substantial lessons of Minamata disease, in particular the serious health and environmental effects resulting from the mercury pollution," but also the need to prevent similar events in the future.

Minamata disease remains a politically contentious issue in Japan, as does the government's support for naming the convention after Minamata. Some victims opposed using the name Minamata for the treaty because they believed that the Japanese government had not done enough—either to assess the damage mercury has caused to humans and the environment or to clean up the pollution—and that Chisso has not been sufficiently held accountable. One Japanese activist went so far as to argue that the naming "profanes the honor of the victims" (Kessler 2013). The Minamata victims have nevertheless had an international influence through their storytelling. Shinobu Sakamoto, who contracted Minamata disease in utero, traveled to Stockholm, Sweden, in 1972 (when she was 15 years old) to share her story at the United Nations Conference on the Human Environment, the first global political conference addressing relationships between people and the environment. A major outcome of that conference was the establishment of UNEP. In 2017, UNEP's deputy executive director Ibrahim Thiaw met with Shinobu Sakamoto after she spoke to the delegates at the first Conference of the Parties (COP) to the Minamata Convention. There he delivered a message that, in effect, dedicated the convention to her and all other victims of mercury poisoning.

Through their storytelling, Minamata disease victims tell of their own and their families' exposure to methylmercury, and the ostracism they faced after being afflicted with the disease. The Minamata stories, however, reflect much more than the human health effects from a local case of methylmercury poisoning: they address broader issues of human dignity, the importance of family and local community, responsibilities of governments and the private sector to address dangerous pollution, people's relationship with the environment, implications of modernity and industrialization, and the role of individual and collective efforts to move forward toward a better and safer future for all of humanity. In these respects, the Minamata stories are not just mercury stories: they are sustainability stories. They are tales of human struggles to live and thrive on a shared and finite planet that has been fundamentally transformed by humanity's interactions with the environmental processes essential for human survival and prosperity.

Stories like those told by the Minamata victims exemplify the ways in which people listen to, and learn from, the past. Oral and written storytelling can effectively raise public awareness and advocate for change. Rachel Carson (1962) caught the world's attention when she wrote *Silent Spring*, the story of the harmful effects of dichlorodiphenyltrichloroethane (DDT) and other similar pesticides, and linked their agricultural use to a decline in bird populations. Her widely read story contributed to the launch of the modern environmental movement in the United States and elsewhere.

Stories remain pervasive in much contemporary discourse around environmentalism (Harré et al. 1999). The telling of stories can also be critical to help envision, support, and guide transitions toward greater sustainability (Chabay 2015; Veland et al. 2018; Kuenkel 2019). In this respect, storytelling not only provides historical information and context, but also informs thinking about what a more equitable and sustainable world may look like.

The courage of the Minamata storytellers, and other victims of mercury pollution who have struggled to be heard and recognized, has inspired this book. The Minamata stories connect to a broader set of mercury stories across the world and through time. Coal burning and industrial manufacturing emits mercury while producing energy and goods. Millions of gold miners in Africa, Asia, and Latin America use mercury to earn an income, but it also threatens their health and the environment. Mercury from these and other sources travels worldwide through the atmosphere, land, and oceans. Pilot whales in the Atlantic accumulate methylmercury; some of this ends up in children in the Faroe Islands who suffer developmental delays from their mothers' consumption of whale meat and blubber during pregnancy. Many people in other regions of the world, including high-income urban residents as well as subsistence fishers, are also exposed to methylmercury from eating seafood. An estimated 200,000 children were born in 2010 in the United States alone to mothers whose methylmercury levels put these newborns at risk of neurological damages (Driscoll et al. 2018; US Environmental Protection Agency 2013).

Aim of the Book

We intend this book for readers with different backgrounds who share an interest in how an interdisciplinary perspective can inform understanding and action on mercury and other sustainability issues. For social scientists whose specialty is in political science, sociology, geography, history, economics, and other subjects, we illustrate how environmental processes and technologies affect social dynamics and change. For natural scientists studying the environment, we show how technological change and social institutions modify physical, chemical, and biological interactions. For engineers who study technical processes or seek to enhance technological performance, we underscore the need to account for environmental processes and governance arrangements in design and evaluation. For practitioners in the area of sustainability, we show how examining issues from a systems perspective can inform their efforts. For all audiences, we demonstrate the utility of a structured analytical framework by which perspectives and information from different fields can be brought together to better analyze and advance sustainability.

Global political attention to sustainable development can be traced back to discussions that informed the 1972 United Nations Conference on the Human Environment in Stockholm (Linnér and Selin 2005). An oft-quoted definition of sustainable development was set out by the United Nations World Commission on Environment and Development (also known as the Brundtland Commission) in their 1987 report *Our Common Future*: development that "meets the needs of the present without compromising the ability of future generations to meet their own needs" (World Commission on Environment and Development 1987, 8). The current global political approach to sustainable development is embedded in the 17 Sustainable Development Goals (SDGs) intended to address economic, social, and environmental issues in an integrated and holistic way until 2030 (Kanie and Biermann 2017; Kamau et al. 2018). We use the term "sustainability" in this book to refer to a societal goal, and the term "sustainable development" as the process by which societies progress toward greater sustainability.

Academics have defined and analyzed sustainability in different ways (Bettencourt and Kaur 2011). Definitions and measures of sustainable development vary in specifying what should be developed and what should be sustained (Parris and Kates 2003). In this book, we apply an anthropocentric perspective and focus on human well-being, now and in the future, as a central challenge of sustainability. Ensuring that Earth's life support systems remain healthy, while maintaining and improving human livelihoods for both current and future generations, is one of the most critical challenges to ever face humanity. Although a broad array of research from multiple disciplines focuses on mobilizing knowledge to inform societal efforts to promote sustainability, additional in-depth empirical research is necessary—not only to further develop and test theories of how people interact with Earth's life support systems but also to generate knowledge that can inform sustainability transitions (Clark 2015).

We wrote this book with the goal of advancing the knowledge and theory of sustainability through an empirically grounded study of the mercury issue. To this end, we aim to draw lessons relevant to other contemporary sustainability challenges in ways that inform practice and policy-making. The analytical framework that we introduce and deploy in this book centers on the concept of a human-technical-environmental (HTE) system that is embedded in the context of institutions and knowledge. We apply our framework, which we call the HTE framework, to the mercury issue, but, as we discuss further in chapter 8, it can also be used to study other sustainability issues. The HTE framework uses a matrix-based approach to examine interactions in systems and interventions aimed at enhancing sustainability. We introduce the framework in further detail, including the components of the human-technical-environmental system and the matrix-based approach, in chapter 2. In the book as a whole, we pose and address four research questions:

1. What are the main components of systems relevant to sustainability?
2. In what ways do components of these systems interact?
3. How can actors intervene in these systems to effect change?
4. What insights can be drawn from analyzing these systems?

Addressing these four questions requires drawing on approaches and insights from literatures across the natural sciences, the social sciences, and engineering. For too long, disciplinary training and boundaries have prevented integrated studies of complex systems. Yet, such systems can only be fully characterized, examined, and understood through an interdisciplinary approach that bridges spatial and temporal scales. Our own scholarly backgrounds are different but overlapping: Henrik specializes in understanding governance processes that deal with chemical pollution and other environmental issues, whereas Noelle examines the transport and fate of mercury in the atmosphere. Both of us have interdisciplinary academic training, and we frequently apply perspectives from multiple fields to advance understanding of the human influence on the environment. In addition, we share an interest in policy-relevant questions of how to promote and govern sustainability transitions.

The analytical and empirical focus of this book connects with the field of sustainability science (Kates et al. 2001; Komiyama and Takeuchi 2006; Kates 2011; Spangenberg 2011; de Vries 2012; König and Ravetz 2017). This rapidly growing and increasingly diverse set of scholarship is defined by the empirical problems it analyzes, where research is motivated by the need for informing action (Clark 2007). Knowledge of how complex systems function is necessary, but not sufficient, for such action. As succinctly argued by Pamela Matson and colleagues (2016, 12): "It is also necessary to understand how people, as active, committed agents of change, can intervene in those systems to make them work differently. Almost always, such interventions require collaboration to be effective." Choices about such interventions are often contested, and must often be made in the face of incomplete scientific information, diverging political and economic interests, diverse social conditions and cultural traditions, and different levels of access to financial and technical resources.

Mercury Stories as Sustainability Stories

Stories involving mercury—an element that humans have mobilized and used for at least 8,000 years—provide a wealth of empirical material through which to analyze sustainability. We tell the mercury stories in this book and analyze them from a systems perspective—examining system components, interactions, and interventions—as a way to help scholars interested in sustainability better understand how human actions have engaged with physical quantities of matter in ways that simultaneously support and are harmful to human well-being over generations. Many cases previously analyzed in the sustainability science literature involve human use of ecosystem resources, such as water and forests (e.g., Carpenter et al. 2009; Gleick 2018; Young et al. 2018). We add to this literature by looking at mercury as a commercial product and a pollutant. This dual focus brings into play characteristics of sustainability issues that are different from those addressed in many previously studied cases.

Humans have known about several of mercury's unique and useful properties since antiquity (Goldwater 1972). Mercury, a naturally occurring element in the Earth's crust, gets its name from the Roman god who was known as a winged messenger. The connection to a messenger god makes mercury a particularly appropriate focus for telling stories relevant to sustainability. Newspapers and magazines in several languages, including English, French, German, Italian, and Spanish, have the word "mercury" in their titles as a nod to the winged messenger. Mercury was also the god of merchants, travelers, and tricksters. The word "mercurial," relating to sudden and unpredictable changes in people's moods and behavior, refers to this deity. The British Broadcasting Corporation (BBC), building off this meaning, referred to mercury in 2013 as "the quixotic bad boy of the periodic table" (Rowlatt 2013). This is a suitable moniker for a substance that is attractive, volatile, and dangerous.

Mercury's ubiquity has made it part of popular culture. Farrokh Bulsara, the dynamic lead singer for the band Queen in the 1970s and 1980s, went by the name Freddie Mercury, a fitting choice for his charismatic and fluid persona. A 1981 recording with Freddie Mercury and David Bowie performing Queen's "Under Pressure"—a song about high-stakes stresses in today's world and hopes for a better future—was played during the Minamata Convention negotiations as delegates broke into smaller groups to address contentious issues. When the negotiations concluded at 6:59 a.m. on Saturday, January 19, 2013, in the Geneva International Conference Center, the organizers instead played another Queen song, "We Are the Champions." The song delivered a congratulatory message to the delegates as well as an exhortation to be the Minamata Convention's chief champions going forward. At the start of the first Minamata Convention COP, "We Are the Champions" played again in the same room, beginning the formal treaty implementation process; the soundtrack quickly switched back to "Under Pressure" as the discussions turned to more politically sensitive issues. Given Queen's "presence" at these events, it seems appropriate that the stylized fish logo for the Minamata Convention is affectionately nicknamed Freddie.

Mercury is mobilized from the Earth's crust by natural processes and by human activities. These natural processes include volcanic eruptions and the weathering of rocks. People have moved large quantities of mercury from geological storage into the atmosphere, land, and oceans during both preindustrial and industrial eras. This human influence has enhanced the amount of mercury cycling through the environment several-fold (Selin 2009; Amos et al. 2013). Consistent with the Minamata Convention, we use the term "emissions" in this book to refer to mercury entering the atmosphere, and the term "releases" to refer to mercury entering land and water. The term "discharges" covers both emissions and releases. Today, every ecosystem on the planet, no matter how remote, contains mercury discharged by humans. Additional mercury is in stockpiles, contaminated sites, and landfills. Current discharges of mercury add to this burden every day, and because of mercury's slippery tendency, it cycles back and forth through the land, atmosphere, and oceans for centuries (Selin 2009). Some mercury that deposits to ecosystems is converted into methylmercury, which then increases in concentration as it moves up food webs.

Mercury is sometimes referred to as quicksilver in English and several other languages because of its slippery and changeable nature. It is the only metal in the periodic table that is liquid at room temperature. But elemental mercury (the silvery liquid form) and methylmercury (the particularly toxic form that caused Minamata disease) are only two of the element's several forms. There are also many other types of mercury compounds. People have used different forms of mercury in a variety of products for thousands of years. Some of mercury's earliest uses were in religious ceremonies, traditional medicines, and paints. Miners used mercury to extract gold and silver from ore for centuries, in a simple amalgamation process that workers in artisanal and small-scale gold mining (ASGM) in many developing countries continue to use to this day. More recently, mercury has been used in a wide variety of products, including dental fillings, pesticides, batteries, light bulbs, and electrical equipment, and as a key element in industrial processes to produce a variety of chemicals and plastics.

Exposure to different forms of mercury can cause serious health problems. Inhaling high concentrations of elemental mercury vapor may lead to severe impacts, especially in occupational settings. Ingesting high doses of methylmercury can be lethal, as seen in Minamata. Lower doses can cause decreases in IQ and other neurological damage in children exposed in utero, and result in cardiovascular and other health problems in adults (Eagles-Smith et al. 2018). Because mercury is long-lived and ubiquitous in the environment, a piece of tuna on a dinner plate anywhere in the world contains methylmercury that may have come from mercury emitted by a currently operating coal-fired power plant and from mercury mined in the 1500s. Multiple factors influence how people are exposed to and affected by methylmercury, whether through individual food consumption patterns and genetics, or broader issues, such as land use change, that affect how much methylmercury is formed in ecosystems. Scientists increasingly realize that examining mercury's effects requires understanding not only its behavior in the environment, but also the influence of societal factors and human-induced changes to ecosystems and the climate (Obrist et al. 2018).

Many societies initially saw mercury as a valuable resource. Over time, with improved scientific knowledge of its environmental and human health impacts, mercury came to be perceived as a dangerous pollutant, a concept that echoes prior descriptions of pollution as "matter out of place" (Douglas 1966). The societal movement of mercury and many other materials that created pollution problems was initially intentional. Metals including lead and cadmium as well as organic chemicals such as the pesticide DDT, the industrial chemicals polychlorinated biphenyls (PCBs), and the stratospheric ozone-depleting chlorofluorocarbons (CFCs) were distributed in commerce before it was discovered that their use and dispersal posed dangers to the environment and human health. Unintentional emission of large amounts of mercury from combustion of coal also links it to other air pollutants, such as sulfur dioxide and carbon dioxide. Technology and engineering play a prominent role in the use of mercury and in mitigation of its harms, drawing attention to the role of industrial processes as they interact with ecosystems and environmental phenomena.

Outline of the Book

In this book, we relate intersecting stories about mercury at different times and in different places. We divide the book into three parts.

In part I, "A Framework for Sustainability Analysis," we devote a single chapter (chapter 2, "Analyzing Human-Technical-Environmental Systems") to presenting our systems-oriented HTE framework, together with the matrix-based approach. We explain how this framework is tied to the four research questions that we listed above by a four-step process that involves (1) identifying and classifying system components, (2) identifying and selecting several system interactions, and tracing the pathways of other interactions that affect them, (3) identifying actors and exploring interventions in the system, and (4) synthesizing insights from the previous three steps. Those insights relate to three thematic areas that shape our discussions and are integral to the structure of this book: systems analysis for sustainability, sustainability definitions and transitions, and sustainability governance.

Part II, "Sustainability Stories about Mercury," comprises chapters 3 through 7. Each of these five chapters focuses on an individual topical mercury system that illustrates major issues and themes in mercury science and governance. We begin each chapter with a short italicized paragraph that describes how the mercury system relates to broader issues of sustainability. We then relate a brief story about mercury to help communicate the complexity of the specific mercury system to readers who approach the book with different backgrounds and interests. Following this introduction, we structure the five chapters identically, based on the four-step process outlined in part I, to make it easier for readers to track and compare system components, interactions, interventions, and insights across the mercury systems. We end each chapter with another italicized paragraph that summarizes its conclusions.

In chapter 3, "Global Human-Technical-Environmental Cycling: Chasing Quicksilver," we address mercury as a pollution and management issue by examining the global cycling of mercury through the atmosphere, land, and oceans, as well as through society. We discuss the increasing analytical and practical difficulties of drawing a clear distinction between natural and anthropogenic sources of the mercury that travels through the environment, in part as a result of the growing influence of human-induced land use and climate changes on the cycling of mercury. We argue that fully evaluating global scale flows of mercury requires that knowledge about mercury's biogeochemical cycling in the environment be better integrated with data on its societal use and transport. This is particularly important when looking at the global mercury cycling system from a sustainability perspective. In addition, we detail the design and content of the Minamata Convention as a primary global-scale intervention to address the full lifecycle of the mercury problem, thus affecting both the environmental and societal cycling of mercury.

In chapter 4, "Human Health: Mercury's Caduceus," we explore the human health system for mercury, examining how people have been affected by mercury through occupational exposure, the use of mercury in medical treatments, and the presence of methylmercury in food. Mercury was extensively used in workplaces and in medicines for centuries after negative effects on humans were documented. Efforts to mitigate health damages from mercury relied on incomplete and varying scientific and medical knowledge of its hazards—many people who suffered from mercury poisoning did not know what caused their symptoms. We show that many early strategies to alleviate human health problems from mercury in workplaces and medical treatments relied on minimizing risk, and that more comprehensive efforts to phase out mercury use did not start until the twentieth century. Although some remaining mercury uses in dental amalgam and vaccines are still judged to have net positive consequences for human well-being, methylmercury exposure from food continues to harm human health worldwide in the twenty-first century, and its risks will continue for generations.

We focus in chapter 5, "Energy, Industry, and Pollution: Mercury, Winged Messenger," on the atmospheric system for mercury, particularly on the origins and consequences of atmospheric mercury emissions from large industrial sources. Mercury emissions from point sources, especially those that burn coal, have added a large amount of mercury to the atmosphere since the start of the Industrial Revolution. Efforts to address mercury emissions from point sources have largely consisted of end-of-pipe controls, and a growing number of countries are adopting technology-based regulations. We show that the incremental application of pollution-abatement technology had substantial positive impacts on reducing mercury emissions in North America and Europe. However, emissions remain high, and in some cases are increasing, in Asia and other regions. Although countries have shown limited willingness to address some of the underlying activities that lead to mercury emissions (including coal burning), incremental actions to control pollution can also affect fossil fuel use and thus the global challenge of climate change.

In chapter 6, "Assets and Liabilities: Mercury, God of Commerce," we discuss the products and processes system for mercury. The use of mercury extends across a broad array of commercial products—including thermometers and other measuring instruments, pesticides, paints, light bulbs, and batteries—as well as in industrial processes, including chemicals manufacturing. We explore how commercial mercury uses had many societal benefits, even though the mercury in these products and processes also harmed human health and the environment. Technology developments in the private sector sparked many early applications as well as later phaseouts of mercury use in consumer goods and manufacturing processes. In contrast, concerns about mercury's environmental and human health dangers influenced the adoption of only some of the technology that made mercury phaseouts possible. Since the late 1900s, when governmental and public pressures on industry to stop using mercury increased, the commercial demand for mercury has declined substantially. Damages from past uses continue to pose challenges because mercury remains in wastes, in contaminated sites, and in the environment worldwide.

We devote chapter 7, "Mining and Sustainable Livelihoods: Mercury, God of Finance," to the ASGM and mercury system. Mercury use in ASGM is a leading global source of mercury discharges to the environment and of mercury exposure for people. We discuss how ASGM miners in developing countries use large quantities of mercury to separate gold from ore. Many ASGM miners depend on mining and mercury use as a source of basic income, but not without cost: their use of mercury harms their health and the local environment and adds to the amount of mercury that cycles through the environment globally. ASGM is influenced by local conditions, but is also linked to global-scale mercury and gold supply chains. Mining often takes place without official permits, affecting efforts to address mercury use and a wide range of other environmental and social problems associated with ASGM. We explore how initiatives targeting mercury use in ASGM involve efforts to reduce mercury use, exposure, and discharges, with a simultaneous goal of phasing out mercury use altogether in gold mining.

Part III, "Lessons for Sustainability," contains the final three chapters of the book. These chapters build on the analysis of empirical material from the second part of the book to synthesize findings related to our four research questions. In these chapters we also propose ways to apply our analytical framework to other sustainability issues, and discuss lessons for further action on the mercury issue targeted toward different audiences.

In chapter 8, "Sustainability Systems: Seeing the Matrix," we look across the systems-oriented analyses of the mercury systems by returning to our first three research questions on system components, interactions, and interventions. We discuss how a limited number of human, technical, environmental, institutional, and knowledge components can be used to describe and analyze the mercury systems, and how some of these components may also be relevant to other sustainability issues. We then illustrate how shorter and longer pathways involving interactions across human, technical, and environmental components coexist, with many interaction pathways crossing spatial and temporal scales. A broad range of actors with different levels of power and influence initiated interventions to change interactions among system components; these interventions had different goals, influenced systems across space and time, and targeted various leverage points. The chapter ends with a discussion of ways that the HTE framework can be further developed and applied to study other sustainability issues.

We come back to Queen and Freddie Mercury as the inspiration for the title of chapter 9, "Sustainability Insights: Earth 'Under Pressure.'" In this chapter, we integrate the major insights from the mercury systems as they connect to the three thematic areas of systems analysis for sustainability, sustainability definitions and transitions, and sustainability governance. First, we discuss the need to account for human, technical, environmental, institutional, and knowledge components together in sustainability analyses. We highlight how the ability of the mercury systems to adapt has affected human well-being both positively and negatively, and stress the importance of looking across temporal and spatial scales when examining systems of relevance to sustainability. Next, concerning sustainability transitions, we argue that variations in values attributed to the benefits and risks of mercury make it difficult to define sustainability, and to evaluate changes to human well-being across populations and over time. We also detail how incremental transitions were important to human well-being. We then turn to sustainability governance, and the importance of ensuring that institutions fit the physical problem they are set up to influence. We discuss how institutional design can help address multiple sustainability issues simultaneously, and argue that environmental and societal factors should both be considered when evaluating whether institutions such as the Minamata Convention are effective.

For the subtitle of chapter 10, "Sustainability Champions: 'We'll Keep on Fighting ... ,'" we quote a line from the Queen song "We Are the Champions." We conclude the book by drawing practical lessons for "champions" who are researchers, policy-makers, and thoughtful citizens by reflecting on ways to move forward in the face of challenges posed by mercury pollution, from local to global scales. We call on mercury researchers to consider mercury in a larger context of other sustainability issues, to expand their work across disciplinary boundaries, and to develop and communicate useable knowledge in collaboration with stakeholders. For decision-makers addressing mercury-related challenges, we suggest looking for different kinds of interventions across multiple geographical scales, focusing attention on high-impact actions, and considering long-term impacts of different intervention strategies. In the final section, we urge thoughtful citizens who are concerned about mercury use and exposure to consider the implications of their consumption choices, to work together to push for efforts to further address mercury related risks and problems, and to share their and others' sustainability stories.

I A Framework for Sustainability Analysis

2   Analyzing Human-Technical-Environmental Systems

Systems-oriented approaches are common in many fields of study, and they are also appropriate for studying sustainability issues. Analytical frameworks provide tools to help identify components of individual systems, examine how these components interact, and explore how these interactions can be changed over time, including toward greater sustainability. We believe that a new analytical framework can help readers with different backgrounds better examine and understand complex sustainability issues from an interdisciplinary systems perspective. In this book, we develop such a framework, applying the perspective of a human-technical-environmental (HTE) system in the context of institutions and knowledge, together with a matrix-based approach. In this chapter, we outline our analytical framework, which we call the HTE framework, and describe how we use it in part II to examine five topical systems involving mercury.

The extraction and mobilization of large amounts of mercury through society and the environment over millennia reflect a broader pattern of how humans have interacted with and depended on the natural world. For much of human history, people's interactions with the environment (and the consequences of those interactions) were largely local, as societies remained small and human mobility was limited. In turn, local environmental conditions shaped much of the early human development that relied on the availability of food, shelter, and energy, and on the ability to ward off predators and pests. Human ingenuity and population growth gradually expanded the scope and depth of these interactions over centuries, in the process fundamentally reshaping relationships between human societies and their surrounding environments. The collective scope of these interactions increased dramatically when the Industrial Revolution began in Europe and North America in the second half of the 1700s, and has further accelerated since the mid-twentieth century (Turner et al. 1990; Steffen et al. 2007).

Advances in knowledge, especially over the past three hundred years, led to the development and application of technologies with far-reaching social, economic, and environmental consequences. These technologies were introduced alongside new institutions that influenced human connections and decisions at an increasingly global scale. Many of these developments in knowledge, technology, and institutions had profoundly positive effects on human prosperity. Most humans live dramatically longer, healthier, and more productive lives than their ancestors did several generations ago. Improvements in production techniques and transportation led to the manufacturing and trade of more and cheaper goods, increasing material standards of living to previously unmatched levels. These benefits are widespread, but they vary sharply, both within and across societies, among people at different levels of income and wealth. In addition, some technologies and institutions had vastly negative environmental consequences, many of which were both unintended and unanticipated.

There is no place on the planet that remains truly unmodified by people. Experts argue that human pressures have replaced natural factors as the main drivers of environmental change, and that the magnitude of these pressures have become so pervasive and profound that they "are pushing the Earth into planetary *terra incognita*" (Steffen et al. 2007, 614). Some analysts suggest that the overwhelming influence of people on the environment represents a new geological epoch, called the Anthropocene (Crutzen and Stoermer 2000). The Anthropocene is viewed as fundamentally distinct from the Holocene ("recent whole"), the postglacial geological epoch of the past 11,700 years (Malhi 2017). The argument that Earth is a human-dominated planet draws attention to the importance of protecting the interacting physical, chemical, and biological cycles and energy fluxes that support life on the planet (Steffen et al. 2007). On a finite, bounded Earth, "everything is connected to everything else" (Sterman 2011, 23). This means that fully understanding social, technological, or environmental factors cannot be accomplished by examining them in isolation.

Researchers from several academic fields have developed analytical frameworks to characterize and examine interactions between people and the environment. Many of these frameworks apply a systems perspective (e.g. Schlüter et al. 2012; Liu et al. 2015). A system is a connection of individual components that together produce results unobtainable by the components alone (Sage and Rouse 2009). Systems, then, are more than the sum of their parts (Bar-Yam 1997). Systems relevant to sustainability are often complex adaptive systems, and include multiple feedbacks, time delays, and nonlinearities. The important dynamics for many of these systems are poorly understood and inadequately conceptualized (Sterman 2011; Levin et al. 2013). To foster a better understanding of complexity, and thus to facilitate the design of new policies and catalyze change in support of sustainable development, John D. Sterman (2011, 21) has called for the launch of a new "systems science of sustainability."

In this chapter, we outline an analytical framework that we refer to as the HTE framework, using the perspective of a human-technical-environmental system coupled with a matrix-based approach. Our framework is designed so that readers from different backgrounds can use a common language and structure to identify and examine systems of relevance to sustainability, without prioritizing the terminology and concepts used by any one particular discipline. We apply our framework to examine five individual mercury systems in part II. The application of the HTE framework involves four steps, following our four research questions from chapter 1: first, cataloging system components; second, mapping interactions among the system components in an interactions matrix; third, identifying past interventions into system components and interactions using an intervention matrix; and fourth, drawing insights from the system analysis. For the fourth step, we identify three thematic areas for further discussion: systems analysis for sustainability; sustainability definitions and transitions; and sustainability governance.

A Matrix-Based Approach to Analyzing
Human-Technical-Environmental Systems

For the first three of the four analytical steps in the HTE framework, we apply a matrix-based approach. Under the three subheadings below, we describe these steps—identifying system components, tracing interactions, and identifying interventions—and explain our matrix-based approach using the example of Minamata disease from chapter 1 and with the help of illustrative figures. The components identified in step one are used to build matrices for further analysis of interactions and interventions in steps two and three. These three steps address our first three research questions, respectively: (1) What are the main components of systems relevant to sustainability?; (2) In what ways do the components of these systems interact?; and (3) How can actors intervene in these systems to effect change? In chapter 8, based on our analysis of the mercury systems in part II, we synthesize results related to the matrix-based approach.

Components: Building Blocks of a System

The first analytical step is to identify the most important components for understanding system operations and dynamics. Analyzing a problem from a systems perspective requires deciding which components to include and which to leave out of the system description. Components are the elements or variables that exist within the boundaries of a defined system; system boundaries can be set at different geographical scales and can include or exclude specific sectors or topics. The identified components need to capture important system behavior, yet be few enough to allow for practical analysis. If everything is described as linking to everything else, identifying the system components and examining their interactions can quickly devolve into an intractable analytical problem, where the selection of overly broad system boundaries prevents the researcher from conducting a meaningful empirically grounded analysis. Because the most important components may change with time, a full description of a system may require a longer-term historical perspective.

When analyzing mercury, it is at least theoretically possible to include in a single system all the major components relating to its extraction, uses, discharges, exposure, and effects on the environment and human health, but such an approach would create a very large and difficult-to-analyze system structure. We therefore separate the mercury issue into five topical systems that correspond with major empirical issues and themes in mercury science and governance. Our selection of topical mercury systems is similar to the focus on an "action situation" within the Institutional Analysis Development framework developed by Elinor Ostrom and colleagues (Kiser and Ostrom 1982; Ostrom 2005; Ostrom 2011). Each topical mercury system has varying spatial and temporal dynamics. Some components are unique to each system, and some are common across two or more mercury systems.

We characterize systems of relevance to sustainability, such as the topical mercury systems, as comprising five different sets of components. Three sets are material, in the form of human, technical, and environmental components. The other two are the non-material institutional and knowledge components; they provide the context within which the human, technical, and environmental components interact. Each system component has a set of attributes that can be defined at a specific time, and many of these attributes change over time. Figure 2.1 shows how individual system components for the five mercury systems will be identified in each chapter in part II; we include in this figure a few illustrative components that are relevant to the Minamata story, which we discuss further below. We use an italicized typeface to emphasize the individual components from figure 2.1, and do the same when we describe each mercury system in its respective chapter.

Human components are people who live in different places and under different circumstances. Attributes of human components include social characteristics such as occupation, education, and level of income. Other attributes may be physical characteristics such as residence or location, or biological factors that influence health, including genetic conditions, age, and mercury levels in the body. People have different concentrations of mercury in their blood, hair, and urine as a result of their individual exposures. It is often analytically useful to consider individual humans as part of larger groups who share common characteristics and engage in similar behaviors. Using the Minamata story as an example, human components may include groups of people such as *factory workers* and *factory owners*. Other possible human components are fishers, pregnant women, children exposed in utero, and other categories of community members.

Technical components take the form of infrastructure and other material artifacts of human society. Collectively, these components have been

Human components	Technical components	Environmental components
Factory workers Factory owners ...	Manufacturing technology ...	Minamata Bay Fish ...

Institutional components		Knowledge components
Markets National pollution laws ...		Techniques for mercury use Health dangers of mercury ...

Figure 2.1
Illustrative system components for the Minamata story.

referred to as the technosphere; the total physical mass of the current technosphere, including such things as buildings, roads, and consumer goods, has been estimated at approximately 30 trillion tonnes (note that throughout the book our use of "tonnes" refers to metric tons). This is about five times larger than the total mass of humans (Zalasiewicz et al. 2017). Attributes of technical components include their mass, quantity, performance characteristics, or concentration of mercury. Substantial quantities of mercury are present in a wide range of technical components; those related to the Minamata story include *manufacturing technology*, mercury used in chemicals production, and equipment that disposed of waste products. Other examples of technical components beyond the Minamata example include mercury in stocks and storage as a raw material, mercury in products such as light bulbs and batteries, pollution control technology, and landfills.

Environmental components consist of the Earth's life support systems and components of the biosphere—including all non-human living organisms in aquatic and terrestrial ecosystems. These range from large-scale systems, such as geological reservoirs, land biomes, the atmosphere, and the oceans, to more local systems such as rivers, lakes, forests, and coral reefs. Attributes of environmental components include physical properties such as wind speed, temperature, and depth, or biological information about organisms such as species, sex, or concentrations of mercury. Different forms of mercury are ubiquitous in environmental components, as they are found at different levels in ecosystems and wildlife in all regions of the world. Environmental components in the Minamata story include *Minamata Bay* as well as the *fish* and other aquatic organisms in which methylmercury accumulated. Mercury also travels long distances via environmental components as it cycles through air, water, and land on both shorter and longer time scales.

Institutional components are social structures outlining rules, norms, and shared expectations that define acceptable or legitimate behavior (Keohane 1989; Young 2002). As such, institutional components are distinct from actors such as international organizations, states, and other stakeholders (which we treat as potential interveners, discussed below). Institutions exist at local, national, regional, and global scales, and their specific rules, norms, and expectations may change over time. Some institutions set standards for human handling of (and exposure to) mercury. In addition, institutions mandate controls on emissions and releases of mercury to air, water, and land. Attributes of institutional components include their membership, scope, and stringency. In the Minamata story, *markets* facilitated the supply of mercury to the factory, but no domestic laws initially controlled the use of mercury in chemicals manufacturing. The outbreak of Minamata disease and other pollution problems, however, triggered the adoption of *national pollution laws*. Other institutions included the legal decisions holding Chisso responsible for releasing methylmercury into Minamata Bay and for providing compensation to Minamata disease victims.

Knowledge components incorporate information about human, technical, environmental, and institutional components and their connections. Attributes of knowledge components—related to specific locations and contexts—include awareness of, or the degree of certainty or uncertainty about, specific data and information. In the Minamata story, knowledge of *techniques for mercury use* influenced production in the Chisso factory. Knowledge of the *health dangers of mercury* and the fact that the disease was caused by methylmercury were unknown to local doctors and researchers until the relevant scientific information was identified and disseminated. The state of knowledge about how mercury affects human health can influence whether or not workers (human components) take safety measures when they handle mercury. Knowledge about the dangers of mercury can also influence uses of mercury in the industrial manufacturing of goods (technical components), the release of mercury into waterways (environmental components), and the formulation of pollution prevention standards and laws (institutional components).

Analysts may choose to define varying numbers of human, technical, environmental, institutional, and knowledge components within a system and at different levels of detail. The most appropriate level of detail for an individual component is largely an empirical question that is heavily influenced by the basic purpose of the system description and analysis. In the Minamata story, individual fishers can be identified as separate human components, or all fishers can be aggregated into a single human component for greater simplicity. Different types of mercury-using technologies can be distinguished as individual technical components, but they can be combined into one technical component if there is no analytical need to keep them separate. Environmental components can be individual species of fish, or all fish can be treated as a single component. Different national mercury laws can be identified as separate institutional components, or as just one collective component. Information about the toxicity of individual mercury compounds can be treated as separate knowledge components, or be aggregated into one.

The attributes of each of the five sets of system components convey information that helps to identify a component's location in time and space. Temporal attributes are defined with reference to a specific point in time, and could include the age of individual people (human components), infrastructure (technical components), and wildlife (environmental components) as well as the dates of laws (institutional components) and scientific discoveries (knowledge components). Spatial attributes can be measured relative to geographical distances or political boundaries on scales ranging from local to global. Some human components like Minamata fishers live within the same municipality, but others, such as commercial fish consumers, are spread across national jurisdictions and geographical regions. A technical component can be a pollution control device that is installed in a specific point source in a set location, or mercury that is traded across borders. An environmental component can be fish in a small lake or in a major ocean such as the Atlantic. An institution may be a law that applies only to a sub-national jurisdiction, or may be the global Minamata Convention. Knowledge about mercury's properties can be highly localized as well as diffused all over the world.

Our categories of system components illustrate that our concept of the HTE system is related to, but distinct from, other system descriptions, including social-ecological systems, social-environmental systems, coupled human-natural systems, human-environmental systems, socio-technical systems, production-consumption systems, and engineering systems (Liu et al. 2007; Ostrom 2009; de Weck et al. 2011; Markard et al. 2012; Selin and Friedman 2012; Levin et al. 2013; Chen 2015; Matson et al. 2016; Colding and Barthel 2019). These different system labels reflect a varying focus on distinct types of system components that are understood to interact in different ways. Each of these system descriptions emphasizes the importance of partially different components, but overlaps inevitably occur as components are conceptualized. In defining the human-technical-environmental system as used in this book, we aim to integrate and give equivalent attention to human-natural and socio-technical perspectives drawn from different literatures. We elaborate on further differences between the HTE framework and other system descriptions in chapter 8.

Interactions: Components Influence Each Other

The second analytical step uses a matrix as a heuristic tool to document and examine material interactions among human, technical, and environmental components, in the context of non-material institutions and knowledge components. The interaction matrix, as we call it, captures the behavior of the system as it changes through time. It documents which specific human, technical, and environmental components interact, how they do so, and the direction in which these interactions take place—that is, which material components influence others. The matrix format is useful to classify one-way as well as two-way interactions. In other literature, conceptual diagrams in which different aspects of systems are connected with boxes and arrows are commonly used to visualize human-natural or social-ecological interactions or networks (e.g., Ostrom 2009; Bodin et al. 2019). Matrices provide the same information as box-and-arrow or network diagrams, but it can be easier to visually compare interactions in different systems within a common matrix structure. Matrices are also used in the engineering systems literature to examine systems and their functions (de Weck et al. 2011; Eppinger and Browning 2012).

Interaction matrices can be presented in different ways. The illustrative matrix in figure 2.2 shows interactions that occur among the human, technical, and environmental components that we previously identified (following the first step of our analytical process) for the Minamata story in figure 2.1. We do not present a detailed matrix like this one in the individual chapters of part II, but we base our analysis in each chapter on such a detailed matrix. The existence of institutions and knowledge, as the non-material components that set the rules and parameters for material interactions, are indicated by large, shaded background rectangles. For purely illustrative purposes, figure 2.2 has four human components, eight technical components, and six environmental components. In the figure, these are indexed by the letters "i," "j," and "k," respectively. Each individual material component of the system occupies both a row and a corresponding column of the matrix. The matrix is read row first and column second.

A shaded cell in figure 2.2 indicates that the component in the row is influencing the component in the column. A shaded cell that lies along the diagonal shows a component interacting with itself. This represents a component's internal dynamics: for example, the deteriorating health of factory workers during the outbreak of Minamata disease (cell A). Interactions can

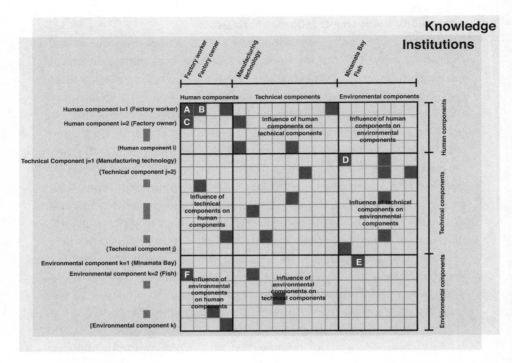

Figure 2.2
Illustrative matrix approach. The interaction matrix shows material interactions taking place in the context of institutions and knowledge. Shaded squares identify where interactions are occurring among material components.

occur between different components within the same set. The interaction between factory workers and factory owners, where workers provide labor to factory owners, is illustrated where the first row and second column intersect (cell B). The reciprocal interaction also occurs: factory owners provide wages to factory workers (cell C). Interactions can also take place between components from different sets. Manufacturing technology, a technical component, affects Minamata Bay, an environmental component, by releasing mercury into it (cell D). To illustrate what a more developed interaction matrix may look like, additional interactions within and across the same set of components are shown by the presence of other shaded but unlabeled cells.

Institutions include rules and other non-material structures that influence identified interactions among the human, technical, and environmental components, and the status of knowledge both facilitates and can reveal such interactions, as we discussed above. The detailed matrix in figure 2.2 identifies whether an interaction occurs among human, technical, and environmental components, and one or several institutional or knowledge components identified during the first analytical step may affect these interactions. For example, the interaction illustrated by cell D—whereby manufacturing technology releases methylmercury into Minamata Bay—is influenced by the existence of markets for both mercury and chemicals made from the mercury-based production process, and it may also be shaped by the adoption of a national law controlling the industrial use and environmental discharges of methylmercury. Access to knowledge about techniques for mercury use also affects whether the interaction in cell D occurs or not.

We trace interactions among human, technical, and environmental components through the matrix by identifying interaction pathways. We do this by first selecting a specific interaction to focus our analysis. We then identify the components that affect this interaction and those that in turn are affected by it. A simple interaction pathway involves cells B and C in figure 2.2, where a factory worker provides labor to a factory owner who in turn provides the worker with wages. In a longer pathway that can also be traced through figure 2.2, manufacturing technology releases mercury into Minamata Bay, illustrated in cell D. Minamata Bay then affects the fish by introducing methylmercury into the food web, illustrated by cell E. The fish in turn affect the factory workers who consume them and begin building up concentrations of methylmercury in their bodies, shown in cell F and cell A. This D-E-F-A pathway also connects to the earlier B-C pathway through the factory worker, and these two pathways could be considered together as one combined pathway if that is analytically useful.

Analysts can use the matrix approach to identify causal links among components by tracing pathways from a selected interaction either forward (to determine potential influences prospectively) or backward (to identify causal factors). In the pathway described in the previous paragraph, going forward from cell D to E to F in figure 2.2, causal links are traced across technological, environmental, and human components to identify how mercury discharges affect the environment and then people. If, in the same example, the purpose is to examine what led to the accumulation of methylmercury in factory workers, the pathway can be traced backward from cell F to E to D. This exercise would help identify the adoption of a local or national pollution law as a potential solution to the risks of methylmercury exposure in factory workers who eat contaminated fish (as well as to other consumers of fish from Minamata Bay).

Changes in the attributes of all five types of system components can provide information about spatial and temporal dynamics. These changes may occur in self-interactions (for example, the aging of populations and infrastructure) or via interactions with other components. Comparing the spatial attributes of the different interacting components in the Minamata story reveals that many interactions between material components were local. In cell D, the individual interaction occurred between a local point source, in the form of a manufacturing plant that discharges methylmercury, and Minamata Bay, which received this methylmercury locally. The methylmercury in Minamata Bay accumulated mainly in local fish (cell E), which in turn was consumed mostly by nearby populations (cell F). At the same time, many of the market-based interactions that involved buying and selling mercury (and the chemicals produced by the factory) were national and international. Knowledge about techniques for mercury use as well as the health dangers of mercury also diffused across international borders.

Using the matrix approach to trace interactions across temporal scales in the Minamata story shows that interactions among material components lasted for both longer and shorter periods of time. The chemical factory in Minamata used commercial mercury in two production processes and released methylmercury into Minamata Bay for several decades starting in the 1930s, and it took nearly 70 years after the methylmercury was first discharged from the factory to clean up Minamata Bay (cell D). Methylmercury discharged into Minamata Bay began to quickly accumulate in local fish as they grew and matured (cell E). After the discharges of methylmercury began, it took more than two decades before the local doctor identified the first patient who had contracted Minamata disease from eating contaminated fish (cell F). Although many fishers and other people died a few years after contracting Minamata disease, some people who suffered irreparable damages to their nervous systems were alive and still affected by Minamata disease well into the early decades of the twenty-first century.

We have taken the illustrative interactions among human, technical, and environmental components related to the Minamata story from figure 2.2, described them in qualitative terms, and aggregated them in their respective sets in figure 2.3, which appears below. In part II, we present interaction matrices in the format of figure 2.3, using mainly qualitative data, as

	1. Human	2. Technical	3. Environmental
1. Human	(1-1) Factory owners employ factory workers	(1-2)	(1-3)
2. Technical	(2-1)	(2-2)	(2-3) Manufacturing technology releases methylmercury into Minamata Bay
3. Environmental	(3-1) **Fish contaminated with methylmercury cause health damages to fish consumers**	(3-2)	(3-3) Minamata Bay fish bioaccumulate methylmercury

Knowledge Institutions

Figure 2.3
Illustrative interaction matrix for the Minamata story.

we examine interactions and provide system-level insights for the mercury systems discussed in each chapter. This allows us to present necessary context and detail for the respective interactions. If sufficient quantitative data were available, and if it were analytically feasible and appropriate, it would be possible to construct a quantitative model. A quantitative model based on the Minamata story could simulate mercury in the fish in Minamata Bay and in the bodies of the people who consume that fish. The interaction matrix could then be used to calculate and describe the rate of change at which mercury builds up in the fish, as well as the rates at which mercury accumulates in (and discharges from) human bodies over time. This calculation is similar to what is known as a first derivative, which identifies whether a mathematical function is increasing or decreasing, as well as its instantaneous rate of change.

The diagonal boxes of figure 2.3 (e.g., boxes 1-1, 2-2, and 3-3) represent interactions among material components in the same set, including where an individual component interacts with itself (the diagonal cells in figure 2.2). The boxes that do not fall along the diagonal represent interactions that involve different sets of material components, where the first number indicates the row and the second number indicates the column of the aggregated matrix (box 1-2, etc.). For example, as noted above in figure 2.2, manufacturing technology affects Minamata Bay in cell D. In figure 2.3, we describe this in general terms as "manufacturing technology releases methylmercury to Minamata Bay" in box 2-3.

Our choice to identify a few particular interactions to focus our analysis for each topical system in part II is based on their prevalence and/or their importance to human well-being. We then trace the pathways that involve those interactions through the matrix. For the Minamata story, we might have chosen to focus on the interaction in cell F, identified by bolded text in box 3-1 in figure 2.3, where fish contaminated with methylmercury cause health damages to fish consumers. We could then analyze the pathway from cell D to E to F, in which pollutant discharges affect ecosystems and ultimately factory workers, by identifying and discussing the pathway in which manufacturing technology releases methylmercury to Minamata Bay (box 2-3), the Minamata Bay fish bioaccumulate methylmercury (box 3-3), and the methylmercury-contaminated fish in Minamata Bay subsequently cause health damages to fish consumers (box 3-1).

In the system interaction section of each chapter in part II, we illustrate how interactions form pathways using box-and-arrow diagrams as shown in figure 2.4. We discuss each identified pathway in greater detail in narrative form. It is important to note that the boxes in these pathway diagrams represent the interactions, and thus capture the way in which two material components influence each other, while the arrows connect two interactions that involve the same individual component. That is, the first arrow in figure 2.4 connects the interaction where manufacturing technology affects Minamata Bay (by releasing methylmercury into it) with the interaction where Minamata Bay affects the fish. The second arrow further connects the interaction in which Minamata Bay affects the fish (by causing methylmercury accumulation) with the interaction where that fish causes health damages to fish consumers. Thus, these diagrams are different from other box-and-arrow diagrams in which boxes

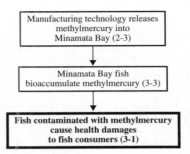

Figure 2.4
Illustrative box-and-arrow diagram for the Minamata story (pathway D-E-F from figure 2.2).

would indicate components and arrows would capture causal connections in the form of interactions.

Interventions: Actors Changing System Interactions

The third analytical step is to identify interveners—the actors who have agency to modify a system—and examine past interventions that have changed the ways material components function and interact. Interveners in the mercury systems include those who use mercury, those who are affected by mercury, those who develop new technology, and those who are engaged in mercury governance. Sometimes these are individuals, including business owners, workers, consumers, researchers, public officials, and members of the general public. In addition, groups of individuals operate in collective entities, such as networks, organizations, and governments, and act across public, private, and civil society sectors. Individuals and groups of actors engage with mercury issues in different ways from local to global levels. Interveners in the Minamata story include doctors and local and national governments. Interveners have differing levels of power and influence, which we evaluate by assessing the degree to which they are able to effect change within a system.

To examine how interveners can influence system interactions by adding or subtracting components or changing their attributes, we use an intervention matrix in every chapter of part II, similar to the one in figure 2.5. We construct the intervention matrix by identifying the components and interactions that the intervener targets, and then describing the intervention in the box that it relates to. In figure 2.5, where we list local and

**Knowledge
Institutions**

	1. Human	2. Technical	3. Environmental
1. Human	(1-1)	(1-2)	(1-3)
2. Technical	(2-1)	(2-2)	(2-3) National government creates new pollution laws
3. Environmental	(3-1) National government sets methylmercury limits for fish sales and consumption; Doctors share knowledge about health damages	(3-2)	(3-3) Local government cleans up Minamata Bay
Interveners			
Local and national governments; Doctors			

Figure 2.5
Illustrative intervention matrix for Minamata.

national governments and doctors as examples of interveners in Minamata, box 2-3 represents how the Japanese national government tried to prevent future damage associated with environmental releases by creating new pollution laws. The local government also cleaned up Minamata Bay, thereby changing how ecosystem components interact with each other (box 3-3). In addition, the national government set methylmercury limits for fish sales and consumption and doctors shared knowledge about health damages (box 3-1) to prevent people from eating fish contaminated with methylmercury.

The intervention matrices in part II identify the main interveners and interventions that have directly or indirectly modified mercury-related interactions among human, technical, and environmental components of each mercury system. We examine interveners and interventions that resulted in changes in levels of mercury discharge or mercury use in particular applications, as well as those that involved initiatives to protect human health and the environment from mercury exposure and pollution (including, but not limited to, national laws and the Minamata Convention). We identify these interveners and interventions through an analysis of the empirical material, building on the previous matrix-based examination of system components and interactions.

Distinguishing interactions from interventions depends on the moment in time that a system is analyzed. Some interactions can be affected by past interventions. In the Minamata story, for instance, no national pollution law existed when Chisso began to discharge methylmercury into Minamata Bay. The subsequent passage of such a law was an intervention that changed the way system components interacted. An examination of the present-day Minamata system would treat the national pollution law as an existing institutional component. Our analysis in part II takes a historical perspective, so we do not explicitly identify a moment in time at which the interaction and intervention matrices are constructed. We choose to do this for clarity—presenting a separate set of system components and related interaction and intervention matrices for multiple points in time for each mercury system would make the system presentations and discussions exceedingly complex. This means, however, that in part II there is some overlap between what we discuss as an interaction and what we treat as an intervention. In the interventions sections, we identify interventions that we believe are most relevant to understand and learn from with respect to sustainability.

We describe the interventions summarized in figure 2.5 qualitatively (as we do for all the interventions we discuss in part II). This is similar to the way we analyze interactions. The intervention matrix can also be used as a basis for quantitative systems modeling—just like the interaction matrix can—if the necessary data are available. In such an approach, where the interaction matrix quantifies the rate at which system attributes change through time, the intervention matrix would describe alterations in that rate. A quantitative intervention matrix might be used to calculate how quickly the accumulation of mercury would change, not only in fish in Minamata Bay but also in the people who eat that fish, once the factory stops releasing any more methylmercury. This corresponds in mathematical terms to a type of second derivative. Analysis of a second derivative matrix can reveal characteristic timescales and identify whether a system exhibits small or large, or stable or unstable, responses to perturbations.

Insights: Lessons about Sustainability

The fourth analytical step, related to our fourth research question (What insights can be drawn from analyzing these systems?), looks across the components and matrices to synthesize insights about how the mercury systems have operated in a broader context of sustainability. We chose three thematic areas to help guide readers from different disciplines toward insights particularly relevant to the types of questions that interest them, and we present these as separate sections in each chapter in part II. First, in the sections titled "Systems Analysis for Sustainability," we focus on insights relevant to those who study environmental or engineered systems and complex adaptive systems more generally. Second, in the sections titled "Sustainability Definitions and Transitions," we center on issues of concern to researchers interested not only in concepts of sustainability but also in how societies can move toward greater sustainability. Third, in the sections titled "Sustainability Governance," we address topics that are of particular interest to scholars who study policy-making and the role of institutions. Following our analysis of the mercury systems in part II, we return to these thematic areas in chapter 9.

Systems Analysis for Sustainability
Many researchers increasingly acknowledge that systems perspectives are necessary to describe and analyze environmental processes influenced by humans. These include those who study mercury and its environmental behavior. Biogeochemical cycle analyses quantify how mercury and other elements such as carbon and nitrogen move and change forms through biological systems, geological processes, and chemical reactions, and quantify their disruptions due to human activities (Klee and Graedel 2004). Other systems research aims to trace the flow of materials such as minerals or fossil fuels through societies, economies, and regions (Erkman 1997;

Fischer-Kowalski et al. 2011). A growing community frames its research as Earth system science, where human activities are seen as a fundamental part of Earth systems (Reid et al. 2010). Some researchers focus on understanding interactions as part of coupled human and natural systems (Liu et al. 2007), or integrated social-ecological systems (Berkes 2017). In addition, some engineering systems literature treats sustainability as a design problem, and researchers work to develop better methods to guide design decisions (Cutcher-Gershenfeld et al. 2004).

Systems research relevant to sustainability often focuses on understanding the properties and behavior of complex adaptive systems. The components of any system have individual attributes such as those we described earlier for human, technical, environmental, institutional, and knowledge components. The system itself may have further properties that are more than the sum of its parts. For example, each fish in a school has its own individual position and velocity, but the shape of the entire school of fish is a property of the system as a whole—the fish swimming together—and is not predictable from individual fish motions. System-level characteristics that emerge from interactions among system components are sometimes referred to as the emergent properties of a complex system (Johnson 2006). Systems relevant to sustainability are also typically adaptive—the patterns that emerge from system interactions may feed back and influence future interactions (Holland 2006; Levin et al. 2013).

Previous work suggests that a particularly important system-level emergent property involves the degree to which systems can change and remain functional when they experience shocks or perturbations (Ross et al. 2007). This is sometimes referred to as a system's adaptive capacity (Smit and Wandel 2006). Similarly, resilience is seen as the capacity of a system to absorb a disturbance and reorganize while retaining its function and structure (Walker et al. 2004; Folke 2016). Adaptation can occur over shorter and longer timescales, and may involve advances in knowledge and innovations in technology and institutions. Whereas the early resilience literature, with its roots in ecology, focused on the behavior of ecosystems (Holling 1973), later analysts have also applied the resilience concept to social systems (Adger 2000; Hall and Lamont 2013). Resilience in this context has been described as the capacity for social-ecological systems to sustain the desired benefits humans gain from ecosystems in the face of disturbances and changes (Biggs et al. 2012).

Much research on complex adaptive systems related to sustainability is shaped by the fact that humans have become a dominating force at planetary scale, as captured in the concept of the Anthropocene. Debates about whether the Anthropocene concept is useful involve controversies over selecting an appropriate start date for this potentially new epoch. This is related to the task of identifying changes in a system's state—in this case, when people became an important driver of changes in the Earth system. For some Earth system processes, changes may be fairly linear toward a threshold, but others can be nonlinear and involve tipping points in the Earth's life support systems (Lenton et al. 2008). Some researchers have argued that crossing certain planetary-scale boundaries, in areas such as biodiversity loss or climate change, can have detrimental or even catastrophic consequences for humanity (Rockström 2009; Rockström et al. 2009). Others have criticized the planetary boundaries idea as a poor basis for conceptualizing environmental challenges (Nordhaus et al. 2012). Chemicals pollution at the global scale is one area in which researchers have nevertheless applied the planetary boundary concept to help assess disruptive effects on vital Earth systems (MacLeod et al. 2014).

Analyzing the mercury systems provides insights both for those who study mercury and for systems analysis more generally. For those who are interested in better tracing how mercury travels through the environment and society, examining the mercury systems can reveal components and processes that are often overlooked in disciplinary analysis, help identify causes of observed changes, or suggest levers for mitigating harms posed by mercury to human health and the environment. For systems analysts, mercury provides an empirical case from which to draw further insights into system operations. Engineering systems researchers may ask how environmental components impact efforts to study socio-technical dynamics. Those interested in adaptation and resilience may consider whether and how resilience thinking can be applied to the management of pollutants. For scientists interested in systems approaches to planetary-scale environmental processes, mercury offers a test of the utility of related perspectives and concepts such as the Anthropocene or planetary boundaries.

Sustainability Definitions and Transitions
The academic literature contains several different definitions of sustainability. Some analysts define sustainability in terms of maintaining natural resources

at a level that does not exceed their ability to be renewed (Daly 1990). This idea, resting on the foundation that there are no substitutes for some forms of natural resources, is sometimes referred to as "strong" sustainability (Neumayer 2003). Another definition of sustainability allows for depleting natural resources as long as other resources that maintain human well-being can substitute for them in the longer term. From this perspective, it is consistent with sustainability to deplete a non-renewable resource to a certain extent, provided that other investments can compensate for its functions. Applying this idea, however, requires understanding how such tradeoffs are valued both in theory and in practice. This idea is reflected in literature that assesses whether fundamental stocks of capital that human well-being depends on—natural capital as well as other types such as manufactured and human capital—are maintained through time (Polasky et al. 2015).

Analysts apply different perspectives on how societies can make progress toward sustainability. From the perspective of ecological modernization, environmental protection and economic growth are regarded as mutually supportive, and the development of new and more environmentally friendly technology is seen as critical to making progress on sustainability (Spaargaren and Mol 1992; Mol 2003). This perspective emphasizes that contemporary societies can be "greened" through new technology without fundamentally changing the basic principles of production, consumption, and trade that are embedded in contemporary capitalism (Neumayer 2003; Meadowcroft 2012; Bulkeley et al. 2013). In addition, the idea of a circular economy suggests that improved recycling and reuse within the existing economy, with the ultimate goal of a closed-loop system, can contribute to greater sustainability (Ghisellini et al. 2016). However, some analysts argue that progress toward sustainability requires much more profound changes in human consciousness and behavior, together with deep alterations to dominant production, consumption, and trade patterns (Princen 2005; Speth 2008; Dryzek 2013).

A growing number of analysts focus on better understanding past and present sustainability transitions, often with an eye toward supporting future transitions (Markard et al. 2012; Feola 2015; Loorbach et al. 2017). Some of these analysts make a conceptual distinction between transitions and transformations (Pelling 2010; Linnér and Wibeck 2019). A transition is seen by these analysts as involving a largely incremental and step-wise change process away from the status quo. In contrast, they view a transformation as a more fundamental and wide-ranging departure from business as usual toward something intrinsically different. Yet, authors in the transition and transformation literature are not consistent in their definition and use of the two terms (Patterson et al. 2017). Both terms—"transition" and "transformation"—nevertheless embody the basic idea that different forms of change are necessary for societies to move to a more sustainable trajectory. We use "transition" as an umbrella term to describe multidimensional change processes to a more sustainable state, as other researchers have done previously (e.g., Markard et al. 2012; Loorbach et al. 2012).

Our analysis of the mercury systems contributes to debates on sustainability concepts and transitions in several ways. We chose in this book to define sustainability, as noted in chapter 1, as centered on human well-being. Those who focus on different ways to define sustainability can use the case of mercury to ask whether any use of mercury could ever have been considered sustainable, and under what conditions. Those who apply ideas like ecological modernization or the circular economy can draw insights from trends in mercury use and discharges over time. For analysts who are interested in better understanding and informing sustainability transitions, the long history of human interactions with mercury offers empirically rich information, focusing on an issue that involves humans, technology, and the environment simultaneously. Analysts who focus on transitions may be particularly interested in the interacting temporal and spatial dynamics of change processes in the mercury systems, where immediate impacts occur simultaneously with much longer, remote feedbacks.

Sustainability Governance
Governance for sustainability requires simultaneously addressing many socio-economic, technical, and environmental issues. Governance structures involving one or multiple institutions can be seen as complex systems that have their own thresholds and tipping points (Young 2017). These institutions are created through collective action, and in turn many of their formal and informal rules shape human activities in a process of mutual co-construction. Governance for greater sustainability requires both reformed and new institutions and networks that can meet governance challenges on a human-dominated planet (Biermann 2014). These can be created through top-down and bottom-up change processes. Governments play many central roles in sustainability governance: they are the negotiators and implementers of international environmental agreements like the Minamata

Convention, and they have the ability to pass domestic legally binding rules and standards. International organizations, private sector actors, market participants, civil society organizations, and individuals also shape governance in a variety of ways.

Many current international and domestic legal, political, and economic institutions are largely ill equipped and inadequate to manage a transition toward greater sustainability (Biermann et al. 2012). This is in part due to a frequent lack of match—or fit—between the scope of these institutions and the biophysical and socio-economic systems that they are designed to govern (Young 2002; Folke et al. 2007; Epstein et al. 2015). A major strand of the governance literature focuses on how to design new and modified institutions to more effectively address sustainability problems, often centering on the importance of paying careful attention to the underlying physical characteristics of the problems that they address (Mitchell 2006). Some of this literature stresses that polycentric and multilayered institutions can improve the fit between institutional scope and properties of biophysical and socio-economic systems (Young 2002). These types of institutions also allow for a large number of actors (including possible interveners) to be involved in different forums and across governance scales.

Social scientists view governance as an inherently social process, whereby actors intentionally seek to steer individuals, groups, and societies toward a collective outcome, such as greater sustainability. Stakeholders may have very different views on how to do that, and even on how to define sustainability. These views are shaped by many political, economic, social, cultural, and environmental factors. In addressing mercury and other sustainability issues, societies have to make normative decisions among a multitude of possible transition pathways as part of any governance process (Meadowcroft 2011; Patterson et al. 2017). The importance of stakeholder involvement in describing a sustainability problem, and identifying and dealing with trade-offs of different options for addressing that problem, is stressed in the sustainability science and governance literatures (Brandt et al. 2013). The importance of broad participation in sustainability governance is also related to issues of equity and justice, as different stakeholders may have very different levels of influence and power in shaping decisions that affect societies and, ultimately, the planet.

Analysts of governance and policy-making processes, and practitioners who are engaged in the creation and implementation of domestic laws and standards or international institutions like the Minamata Convention, will find in part II many examples of how efforts to govern different aspects of the mercury issue evolved over time, with many intended as well as unintended results for human well-being. Those who are interested in how to design more effective revised or new governance structures may be particularly interested in how institutions fit with different material components and characteristics of the mercury systems. The overlapping nature of mercury governance at multiple geographical scales also offers a comparative perspective on how local, national, regional, and global efforts and institutions addressed multifaceted mercury issues of much importance to human well-being. In addition, governance scholars may be interested in linkages between mercury and other sustainability issues.

The HTE framework that we introduce in this chapter forms the structure for examining the topical mercury systems in the five chapters of part II. These chapters are organized based on a common four-step approach associated with the book's four research questions. First, they describe the human, technical, environmental, institutional, and knowledge components (research question 1). Second, they examine interactions between components (research question 2). Third, they look at system interventions focused on sustainability (research question 3). Fourth, they draw insights relevant to the three areas of systems analysis for sustainability, sustainability definitions and transitions, and sustainability governance (research question 4). In the three chapters in part III, we synthesize this empirical material. Chapter 8 returns to the first three research questions, whereas chapter 9 addresses research question 4. Chapter 10 concludes the book by drawing lessons for future efforts to further address the mercury problem, targeted toward researchers, decision-makers, and thoughtful citizens.

II Sustainability Stories about Mercury

3 Global Human-Technical-Environmental Cycling: Chasing Quicksilver

Human activities have dispersed substances such as mercury, lead, and carbon in the atmosphere, ocean, and land, where they are now present in quantities much higher than before there was an anthropogenic influence. These substances spread across the Earth through winds and water currents. Many scientific studies of how people have altered the environment begin with the perspective of biogeochemical cycling, which accounts for how substances move and change form in the environment. However, biogeochemical cycling processes are not the only global-scale flows of materials relevant to sustainability. Substances in commerce also cross the globe through trade, as do products and wastes that contain them. A global human-technical-environmental cycling perspective provides an expanded view of the transport and transformation of substances relative to that of global biogeochemical cycling analysis by encompassing environmental and societal flows together.

While living in New Spain (now Mexico) in 1554, the Spanish merchant Bartolomé de Medina developed the patio process of silver amalgamation, which uses mercury in the silver mining process. The high value of silver in the 1500s and 1600s helped power international trade and commerce, connecting Europe, the Americas, and Asia (Flynn and Giráldez 1995). In the Americas, where mines contained large quantities of low-quality silver-containing ore, the patio process made extracting the valuable metal easier for the Spanish colonialists. It works by allowing an amalgamation of ore and mercury to mix in a shallow, open-air courtyard for several weeks. The large quantities of mercury required for this process originated mainly from mines in Almadén in Spain, Idrija in modern Slovenia, and Huancavelica in Peru. The "loss" of mercury to the environment in silver production could range from 0.85 to more than 4 times the amount of silver produced (Nriagu 1993). The total amount of mercury "lost" in South

America between 1570 and 1820 may have been as high as 126,000 tonnes (Nriagu 1994). That is more than 50 times the annual present-day anthropogenic emission of mercury to the atmosphere (UNEP 2019).

Centuries after the colonial South American mining boom, scientists are trying to determine where all of the mercury that was extracted and used ended up, and whether it remained locally or cycled globally through the environment. Scientists attempt to answer this key question by examining environmental archives such as lake sediments, bogs, and ice cores that record mercury deposition over time. Mercury pollution from colonial mining, for instance, can be found in sediments in South American lakes (Cooke et al. 2009). However, deposition recorded in other environmental archives, particularly in the Northern Hemisphere, was not elevated around the time that mercury was used during colonial mining in South America (Engstrom et al. 2014). Deep ocean sediments around Antarctica do not show elevated deposition from this silver mining activity either (Zaferani et al. 2018). Some modeling studies, in contrast, suggest that a substantial quantity of mercury from silver mining traveled globally via environmental processes, and that it continues to circulate in the environment worldwide (Amos et al. 2015). Other researchers who have attempted to synthesize measurement and model data argue that emissions from historic mining only had a moderate influence on current global mercury cycling (Outridge et al. 2018).

Identifying the time when human activities first distributed mercury globally in the environment, and thus the relative quantities of human-induced preindustrial and industrial mercury pollution, is not just of interest to scientists. It also provides policy analysts an environmental benchmark against which to compare the effectiveness of strategies to reduce mercury discharges and concentrations. Environmental records of pollutant deposition that predate the Industrial Revolution are often considered to be a proxy for natural levels absent human influence. Different environmental archives provide a remarkably consistent picture of a dramatic global-scale increase in mercury deposition since the beginning of the nineteenth century. The human influence is even larger if preindustrial mercury deposition was also elevated due to earlier anthropogenic activities, such as colonial mining in South America going back to the 1500s. This would also imply that present-day human influence on the environmental cycling of mercury could persist for many centuries in the future.

Studies of different environmental archives and models of biogeochemical cycling tell a mixed story of mercury mobilization, but they all focus on the environmental aspects of when mercury became a global pollutant. Mercury, however, also travels through societies. As we indicate in the chapter title, when examining the international distribution of mercury, it is also necessary to chase the quicksilver that travels across borders—as a commodity, in mercury-added products, and as a contaminant in food and biota. People carried mercury across the Atlantic Ocean in large quantities starting in the 1500s for use in South American silver and gold mining. The social and cultural consequences of the economic processes that this mercury fueled were also international in scope: the silver trade and its distribution of resources and capital was a driving factor in the growth of international economic systems of production and consumption starting in the sixteenth century (Moore 2003). The term "quicksilver," a synonym for mercury in many languages as noted in chapter 1, is an adjective that means moving or changing rapidly and unpredictably—a fitting description of many of the dynamics of mercury's travels in both the environment and society.

This chapter focuses on mercury as a global pollutant, and the role of the Minamata Convention in addressing activities that contribute to the global cycling of mercury. We use the perspective of global human-technical-environmental cycling to contrast with the narrower perspective typical of global biogeochemical cycling analyses. In the section on system components, we identify where mercury is present in the environment and society, and the institutions and knowledge that provide the context for its global-scale material cycling. We address processes that influence the distribution and fate of mercury in the section on interactions. In the section on interventions, we give an overview of the types of actions designed to address mercury at a global scale by discussing the structure and the main provisions of the Minamata Convention. In the final section on insights, we highlight the importance of accounting for a broad range of system dynamics in understanding the global cycling of mercury, how variations in timescales of societal and environmental cycling of mercury affect transitions toward sustainability, and factors that affect the implementation and effectiveness of the Minamata Convention.

System Components

The total quantity of mercury present on Earth is fixed, but human activities have moved large quantities of this mercury to different locations. Much mercury has dispersed throughout the environment, where it is present at varying concentrations. Additional mercury remains in places controlled by people as stockpiles, in products, or in landfills. A small amount of mercury is also present in wildlife and in human bodies. Figure 3.1 shows the human, technical, environmental, institutional, and knowledge components relevant to the global mercury cycling system.

Mercury is present in *geological reservoirs* in the Earth's "mercuriferous belts"—areas associated with volcanic activities and plate tectonic boundaries (Gustin et al. 2000). The highest concentration of mercury in these belts is typically in cinnabar (mercuric sulfide, or HgS), but mercury also exists in small amounts in other minerals and in fossil fuels such as coal. Mercury is emitted to the *atmosphere*, and circulates among *land* and *oceans*, as a result of natural processes (such as volcanic eruptions and the weathering of rocks) as well as human activities. Archeological data suggest that *miners* produced *extracted mercury* from cinnabar as early as 6300 BCE in the Sizma district in southwestern Turkey, which may have been the world's first underground mine (Brooks 2012). Early written mentions of cinnabar mines date back to the work of Theophrastus of Eresus in the fourth century BCE (Goldwater 1972). Much of the mercury mined over human history came from the historically large mines in Almadén, Idrija, and Huancavelica, as well as from mines in China (Wanshan), Italy (Monte Amiata), and the United States. The annual minerals commodity survey issued by the US Geological Survey (2017) reports that 600,000 tonnes of

Human components	Technical components	Environmental components

Miners
Producers and consumers of
  goods
Producers and consumers of
  energy
	Extracted mercury
	Extracted geological materials
	  containing mercury
	Mercury in commerce
	Mercury in stockpiles and
	  landfills
	Mercury-added products
	Mercury in production
	  processes
		Geological reservoirs
		Atmosphere
		Land
		Oceans
		Terrestrial and aquatic
		  ecosystems
		Living organisms

Institutional components	Knowledge components
Mercury markets Regional treaties Global Mercury Partnership Minamata Convention Trade controls	Forms of mercury Properties of mercury Long-range transport Mercury concentrations in the environment Quantities of mercury in stockpiles and trade

Figure 3.1
Components in the global mercury cycling system (referenced in the text in italic type).

mercury remain in the Earth's crust globally in places where extraction is currently or potentially feasible.

Elemental mercury is extracted from cinnabar through a relatively simple process of heating the crushed ore in a furnace. This process remained more or less the same for thousands of years. When heated, the mercuric sulfide reacts with oxygen from the air to form sulfur dioxide, and the mercury vaporizes. Mercury, which condenses at 357 degrees Celsius (675 degrees Fahrenheit), a lower temperature compared with the other gases, is then captured and cooled into liquid form before it is stored, traded, and used. Extracting mercury from cinnabar that often contains less than 1 percent of mercury has not only required vast amounts of ore, but also large quantities of biomass fuel from areas around the largest mercury mines. Data on mercury mining are uncertain; Lars Hylander and Markus Meili (2003) estimate that nearly 1 million tonnes of mercury have been extracted throughout history.

People have moved mercury from geological reservoirs both intentionally and unintentionally, and for many different reasons. Much of the early mercury extracted from cinnabar was used in other mining operations to produce gold and silver; mercury is still used in the artisanal and small-scale gold mining (ASGM) sector today (see chapter 7). *Producers and consumers of goods* relied on mined and recycled mercury for millennia: in ancient to modern medicines (see chapter 4); in a variety of products such as thermometers, paints, pesticides, and batteries; and in several mercury-based manufacturing techniques to help produce other goods such as mirrors, hats, and chemicals (see chapter 6). *Producers and consumers of energy* have unintentionally discharged mercury present in trace amounts in other *extracted geological materials containing mercury* into the environment. A substantial fraction of mercury emissions from such geological material has come from burning coal, but industrial processes such as metal smelting and cement production have also resulted in mercury emissions and releases (see chapter 5).

Mercury in commerce has been traded since at least Roman times. Elemental mercury has been measured and priced on *mercury markets* since 1927 using a common but unique unit—a 76-pound flask. The exact origin of this specific measurement unit is unclear, but mercury sellers and buyers likely first used this distinctive flask in the Almadén mine in Spain (Myers 1951). Mercury flasks come in different shapes, but they are typically made of welded steel, have a screw cap, and are roughly the size of a two-liter container. In metric units, a 76-pound flask equals 34.5 kilograms, and 29 of those flasks make up a tonne (Brooks 2012). The total quantity of mined and sold mercury that is under human control today is unknown. A report by the United Nations Environment Programme (UNEP) that quantified best estimates of the total amount of mercury supply, demand, and trade in 2015 did not include statistics for *mercury in stockpiles and landfills* or mercury accumulated in society (comprising the amount of mercury in *mercury-added products* and *mercury in production processes*) because reliable data were unavailable (UNEP 2017).

Knowledge of the different *forms of mercury* and their behavior is relevant to understanding mercury cycling in both the environment and in commerce. *Properties of mercury* have been discovered over time, and have important implications for how different forms can be used and for analyzing environmental flows and impacts on living organisms. Elemental mercury exists on its own, both in liquid form and as a gas. It is the only metal in the periodic table that is liquid at room temperature. This unique property gave mercury its moniker "quicksilver" as well as its chemical symbol Hg, which refers to hydrargyrum, or "water-silver" in Greek. Elemental liquid mercury is highly volatile. Left on its own in open air, it readily evaporates into the gas phase. Gaseous elemental mercury, the predominant form of mercury found in the atmosphere, is both colorless and odorless, and therefore difficult to detect without advanced measuring equipment. In aquatic systems, mercury can be present as dissolved gaseous mercury, which is predominantly composed of elemental mercury.

Mercury compounds involve atoms of mercury that are bound to atoms of other chemical elements that can only be separated by chemical reactions. Mercury compounds exist in two categories, inorganic (not containing carbon) and organic (carbon-containing). Elemental mercury is also inorganic, as it does not contain carbon. Some mercury compounds exist naturally in the environment, but scientists in laboratories have synthesized additional ones. Cinnabar, or mercuric sulfide, from which most commercially traded elemental mercury is extracted, is a solid and relatively stable inorganic mercury compound in the Earth's crust. Mercurous chloride (Hg_2Cl_2), another naturally occurring solid inorganic mercury compound, is sometimes referred to as calomel. Various other inorganic mercury compounds can be found in freshwater, oceans, and soils. Organic mercury compounds include methylmercury ($[CH_3Hg]^+$) and dimethylmercury (($CH_3)_2Hg$). These are present in aquatic systems and in biota. Other organic forms, such as ethylmercury ($C_2H_5Hg^+$) and phenylmercury (C_6H_5Hg), are largely commercially produced.

Levels of mercury in the environment, in animals, and in people increased rapidly following the onset of industrialization (AMAP 2011). Estimates of the total amount of mercury present in the global atmosphere, oceans, and soils today vary dramatically: a 2018 global mercury assessment conducted by UNEP estimated this number at just under 500,000 tonnes (UNEP 2019), but other estimates range as high as 1.5 million tonnes (Obrist et al. 2018). Much of this mercury originated from anthropogenic sources (Amos et al. 2013; N. E. Selin 2014). The amount of mercury in the atmosphere (4,400 to 5,300 tonnes) is reasonably well known; this amount is estimated to have increased, due to pre- and postindustrial human activities, by a factor of 5 or more (Amos et al. 2013; Outridge et al. 2018). Estimates of the total amount of mercury in the oceans range from 270,000 to 450,000 tonnes. This amount has increased by up to a factor of 2.8 due to human activities, and the amount in the surface ocean may have increased by up to a factor of 6 (Amos et al. 2013). Human activities may have increased the amount of mercury in terrestrial soils by 20 to 50 percent; current estimates of total mercury in soil range from 250,000 to 1 million tonnes (Obrist et al. 2018). Mercury is also present in *terrestrial and aquatic ecosystems* and in *living organisms* such as fish, birds, and marine mammals, as well as in people.

Estimates of mercury's global biogeochemical cycle attempt to quantify the amount of mercury that is present in and travels through the global environment, and the relative influence of human and natural activities. Biogeochemical cycle analyses evaluate how substances transport and transform through biological systems (living things, including plants and animals), geological processes (such as fossil fuel formation and plate tectonics), and chemical reactions (changing the chemical composition of specific substances). See figure 3.2 for an example of a global biogeochemical cycle diagram for mercury. Biogeochemical cycle estimates are constructed by combining measurement data and models, but many relevant environmental processes are uncertain, and data are lacking for some time periods and regions. As noted above, a major area of uncertainty is how much anthropogenically discharged mercury cycles globally versus mercury that remains locally. David G. Streets and colleagues (2017) calculated that 1.5

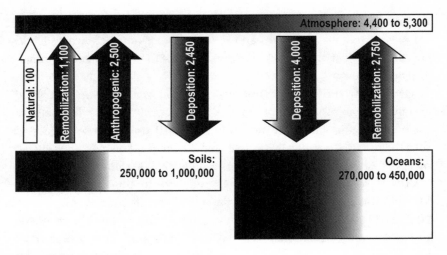

Figure 3.2
Global biogeochemical cycle for mercury. Quantities in boxes (tonnes) and fluxes in arrows (tonnes/year), drawing from Amos et al. (2013), Selin (2014), Obrist et al. (2018), and UNEP (2019). Shaded regions indicate approximate fractions of anthropogenic contribution.

million tonnes of mercury have been discharged into the environment by anthropogenic activities over time, of which 30 percent was emitted to the atmosphere and 70 percent released to land and water. They estimated that 40 percent (390,000 tonnes) of those historical land and water releases are sequestered in sediments or at contaminated sites and do not cycle globally. Other estimates of total mercury entering the global biogeochemical cycle as a result of human activities are much lower: the 2018 global mercury assessment attributes only 79,000 tonnes in the global biogeochemical cycle to anthropogenic activities (Outridge et al. 2018; UNEP 2019).

Scientific knowledge that some forms of mercury undergo *long-range transport* in the environment emerged in the 1970s with the discovery of *mercury concentrations in the environment* in areas far from emission sources. Knowledge about environmental transport and levels of mercury at a global scale evolved somewhat separately from the development of information on the *quantities of mercury in stockpiles and trade*. A recently published study, however, combined information on product flows with environmental releases to quantify the mercury flows addressed by articles of the Minamata Convention (Selin et al. 2018). Other efforts have attempted to assess the quantities of mercury flowing through society, largely with the goal of better quantifying releases to the environment from different sources in selected countries through substance flow analyses (Sznopek and Goonan 2000; Cain et al. 2007). The societal circulation of mercury results from supply and demand dynamics on international and domestic mercury markets as well as trade in mercury-added products. Some work on tracing mercury flows in society has also been carried out at national and sub-national scales (Hui et al. 2016; Svidén and Jonsson 2001).

International institutions began to address mercury as a pollutant in the 1970s (Selin and Selin 2006). The Organisation for Economic Co-operation and Development (OECD) issued a recommendation to its member states in 1973 to reduce anthropogenic discharges of mercury to the lowest levels possible. Countries developed several *regional treaties* starting in the 1970s that addressed releases of mercury into bodies of water in Europe and North America. European and North American countries in 1998 adopted a heavy metals protocol that included mercury under the Convention on Long-Range Transboundary Air Pollution (CLRTAP). Global political efforts to address mercury problems gained momentum in the early 2000s, largely in the form of voluntary collaboration under UNEP's *Global Mercury Partnership* program (Sun 2017). This still-active program brings together national governments and other stakeholders to build and diffuse knowledge about the risks of mercury and the availability of mercury-free alternatives. The negotiations for the *Minamata Convention* began in 2010, and the treaty was adopted at the diplomatic conference in Kumamoto, Japan, in October 2013 (as mentioned in chapter 1). It entered into force in 2017, and by early 2020, 117 countries and the European Union (EU) had become parties. In addition, several countries have instituted *trade controls* on mercury in commerce, regulating exports as well as imports.

Interactions

Much of the mercury that people have extracted from geological storage continues to cycle through society and the environment. Figure 3.3 shows interactions in the system for global mercury cycling: we have selected three interactions in that matrix (the items in bold type in boxes 1-3 and 2-2) to focus on in this section; we then trace the pathways that influence them, which we summarize in figure 3.4 (where the bold boxes correspond

Knowledge

Institutions

	1. Human	2. Technical	3. Environmental
1. Human	(1-1)	(1-2) Producers and consumers affect mercury uses and quantities in commerce	(1-3) **People discharge mercury into ecosystems; People alter ecosystems**
2. Technical	(2-1)	(2-2) **Mercury in commerce is traded, reused, or enters stockpiles**; Stockpiles and mercury-based production techniques influence mercury in commerce	(2-3) Mercury in commerce is emitted and released
3. Environmental	(3-1)	(3-2) Geological reservoirs provide mercury for commercial activities	(3-3) Mercury cycles through atmosphere, land, and oceans and changes form; Ecosystem conditions and processes lead to methylmercury production; Methylmercury in ecosystems adversely affects living organisms

Figure 3.3
Interaction matrix for the global mercury cycling system.

to the selected interactions). First, people discharge mercury into ecosystems (box 1-3), and the mercury then undergoes global biogeochemical cycling through the atmosphere, land, and oceans and changes form (box 3-3). Second, people alter ecosystems (box 1-3), leading to changes in mercury cycling and methylmercury production (box 3-3). Third, mercury in commerce is traded, reused, or enters stockpiles (box 2-2), as its supply is affected by geological availability and market interactions, and this mercury can subsequently be emitted and released to the environment (boxes 3-2, 2-2, and 2-3).

Mercury Discharges and Global Biogeochemical Cycling

People have discharged mercury into ecosystems (box 1-3), thus altering the global biogeochemical cycle of mercury. These mercury discharges

a) Mercury discharges and global biogeochemical cycling: People discharge mercury, which cycles through ecosystems and converts to methylmercury

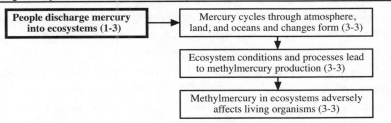

b) Climate and ecosystem changes, mercury cycling, and methylmercury production: People alter ecosystems in ways that lead to changes in mercury cycling and methylmercury production

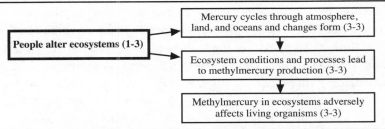

c) Mercury cycling in society: Mercury in commerce is traded across borders and leads to emissions and releases

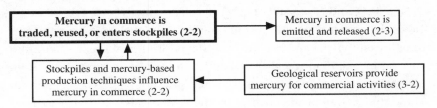

Figure 3.4
Pathways of interactions for the global mercury cycling system. Bold boxes indicate the selected interaction for the subsections below.

have occurred in several different ways, including through primary mercury mining, the use of mercury in gold and silver mining, the burning of coal, the loss of mercury from mercury-added products and industrial production processes, the failure to properly manage mercury in stockpiles and landfills, and the unsafe disposal of mercury and mercury-containing wastes. This anthropogenically discharged mercury, together with mercury that is emitted and released from natural sources, then cycles through the atmosphere, land, and oceans. Some of this mercury changes its form through chemical and biological processes during this cycling (box 3-3).

Scientists initially considered mercury a local pollution and exposure problem. In the 1970s, researchers made the first measurements of mercury in areas remote from anthropogenic point sources. Researchers who measured mercury in Greenland attributed increasing concentrations in glacial ice to human sources (Weiss et al. 1971), although another study conducted in Greenland a few years later found no evidence of anthropogenic increases (Weiss et al. 1975). Early mercury measurements were later found to have major contamination problems (Boutron et al. 1998). Nevertheless, in 1969, Göran Löfroth and Margaret E. Duffy (1969, 17) noted: "Mercury, like lead and the organochlorine pesticides, may turn out to be a global pollutant." This, however, was not a widely accepted conclusion at the time. A 1972 report of the Joint Food and Agriculture Organization/ World Health Organization (FAO/WHO) Expert Committee on Food Additives stated: "The sources of direct pollution due to man's activities can have only local effects on the mercury levels found in fish, e.g., in estuaries and coastal areas. The largest reservoir of mercury is the open seas, and is not appreciably affected by pollution caused by man" (Joint FAO-WHO Expert Committee on Food Additives 1972, 12).

Scientists and policy-makers became increasingly aware by the late 1980s that mercury pollution could travel on a global scale before depositing into terrestrial and aquatic ecosystems. Studies of the human influence on mercury's biogeochemical cycle relied on growing evidence from environmental archives such as ice cores, peat bogs, and lake sediments. These studies were made possible by the development and increased use of ultraclean analytical procedures for trace metal analysis (Vandal et al. 1993; Fitzgerald et al. 1998). The first International Conference on Mercury as a Global Pollutant was held in 1990, in Gävle, Sweden, reflecting growing scientific agreement on mercury's global reach. Yet, some scientists in the late 1990s still debated whether the presence of mercury in remote areas was a result of human activities or of natural geological processes (Fitzgerald et al. 1998; Rasmussen 1998). A 2002 global mercury assessment synthesized the scientific literature and conclusively deemed mercury a matter of global-scale concern as a result of long-range transport and its potential hazardous impacts on the environment and human health (UNEP 2002).

The 2018 global mercury assessment estimated that human activities emitted roughly 2,500 tonnes of mercury to the air in 2015 (UNEP 2019). Global mercury emissions increased by 20 percent between 2010 and 2015 because of a growth in emissions in every region except Europe and North America. The 2,500 tonnes equaled about a third of total mercury emissions to air in 2015. Twice that amount, or about 5,000 tonnes, came from environmental processes that cycle previously released mercury from both anthropogenic and natural sources. These are sometimes referred to as legacy emissions. An additional (small) amount is from present-day natural sources, such as volcanoes and other geological processes. Emissions to air are generally considered to cycle globally.

Releases of mercury to water and land are less well quantified than emissions to air. Releases from ASGM to water and land together were estimated to total 1,220 tonnes in 2015 (UNEP 2019). Other mercury releases to water that researchers are able to quantify are from the mining sector, municipal sewage systems, leaks from mercury-added products, and wastewater from coal-fired power plants and coal washing; these sources totaled 580 tonnes in 2015. This inventory is incomplete, however, as data are lacking for additional sources. There are particularly high uncertainties regarding estimates for mercury releases to land beyond the combined estimate for ASGM to water and land. A very rough estimate put anthropogenic releases of mercury into soils at between 7,000 and 8,000 tonnes in 2015 (UNEP 2019). Large-scale mining and production of minerals such as gold and zinc, which shift large amounts of soil from one place to another, caused much of this release. Mercury-added products going into solid waste streams also add mercury to soils. It is unknown what fraction of present-day releases stays locally in soils compared to what enters waters and the air, and thereby adds to the global biogeochemical cycle.

Mercury in the atmosphere can travel longer or shorter distances, depending on its form. Much mercury is emitted as elemental mercury. Current research estimates that elemental mercury may stay in the atmosphere for an average of six months to a year before it returns to the surface (Horowitz et al. 2017). That is enough time for winds to carry it around the entire globe—a year is the average time that air takes to circulate between the Northern and Southern Hemispheres (Jacob et al. 1987). Elemental mercury lasts so long in the atmosphere mostly because it is relatively insoluble in water and does not get taken up into rainfall. But elemental mercury can settle out of the atmosphere through dry deposition (Jiskra et al. 2018). Some elemental mercury undergoes chemical reactions in the atmosphere that transform it into the more soluble form of gaseous oxidized mercury, which can then either more easily rain out or undergo dry deposition. The specific chemical composition of gaseous oxidized mercury, however, remains unknown (Jaffe et al. 2014). A fraction of the mercury emitted from anthropogenic sources is already in its soluble form, which can be attached to atmospheric particulate matter; these forms of mercury can deposit nearer to sources (tens to hundreds of kilometers).

Mercury continues to cycle among the atmosphere, oceans, and land long after it first enters the environment. Mercury falling out of the atmosphere through wet and dry deposition is taken up by land or water surfaces, where some of this mercury changes its form again in terrestrial and aquatic ecosystems. Because mercury is such a volatile substance, it can return from land and surface waters to the atmosphere as elemental mercury, and start the biogeochemical cycle all over again. In addition to its transport through the atmosphere, mercury moves from place to place with rivers and ocean currents, both across regions and between surface and deeper waters. Vertical transport of mercury to the intermediate and deep ocean occurs on timescales of decades to centuries. Returning to preindustrial levels of mercury in the atmosphere, oceans, and land would take centuries to millennia if anthropogenic emissions and releases stopped completely because the only way environmental processes can truly get rid of this legacy mercury is through burial in sediments, an extremely slow process (Selin 2009).

Specific ecosystem conditions and processes lead to the production of methylmercury (box 3-3). Only a small fraction of the elemental mercury that cycles through the environment is converted into methylmercury. This process largely occurs in aquatic environments lacking oxygen where bacteria transform inorganic mercury into methylmercury. The bacteria that do this share a common gene cluster, and include sulfate-reducing bacteria, iron-reducers, and methanogens (Hsu-Kim et al. 2018). Methylmercury, which is much more toxic than elemental mercury at low concentrations, bioaccumulates and biomagnifies in living organisms. ("Bioaccumulation" refers to the net accumulation of a substance over time in an individual organism from different sources. "Biomagnification" is the progressive buildup of a substance at successively higher levels of a food web.) Concentrations of methylmercury in ecosystems can consequently reach high levels that adversely affect living organisms (box 3-3). Methylmercury is taken up in multiple organs, including the brain, kidneys, and liver (Wolfe et al. 1998; Eagles-Smith et al. 2018). The highest risks are faced by top predators, which in many cases are humans and other large mammals, but can also be predatory birds and fishes (Scheuhammer et al. 2007).

Effects on wildlife are similar to—and maybe even greater than—those seen in humans (European Environment Agency 2018). (We discuss human health impacts in chapter 4.) Researchers base their knowledge of wildlife effects largely on laboratory studies, but have also observed some effects in the wild (Evers et al. 2008). At high exposure levels, fish can die, which happened in Minamata Bay. Non-lethal exposure levels can affect fish growth and their ability to reproduce (Depew et al. 2012). Swedish researchers trying to understand abnormal neurological signs in fish-eating birds in the late 1960s were among the first to identify wildlife impacts of methylmercury (Clarkson and Magos 2006). Exposure remains widespread; for example, 40 percent of the surface water bodies in Europe contain levels of mercury that exceed EU guidance intended to protect fish-eating birds and mammals from adverse effects (European Environment Agency 2018). Many Arctic species have methylmercury concentrations that may damage their health, including polar bears and whales (AMAP 2018). These impacts occur in the context of other stresses, including habitat loss, climate change, hunting, and infectious diseases.

Climate and Ecosystem Changes, Mercury Cycling, and Methylmercury Production

People alter ecosystems in multiple ways (box 1-3). These alterations influence the transport and distribution of mercury and the conditions for methylmercury production. That is, people's interactions with the environment can affect the environmental cycling of mercury even where those interactions do not directly involve mercury use or discharges. Human activities that alter ecosystems change how mercury cycles through the atmosphere, land, and oceans, and also affect the biological and chemical processes that convert mercury between its different forms (box 3-3). This means that even if people stopped discharging mercury into the environment, human activities would continue to influence the ways in which mercury cycles through it. Human impacts on the environment are increasing in the context of industrialization and a rapidly urbanizing population, and are changing landscapes, affecting biodiversity, and altering the global climate system. Human-driven change will remain a major influence on global mercury cycling in the future (Obrist et al. 2018).

Human-induced land use changes such as deforestation and infrastructure development have altered land surfaces and their characteristics, influencing the biogeochemical cycling of mercury (Hsu-Kim et al. 2018). Changing land use influences how, where, and when different forms of mercury, often after long-range atmospheric transport, enter ecosystems, both through wet and dry deposition. Wildfires, which people can set intentionally and unintentionally, move mercury from terrestrial ecosystems into the atmosphere. Changes in the frequency and scope of wildfires can alter not only the amount of mercury that is emitted, but also the movement of mercury within ecosystems that are affected by fire (Kumar et al. 2018). Any process that creates soil erosion, such as vegetation removal and road construction, can lead to runoff of mercury from land, increasing mercury concentrations in streams and rivers (Hsu-Kim et al. 2018). Studies have shown, for example, that converting forested land to agricultural use can cause mercury to be released from soils and enter waterways (Kocman et al. 2017).

People alter environmental processes in ways that make ecosystem conditions and processes more or less favorable to producing methylmercury (box 3-3). In turn, this methylmercury in ecosystems adversely affects living organisms (box 3-3). The building of hydroelectric dams, for instance, creates aquatic environments where biological production of methylmercury can increase (Friedl and Wüest 2002). Burning coal and other fossil fuels increases deposition of sulfates to aquatic ecosystems, fueling the activity of bacteria that convert other forms of mercury into methylmercury (Gilmour et al. 1992). Where human activities affect organisms such as plankton, this can lead to changes in mercury bioaccumulation that propagate through food webs (Krabbenhoft and Sunderland 2013). Changes in the input of terrestrial organic matter and nutrients to aquatic ecosystems, for example, can alter the processes by which methylmercury is taken up by plankton (Jonsson et al. 2017). Other actions such as harvesting of seafood can alter the structure of food webs and affect wildlife and human exposure to methylmercury (Eagles-Smith et al. 2018). One study calculated that because of changes in food webs related to overfishing, Atlantic cod ate a diet higher in methylmercury in the 2000s than they did in the 1970s, resulting in a 23 percent increase in their methylmercury concentrations (Schartup et al. 2019).

Human-induced climate change, a central driver of ecosystem change, leads to changes in mercury cycling and methylmercury production (Obrist et al. 2018). Climate change increases the frequency and intensity of wildfires, which as mentioned above can lead to increased emissions of historically deposited mercury from land. Many other processes that control the remobilization of historically discharged and naturally occurring mercury from land, oceans, and contaminated sites are temperature-sensitive, and are thus affected by rising global, regional, and local temperatures. Permafrost contains a large amount of mercury, some of which may be emitted to the atmosphere when it melts (Schuster et al. 2018). The loss of sea ice in polar regions may increase the amount of mercury that cycles between the oceans and the atmosphere, because ice prevents sea-air exchange of mercury (UNEP 2019). Climate-induced changes also affect ocean ecosystems. For example, changes in temperature and ocean dynamics influence mercury cycling, and changes in the productivity of ecosystems affect methylmercury production. In addition, climate change affects food web structures by changing the presence and health of species, altering patterns of methylmercury production and bioaccumulation.

Mercury Cycling in Society

A biogeochemical cycling perspective focuses on the ways in which mercury transports and transforms in the environment, but mercury in commerce also cycles as it is traded, reused, or enters stockpiles (box 2-2). UNEP estimated that total global demand for commercial mercury in 2015 was 4,720 tonnes. This mercury demand covered a range of uses in products (31 percent), industrial processes (32 percent), and ASGM (37 percent) (UNEP 2017). Some, but not all, of this mercury is included in global inventories of emissions and releases once it enters the environment. Global mercury trade still shifts large quantities of mercury across societies, though this trade has recently declined: exports of mercury reported in 2015 totaled just over 1,300 tonnes, a 2.5-fold decrease since 2010 (UNEP 2017). However, official trade data are incomplete, and some mercury is also traded illegally. The uncertainty in how much mercury is traded or remains in mercury stockpiles underscores the importance of better tracking commercial mercury as well as the difficulty in quantifying its magnitude and impact.

During much of human history, primary mining of cinnabar from geological reservoirs provided mercury for commercial activities (box 3-2). Starting in the 1800s, as mercury use increased with its growing application in products and industrial processes, more commercially available mercury came from reuse. Only about a third of mercury demand in 2015 was sourced from primary mining, with the rest coming from recovery and recycling, the production of mercury as a byproduct of other mining, and the drawdown of stockpiles (UNEP 2017). Roughly half of all of the demand of commercial mercury occurred in East and Southeast Asia. China was the world's largest producer of mined mercury in 2018 with an estimated 3,000 tonnes, representing 88 percent of globally mined mercury (US Geological Survey 2019). Most of the mercury mined in China was used domestically in industry. Primary mercury mining also occurred in the Kyrgyz Republic, largely intended for export to other countries. Previously closed mercury mines were recently illegally reopened in Mexico and Indonesia (UNEP 2017).

Historical changes in the demand and supply of mercury have resulted in large global price fluctuations and changes in the flows of elemental mercury in commerce. Mercury has been traded internationally for millennia, but mercury exports and imports increased sharply together with a growth in its industrial use. During the first half of the twentieth century, the international price of mercury spiked during both World War I and World War II (see figure 3.5). Some mercury uses during these wars were for purposes such as making explosives, but the 1943 US minerals yearbook noted that the principal use of mercury in wartime, like peacetime, was for manufacturing pharmaceuticals, and it reported sharp increases in mercury consumption for this purpose that year due to "the Army's large requirements for prophylactics and antiseptics" (Meyer and Mitchell 1945, 719). Coupled with disruptions in imports from major supply centers in Europe, the increased demand by the US military drove a sharp increase in the price of mercury (Meyer and Mitchell 1941). The modern price of mercury peaked in 1965. Commercial mercury use was at its highest in the 1970s (Horowitz et al. 2014). Reductions in mercury use and growing awareness of mercury as an environmental pollutant led to a strong decline in the global market price of mercury between 1970 and the early 2000s.

Increased demand and trade controls in the early 2000s changed the dynamics of the global mercury market. The price increase starting in 2003 was in large part driven by a growth in ASGM. Anticipation and implementation of bans the EU and the United States placed on elemental mercury exports, which were adopted in 2008 and came into effect in 2011 and

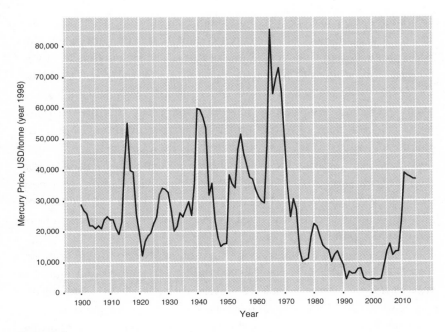

Figure 3.5
Global mercury price, given in USD per tonne in 1998 dollars. *Data source*: US Geological Survey (2014).

2013, respectively, also affected mercury prices (Wilburn 2013). Because of these export bans as well as other trade restrictions instituted by a growing number of countries, there is no longer a globally uniform price for mercury, but rather several different markets. The current price of mercury is relatively low and stable in the EU and the United States, because in both of these closed markets demand is lower than supply. In contrast, mercury prices are higher in other places as a result of restricted supply and continuing demand mainly in the ASGM sector. After the EU and the United States export bans, other places emerged as major mercury trading hubs, including Hong Kong and Singapore (UNEP 2017).

Changes in the dynamics of mercury markets influence the supply and global distribution of mercury. This includes the presence of mercury stockpiles as well as the use of mercury-based production techniques, which influence the amounts of mercury in commerce (box 2-2). Much of the short-term increase in exports from the EU and the United States in the late 2000s and early 2010s—before their export bans on elemental mercury took effect—came from the decommissioning of old chlor-alkali plants that had used large quantities of mercury (see more about the process of chlor-alkali production in chapter 6). Some of this excess mercury made its way, both legally and illegally, to developing countries for use in the ASGM sector, as did the excess mercury from other countries experiencing a simultaneous decline in industrial mercury demand. Many countries also have mercury import bans, or restrict imports to specifically approved uses. The amount of mercury trade that occurs outside legal channels is difficult to document, but has likely increased concurrently with the decrease in legal mercury trade since 2010 (UNEP 2017). Illegal primary mining of mercury, particularly in Indonesia and Mexico, has also increased as a result of the growing demand for mercury in ASGM coupled with tighter restrictions on supply and export in the 2010s.

Much mercury in commerce is emitted and released to the environment (box 2-3). One estimate is that 540,000 tonnes of commercial mercury has entered the environment since 1850 (Horowitz et al. 2014). Some of this mercury was likely in forms that neither cycle further in the global environment nor form methylmercury, but this is still a much larger amount than some estimates of the anthropogenic contribution to the global biogeochemical cycle (underscoring the high degree of uncertainty in much mercury data). Discharges can occur shortly after its initial use, as in the case of ASGM, or after a time lag, for example after a mercury-added product such as a thermometer is discarded at the end of its useful life. Just over half of all global mercury demand in 2015 (2,490 tonnes of the total demand of 4,720 tonnes) was estimated to be discharged into the environment directly after its initial use (UNEP 2017). Nearly 70 percent of this discharged mercury came from ASGM. Of the remainder, 600 tonnes were from mercury used in industrial processes, and 150 tonnes originated from commercial products.

Interventions

The Minamata Convention, the primary global-scale intervention addressing mercury, is the outcome of scientific and political processes that date back to a UNEP Governing Council decision in 2001 to launch the global scientific assessment of mercury completed in 2002. The resulting assessment report published a year later concluded that there "was sufficient evidence of significant global adverse impacts to warrant international action to reduce the risks to human health and/or the environment" (UNEP 2002, paragraph 139). This led Norway, Switzerland, and the EU to call for an international legally binding agreement on mercury (Selin and Selin 2006; Eriksen and Perrez 2014). Other industrialized and major developing countries including the United States, Canada, Japan, Russia, China, India, and Australia opposed treaty negotiations at that time. Several of these countries were reluctant to negotiate yet another global treaty on hazardous substances, having recently adopted the 1998 Rotterdam Convention on the Prior Informed Consent Procedure for Certain Hazardous Chemicals and Pesticides in International Trade and the 2001 Stockholm Convention on Persistent Organic Pollutants (POPs). In addition, some countries opposed international mercury controls impacting mining, industrial manufacturing, and the energy sector (H. Selin 2014).

As a compromise between countries that supported and opposed treaty negotiations, the UNEP Governing Council in 2003 launched an international voluntary program to reduce mercury pollution, and in 2005 created the Global Mercury Partnership program. Countries agreed at the UNEP Governing Council meeting in 2007 to establish a working group to assess options for enhanced voluntary measures and to explore the possible role of existing and new legal instruments. There was enough political support by 2009 to launch treaty negotiations on mercury; in particular, the new Obama administration's support for a new multilateral agreement was a major political trigger, causing first Canada and Australia and later India and China to remove their objections (H. Selin 2014). Treaty negotiations began in Stockholm in June 2010 and concluded in Geneva in January 2013 where, as we mentioned in chapter 1, organizers played Queen's "We Are the Champions" to celebrate the historic moment. Current parties to the Minamata Convention include the EU, larger political powers such as the United States, China, India, Canada, Brazil, Nigeria, Indonesia, and Japan, and many medium-sized and smaller countries in Africa, Asia, Europe, and Latin America.

The Preamble of the Minamata Convention, consistent with the first global mercury assessment as well as with more recent assessment reports, recognizes mercury as a chemical of global concern. As stated in Article I, the overall objective of the Minamata Convention is to "protect the human health and the environment from anthropogenic emissions and releases of mercury and mercury compounds." The treaty specifies a set of legal mandates that cover the entire lifecycle of production, trade, use, emissions, releases, handling, stockpiles, and disposal of mercury. When it comes to implementing its provisions, the individual parties that have joined the treaty hold ultimate responsibility: as with all international treaties under public international law, joining the Minamata Convention is voluntary, but implementation is obligatory for all parties. All parts of the Minamata Convention are legally binding for the parties, but some treaty provisions mandate action (using the directive "shall") whereas other provisions adopt a more persuasive tone (using words such as "should" or "may"). The Minamata Convention includes control provisions that address specific aspects of the mercury issue in Articles 3–12 and enabling provisions in support of treaty implementation in Articles 13–24 (Selin et al. 2018).

Figure 3.6 identifies interventions in the global mercury cycling system, focusing on Articles 3–12 and Article 16 of the Minamata Convention.

	1. Human	**2. Technical**	**3. Environmental**
1. Human	(1-1) Facilitating formalization of ASGM (Art. 7)	(1-2) Measures to address commercial uses of mercury in products and processes (Art. 4–6); Measures to address uses of mercury in ASGM (Art.7)	(1-3) Addressing mercury waste (Art.11)
2. Technical	(2-1) Human health protection from occupational exposure (Art. 16)	(2-2) Mercury reuse and trade restrictions (Art. 3); Application of technology-based control strategies (Art. 8–9); Storage and stockpiles (Art. 10)	(2-3) Provisions controlling emissions and releases (Art. 8–9)
3. Environmental	(3-1) Human health protection from environmental exposure (Art. 16)	(3-2) Primary mining bans and supply-side interventions (Art. 3)	(3-3) Provisions about contaminated sites (Art. 12)

Knowledge Institutions

Interveners

Minamata Convention parties; Minamata Convention bodies; Global Mercury Partnership participants

Figure 3.6
Intervention matrix for the mercury global cycling system.

The figure's bottom row identifies interveners: Minamata Convention parties, Minamata Convention bodies, and Global Mercury Partnership participants. The control provisions in Articles 3–12, together with Article 16, address nearly all aspects of the interaction pathways described in figure 3.4. In the following section, we refer back to figure 3.6, first as we summarize how each of the main treaty provisions applies to the aspects of the mercury issue that we examine in this and the next four chapters, and then as we discuss the provisions in the Minamata Convention on knowledge and institutions. We omit the additional enabling provisions found in Articles 13–15 and 17–24 from the intervention matrix in figure 3.6 because they address knowledge and institutional components that affect the entire mercury global cycling system.

Provisions Addressing Mercury
in Human-Technical-Environmental Systems

The Minamata Convention sets up several supply-oriented controls in Article 3 by addressing primary mining of, reuse of, and trade in mercury (boxes 3-2 and 2-2). The treaty prohibits the opening of new mercury mines, although parties with operating mercury mines may continue mining for up to 15 years after becoming a party. Historically large mercury mines in Idrija and Almadén had been closed in 1995 and 2002, respectively. China, by far the country with the most continuing primary mercury mining, announced in 2017 that it would prohibit primary mercury mining in existing mines in 2032. The only major legal exporting mine still operating when treaty negotiations began was the Khaidarkan mine in the Kyrgyz Republic, which generated roughly 250 tonnes of mercury in 2009 (Brooks 2011). The government of the Kyrgyz Republic in the late 2000s announced its intention to close the mercury mine in return for outside financial and technical assistance (Earth Negotiations Bulletin 2010). The mine, however, still produced an estimated 20 tonnes of mercury in 2018 (US Geological Survey 2019), and the Kyrgyz Republic had not yet become a party to the Minamata Convention by early 2020.

Article 3 mandates that parties identify individual mercury stocks exceeding 50 tonnes, and mercury supply sources generating over 10 tonnes per year. It prohibits the reuse of excess mercury from chlor-alkali facilities, the largest secondary source of mercury. The Minamata Convention bans the export of elemental mercury from one party to another unless it is intended for a use that is allowed under the treaty or for environmentally sound interim storage. The permitted trade in elemental mercury is governed by a prior informed consent (PIC) scheme. Under this PIC scheme, the government of a firm seeking to sell mercury to a buyer in another country must receive written approval from the government of the importing country before the trade can proceed. Parties are only allowed to export elemental mercury to non-parties that have measures in place to protect human health and the environment, and that follow treaty provisions on allowed uses, storage, and disposal. Parties can only import elemental mercury from a non-party if that country provides guarantees that the mercury comes from a source allowed under the treaty.

Minamata Convention parties that have enacted policies going beyond the treaty's trade provisions have further influenced societal flows of mercury. Sweden took on a leading role when it banned elemental mercury exports in 1997. As noted above, the EU and the United States both adopted elemental mercury export bans in 2008. The EU ban entered into force in 2011 and the US ban took effect in 2013. The United States in 2016 and the EU in 2017 expanded their bans to include mercury compounds (not addressed by the Minamata Convention PIC scheme). The United States banned the export of several mercury compounds as of 2020, and the EU's ban was implemented stepwise in 2018 and 2020. Many other parties also strengthened their export provisions in the 2010s as part of implementing the Minamata Convention, but some stopped short of total bans. For example, Switzerland announced in 2017 that it would allow export of elemental mercury for dental amalgam for another decade (Carey 2017). In addition, a growing number of countries have elected to ban all imports of mercury, or restrict it to a few uses such as in dentistry.

The Minamata Convention has a clear focus on human health protection in its overall objective, as well as under Article 16 on health, which deals with both environmental exposure (box 3-1) and occupational exposure (box 2-1). Implementing the treaty provisions would contribute to the protection of human health in varying ways. The preamble notes that mercury poses health concerns for vulnerable populations—especially women, children, and also future generations—particularly in developing countries. Under Article 16, parties are encouraged to promote the development and implementation of strategies to identify populations at risk from mercury exposure based on different exposure routes across different groups of people, countries, and regions. Parties are urged to promote the availability of health care services for the prevention, treatment, and care of populations who are affected by mercury exposure. Parties should establish and strengthen the institutional and professional capacities for the prevention, diagnosis, treatment, and monitoring of health risks from mercury. In chapter 4, we discuss human health aspects of the use of, and exposure to, elemental mercury and mercury compounds and related efforts to mitigate human health risks.

A major part of the Minamata Convention addresses mercury emissions to air and mercury releases to land and water. Articles 8 and 9 of the treaty introduce a set of mandates for parties to control mercury emissions and releases (box 2-3). These rely on technology-based control strategies (box 2-2) and are focused on large industrial point sources. Parties have some flexibility in defining and applying their own specific strategies to control emissions, but they are required to take action to address mercury emissions from both new and existing stationary sources for the purpose of controlling and, where feasible, reducing such emissions. Parties shall also identify and control releases from other relevant point sources that are not covered by the obligations for products, processes, and sources of air emissions. In chapter 5, we discuss mercury emissions to air from coal combustion and other industrial point sources, and examine efforts to address mercury emissions from such sources.

The Minamata Convention focuses on different areas of intentional mercury use that result in direct human exposure and environmental emissions and releases, including in commercial products and production processes, in Articles 4, 5, and 6 (box 1-2). Article 4 prohibits the manufacture, import, and export of nine mercury-added product categories, and states that parties should take steps to phase down the use of mercury in dental amalgam. Article 5 mandates the phase out of mercury use in two industrial processes and imposes mercury use restrictions in three others; parties shall discourage the manufacture and commercial distribution of new mercury-added products and the development of new facilities that use mercury in manufacturing processes. Parties can also apply for exemptions to phaseout dates for both products and processes under Article 6. Article 10 mandates that parties store and manage stockpiles of mercury in an environmentally sound manner (box 2-2). Article 11 addresses the disposal of mercury wastes (box 1-3). Article 12 requires that parties endeavor to develop strategies for identifying and assessing mercury-contaminated sites (box 3-3). We discuss the use of mercury in products and processes, its consequences (including for stockpiles and contaminated sites), and efforts to reduce and phase out such mercury use in chapter 6.

ASGM, an area of growing mercury use that leads to substantial environmental discharges and human health problems, is included in a separate article in the Minamata Convention. Under Article 7, parties with ASGM within their territory shall take steps to reduce (and where feasible eliminate) mercury use (box 1-2). They shall also take steps to reduce the emissions and releases of mercury in such activities. To this end, parties with "more than insignificant" ASGM are required to develop a national action plan. Among other things, such a plan must outline national objectives and reduction targets, actions to eliminate mercury-based amalgamation practices that are particularly damaging to the environment and human health, and strategies for promoting the reduction of emissions, releases, and human exposure to mercury. It must also include steps to facilitate the formalization of the ASGM sector (box 1-1). The parties that develop national action plans must submit them to the Secretariat, and every three years they also need to provide the Secretariat with progress reviews toward meeting their obligations under Article 7. In chapter 7, we discuss the ASGM and mercury issue, including different strategies for reducing mercury use, discharges, and exposure.

Provisions Addressing Institutions and Knowledge

The Minamata Convention identifies the Conference of the Parties (COP) as the treaty's supreme decision-making body (Article 23) and stipulates that the Secretariat, hosted by UNEP Chemicals in Geneva, provide administrative services to all treaty bodies and parties (Article 24). The treaty also establishes a committee to promote the implementation of, and review compliance with, articles and provisions (Article 15). The COP can create additional bodies, and has set up a governing board for the treaty-specific funding mechanism—the Specific International Programme—that provides financial support together with the Global Environment Facility (GEF). The Minamata Convention explicitly identifies the importance of the World Health Organization (WHO) and the International Labour Organization (ILO) in treaty implementation. Other international organizations that support implementation include UNEP, the United Nations Development Programme (UNDP), and the United Nations Institute for Training and Research (UNITAR). Continuing work under the Global Mercury Partnership also involves national governments and many different non-state organizations.

The Minamata Convention contains language on support geared to helping developing countries meet their treaty-based obligations in Articles 13 and 14. Article 13 defines the treaty-based mechanism for providing financial resources to developing countries. Pursuant to Article 14, parties shall cooperate to provide capacity-building, technical assistance, and technology transfer to developing countries. These treaty-based mandates are practically and politically important to developing-country parties, which may struggle with a lack of domestic resources or see the fulfillment of these obligations by wealthier countries as an important indication of their commitment to effective and equal treaty implementation. In addition, during negotiations, developing countries considered the inclusion of strong treaty language on financial support, capacity building, and technical assistance as necessary for agreeing to the establishment of the implementation and compliance committee (H. Selin 2014).

Several Minamata Convention articles relate to knowledge provision. Parties are required to facilitate the exchange of scientific, technical, economic, and legal information on a number of topics (Article 17). These topics include the reduction or elimination of the production, use, trade, emissions, and releases of mercury; on technically and economically viable alternatives for mercury use in products and processes as well as control measures; and on mercury-related health impacts. Parties are also required to promote and facilitate public information, awareness, and education about mercury (Article 18). They must endeavor to cooperate to develop and improve: inventories of use, consumption, emissions, and releases of mercury; scientific information on the environmental behavior and human impacts of mercury; information on commerce and trade in mercury and mercury-added products; and research on control techniques (Article 19). Parties may develop an implementation plan for the entire Minamata Convention if they deem that helpful for domestic purposes, but they must report on national measures and their effectiveness (Articles 20–21). The COP is tasked with carrying out periodic evaluations of the effectiveness of the Minamata Convention, with the first one beginning no later than six years after the treaty entered into force—which is 2023 at the latest (Article 22).

The provisions of the Minamata Convention affect broader efforts to promote sustainability, including those under the Sustainable Development Goals (SDGs) (UNDP 2016). Goal 3 (Good Health and Well-Being) focuses on protecting human health, where mercury is an important element of a suite of pollution problems. Goal 14 (Life below Water) targets marine pollution, as methylmercury threatens the well-being of both marine animals and humans who consume seafood or who are exposed in utero. Food consumption also connects to Goal 2 (Zero Hunger), as many less affluent people rely on fish for much of their food and nutrition intake. Goal 8 (Decent Work and Economic Growth) draws attention to safety in the workplace, where workers who come in contact with mercury are at risk. Goal 1 (No Poverty) is relevant to ASGM miners and others who are driven into mercury-using sectors by poverty. Goal 12 (Responsible Consumption and Production) puts a focus on reducing the negative environmental and human health impacts of the use of hazardous substances, including mercury. Goal 7 (Affordable and Clean Energy) highlights the challenge of coal burning, which leads to mercury emissions, as a source of energy and driver of climate change.

Insights

The South American colonial mining story that began this chapter illustrates how a systems view of global mercury cycling comprises not only the quantities and fluxes incorporated in assessments of mercury's biogeochemical cycle, but also societal flows, institutions, and knowledge. In this section, we examine insights from our analysis of the global mercury cycling system and illustrate some of the difficulties in chasing quicksilver in the environment and society. First, understanding important system dynamics requires considering a broad range of material and non-material connections. Second, it is difficult to define criteria for sustainability with respect to the global human-technical-environmental cycling of mercury, but incremental change is occurring. Third, the implementation and effectiveness of the Minamata Convention are shaped by the global human-technical-environmental cycling of mercury in the context of other efforts to govern sustainability.

Systems Analysis for Sustainability
A systems-level accounting for global-scale mercury cycling needs to consider mercury in the environment and society simultaneously. Focusing only on

the biogeochemical cycle of mercury in the environment misses important societal and technical processes. Accounting for mercury's cycling in society alone, in contrast, omits the extensive cycling of mercury through the atmosphere, land, and oceans. Some studies suggest that the global environmental cycling of mercury from human activities goes back at least five centuries, but most mercury from anthropogenic sources has been discharged during the industrial era. The trade of mercury across societies accelerated with the start of the Spanish colonial mining in Latin America in the 1500s, and industrialization further increased its societal flows. Commercial mercury may be reused multiple times, and excess mercury can be stored before it is traded, used, and discharged into the environment. Discharged mercury connects near and distant ecosystems as well as human health harms through its environmental cycling, but data on the quantities and locations of mercury in society and in the environment are incomplete. The ability to better analyze global mercury cycling would be aided by further and more comparable data on mercury in technical and environmental components.

The global mercury cycling system changed dramatically over time as a result of human activities. Concentrations of mercury in the atmosphere, land, and oceans are now greatly elevated, human-mobilized mercury has reached ecosystems in all regions of the Earth, and some of this mercury has been converted into methylmercury. This increase is consistent with trends in other human-induced environmental pressures, including global population and carbon dioxide emissions, as well as economic indicators such as world gross domestic product (Steffen et al. 2007). Mercury will continue to cycle in the environment over long timescales, but mercury methylation and bioaccumulation processes can change more rapidly along with ecosystem changes. Interventions targeting societal flows of mercury can also have relatively rapid impacts. Yet, people and wildlife are unable to adapt biologically to make methylmercury concentrations less damaging to their bodies. Human forces affecting the supply and demand for mercury have historically shaped its trade patterns as well as its price. The Minamata Convention sets out to further reduce the supply, trade, and use of mercury over the next few decades. It is too soon to say if this will happen, but the rapid decline in mercury use in some regions over the past 30 years shows that change is possible.

Evolving knowledge about the extent to which elemental mercury and mercury compounds cycle in the environment has changed perceptions of the scale of the mercury problem over time. Mercury was thought of as mainly a local, national, and regional pollution problem until the late 1900s based on the common understanding at the time of how mercury was dispersed in the environment. Global concern about mercury emerged with the realization that mercury can travel long distances in the atmosphere. Mercury remains both a local and a global pollution challenge, and its local and global dynamics are linked. Mercury that deposits locally may re-volatilize and begin to cycle in the environment later, posing future risks in other places. Local land use changes and global climate change may alter the form and availability of mercury compounds that cycle in the environment, further complicating systems-focused efforts to analyze the scale and scope of the mercury problem. Changing knowledge about the extent to which mercury travels globally in the environment has also influenced views of the appropriate scale of institutions to address mercury use and pollution. A comprehensive view of the mercury problem should account for its local and long-range aspects simultaneously.

Sustainability Definitions and Transitions
Defining whether the global mercury cycling system is becoming more sustainable is complicated because of the long timescales of mercury cycling in the atmosphere, land, and oceans. Definitions of sustainability focusing on hazardous substances often stress the importance of eliminating uses and discharges, but much mercury from anthropogenic (as well as natural) sources already exists in the environment. Societal measures to reduce discharges from human activities will lower the overall amount of mercury that will cycle in the environment in the future. Other human activities, however, increase the global environmental cycling of mercury. Human-induced trends of land-use changes and climate change contribute to the remobilization from environmental storage of mercury previously discharged from both human and natural activities. This means that merely lowering the levels of mercury uses and discharges will not solve the problem of mercury damaging the environment and human health. Assessing whether the global-scale human-technical-environmental system for mercury is moving toward sustainability will thus require attention to societal as well as environmental trends.

Societal efforts addressing the supply of mercury have gradually become more stringent over time with the closure of large primary mercury mines and tighter controls on the reuse of mercury, especially from industrial sources. Supply of new mined elemental mercury was seen as critical to meeting societal demand well into the second half of the twentieth century. Commercial mercury use only began to decline in the early 1970s, partially as a result of environmental and human health concerns. The adoption of elemental mercury export bans by the United States and the EU in the 2000s further affected the international supply and price of elemental mercury. The Minamata Convention provisions that phase out primary mercury mining, ban the reuse of excess mercury from chlor-alkali production, and permit exports of elemental mercury only for those few uses that are still allowed under the treaty are important global-scale interventions in an ongoing incremental transition process. The closing of mercury mines and expanded introduction of other supply restrictions, including the recent actions by the US and the EU to ban the export of mercury compounds, will reduce the environmental distribution and cycling of mercury and also have positive effects on human exposure and well-being.

Global-scale transitions involving mercury will occur over various timescales moving forward. Environmental concentrations of mercury will not return to preindustrial levels for centuries and perhaps millennia, but changes involving use and discharges will advance over the coming decades. The Minamata Convention's control provisions on mercury use, emissions, and releases vary with respect to their deadlines and stringency. The bans and phase-downs of mercury use in many mercury-added products were relatively rapid, with an initial target date of 2020. Bans on uses of mercury in some industrial production processes are in effect as of 2018 and 2025, but mercury use can continue in other processes. Parties to the convention also have the ability to request extensions from the target dates for products and processes, and the controls on emissions and releases from point sources do not rely on universal numerical reduction targets. Parties must apply technology-based standards, but they are given much freedom in setting their own standards. This will, at least in the short-term, result in much national variation in regulatory stringency and mercury discharges. As a result, trends in environmental discharges may differ across countries and regions. Some environmental compartments will also respond more rapidly than others to varying changes in discharges.

Sustainability Governance

The global scope of the Minamata Convention, together with its lifecycle approach, is designed to fit both the socio-economic and environmental dimensions of the mercury issue. The treaty was adopted based on the scientific and political recognition that mercury transports globally, both in the environment and society. The trade of mercury across continents has been documented for centuries, but only after the release of the first global mercury assessment in 2002 did a global political forum officially recognize that mercury also travels worldwide in the atmosphere. The stated goal of the Minamata Convention to protect human health and the environment from anthropogenic emissions and releases of mercury and mercury compounds puts a clear focus on human well-being in the context of sustainability. To this end, the Minamata Convention sets out a global legal framework explicitly designed to address all aspects of the lifecycle of the mercury issue, including production, trade, use, emissions, releases, handling, stockpiles, and disposal.

Effectiveness evaluation is an important part of implementing the Minamata Convention, as mandated in Article 22, and the ability to conduct such evaluations is shaped by the fact that material components in the global human-technical-environmental system for mercury change over vastly different timescales. The long timescales of global environmental cycling of mercury, and the growing importance of remobilizing legacy emissions from environmental compartments, make it difficult to causally link changes in environmental concentrations with measures taken under the Minamata Convention. An increase in legacy emissions may result in additional mercury entering into global environmental cycling, offsetting some of the anthropogenic mercury emissions and releases that are reduced by measures related to the implementation of the Minamata Convention. A full analysis of the effectiveness of the Minamata Convention should also consider political and technological factors, such as the degree to which parties phase out mercury supply and use, mandate the application of pollution control technology on point sources, and fulfill provisions on financing, capacity building, and technology transfer.

The ability to protect the environment and human health from mercury is increasingly affected by the governance of other sustainability issues. Policies that address coal and other fossil fuel uses in the context of other forms of air pollution and climate change affect levels of mercury emissions from these sources. Material connections with other sustainability issues also influence the global environmental cycling of mercury. Human alterations to the carbon cycle through the burning of fossil fuel and deforestation change the ways in which mercury travels in the atmosphere, land, and oceans as well as the processes it undergoes in the environment. These include processes of methylation and bioaccumulation in aquatic environments, which have important implications for animals and seafood consumers worldwide. The Minamata Convention's links to several SDGs show that mercury abatement is connected with human-centered efforts on addressing poverty, hunger, food safety, good health, work, consumption and production, and energy. In this respect, implementing the Minamata Convention is an important aspect of broader global governance efforts that aim to move societies toward greater sustainability.

The global cycling of mercury makes it necessary to chase quicksilver through both the environment and society. Mercury is discharged into the environment from natural events as well as by human activities. People have dramatically altered the global biogeochemical cycling of mercury in the environment over millennia, but especially during the last few hundred years. The international trade in mercury and mercury-containing goods increased during that same time period, resulting in large amounts of mercury cycling through society. Much of this mercury eventually ended up in the environment. Human-induced land-use and climate changes increasingly contribute to the re-release of historical emissions from both natural and anthropogenic sources, making it more analytically difficult, and less useful for informing sustainability transitions, to separate between natural and anthropogenic sources of mercury cycling in the environment. The Minamata Convention takes a lifecycle approach to mercury, addressing both societal cycling and environmental discharges. The legacy of past mercury discharges, and the impacts of today's discharges, will affect ecosystems and people for generations, as it can take centuries, if not millennia, for discharged mercury to return to geological storage.

4 Human Health: Mercury's Caduceus

Industrialization increased people's exposure to a wide variety of hazardous substances, including mercury. Some adverse human health effects of this exposure, especially those that occur at high doses, are readily apparent to observers and have been known to science and medicine for centuries. Other effects can be more difficult to discern, especially at low levels of exposure, in part because health outcomes can be separated in time and place from the use and release of hazardous substances. Knowledge of the toxicity of specific substances and their human health impacts is distributed unevenly across the world and at different times. Societal risk perceptions and responses also differ: private sector actors and regulatory agencies assess costs and benefits and respond in varying ways to human health dangers. Hazardous substances such as mercury continue to pose a wide range of health threats, involving all parts of the planet that are inhabited by humans. Ensuring good health is a primary goal of sustainability, and addressing and managing health risks from the products of modern society will continue to be a central challenge for ensuring future human well-being.

Dr. Hajime Hosokawa and municipal health official Hasuo Itō classified the first official case of Minamata disease on May 1, 1956, without knowing its cause (George 2001). Local fishers had complained for several years that wastewater dumped into Minamata Bay from a local factory was killing fish, and they observed how cats that had eaten fish would dance crazily, jump into the sea, or die (Harada 1995). The fishers and their family members suffered from numbness, visual and hearing impairments, and loss of motor control, among other symptoms, and some had died (Takeuchi et al. 1959). The local committee to address the unknown problem, known as the Strange Disease Countermeasures Committee, designated the first 52 certified victims on December 1, 1956. By that date, 17 victims had died.

Leaders of the local factory denied all responsibility for causing the illnesses, and held back information on their use and releases of mercury; the factory, owned by the Chisso Corporation, had used mercury as a catalyst to produce acetaldehyde since 1932 and to manufacture vinyl chloride monomer (VCM) since 1941. As local tensions grew, the Minamata-born poet Gan Taginawa suggested a parallel between Dr. Hosokawa and the small-town doctor in Henrik Ibsen's play *An Enemy of the People*, who was attacked for pointing out polluted water in the town spa, an enterprise critical to the tourist industry (George 2001).

City, prefectural, and national officials as well as many local residents who depended economically on the factory defended Chisso. One of those defenders was Hikoshichi Hashimoto, the mayor of Minamata, who had himself developed the mercury-based acetaldehyde production technique when he previously worked at the factory. The Japanese government did not officially acknowledge that pollution from the factory had caused Minamata disease until September 1968, a few months after Chisso had stopped using mercury in acetaldehyde production. The company continued to use mercury in VCM production until 1971. In 1969, 112 people representing 29 families filed a lawsuit against Chisso in Kumamoto District Court. Four years later, the court found Chisso responsible and established that all certified patients had a right to compensation. But many people who fell ill struggled to get certified. The legal process continued until 1988, when the Japanese Supreme Court found a former Chisso president and factory manager guilty of negligent homicide and gave them suspended two-year prison sentences. After an extensive remediation that involved removing contaminated sediments, the governor of the Kumamoto prefecture declared the fish in Minamata Bay safe for consumption in 1997, more than 40 years after the formal identification of the first Minamata disease patient.

The story of methylmercury poisoning in Minamata, where hundreds of people died, illustrates risks from the millennia-long relationship between mercury and human health. That relationship has manifested in two main ways. First, people have used mercury in medicine, with mixed results, in an attempt to improve well-being and longevity. Second, occupational and dietary exposure to mercury has caused severe health damages. In the title of this chapter, we echo the (often ill-advised) use of mercury in many medical therapies by referring to the irony in the mistaken use of the caduceus of Hermes (the Greek equivalent to Mercury) in place of the similar-looking rod of Asclepius as a symbol of the medical profession. Asclepius is the god of medicine, while Hermes/Mercury is the god of commerce and thieves. Asclepius's rod features a serpent entwined around it; the caduceus features two serpents around a winged staff. Hermes's caduceus was adopted in 1902 as the insignia for US Army medical personnel, for example. The mix-up between the two symbols goes beyond visual similarity, however, to suggest the complicated symbolism of the serpent throughout history, as a creature possessed of destructive and healing powers alike (Prakash and Johnny 2015).

Mercury harmed people in both preindustrial and industrial societies, and continues to adversely affect people throughout the world today. Mercury's health impacts vary with both the forms and the amounts of mercury people are exposed to. The challenges of mobilizing actions to address mercury's risks and impacts have changed with time. People adversely affected by mercury pollution have clashed with powerful economic and political interests when looking for support and compensation. In the case of Minamata, it took decades before the Chisso Corporation and relevant authorities took meaningful action to stop discharges of methylmercury into Minamata Bay. Negative health problems from mercury continue to affect populations living far away from any emission source—especially pregnant women and small children—who are nevertheless at risk from long-range transport of emissions and exposure because of the food they eat. Discussions about the burdens of human exposure and the strategies and costs of mitigation, from local to global scales, continue today in the context of the implementation of the Minamata Convention, which also recognizes future generations as a particularly vulnerable group of people.

In this chapter, we examine how human health is affected by mercury and mercury compounds, and we consider measures taken to reduce negative health impacts from mercury. First, in the section on system components, we outline how human, technical, and environmental components influence how people use and are exposed to mercury, and how institutions and knowledge are linked to mercury as a human health issue. In the section on interactions, we detail pathways by which the societal use of mercury and its presence in the environment harms human health through occupational, medical, and dietary exposure. Our discussion of interventions focuses on efforts to address mercury-related health problems, including those that have been taken to address high-dose occupational exposure in particular types of workplaces, to phase out mercury use in medicine, and to mitigate harms from low-dose dietary exposure from seafood. In the final section on insights, we look at how mercury's inherent properties and system interactions influence mercury-related human health problems, how changing perceptions of the way mercury affects people's health have led to largely incremental efforts to mitigate these impacts, and the need for multi-scale governance strategies to effectively address mercury's human health dangers.

System Components

Mercury exposure has posed a threat to human health for millennia. It remains a current problem, and it will continue to be problematic for many generations to come. The forms of elemental mercury and mercury compounds that we discussed in chapter 3 have different levels of toxicity, with important implications for human exposure and health. People are exposed to varying levels of different forms of mercury. Exposure to both inorganic and organic mercury can be short-term or long-term and thus manifest as acute or chronic poisoning. The phrase "the dose makes the poison" applies to all forms of human exposure to mercury. That famous phrase itself is attributed to the Swiss physician Paracelsus (1493–1541), and has its roots in his laboratory work with mercury (Grandjean 2016). Figure 4.1 shows the main human, technical, environmental, institutional, and knowledge components for the mercury and health system.

Human components	Technical components	Environmental components
Workers in mercury-related sectors Employers in mercury-related sectors Medical professionals Medical patients Consumers of food Producers of commercial market food Non-commercial harvesters	Mercury in production processes Mercury-containing medicines and medical treatments	Ecosystems near mercury sources Ecosystems far from mercury sources Fish, shellfish, and marine mammals Rice

Institutional components	Knowledge components
National and local laws and regulations Dietary recommendations Cultural norms Minamata Convention	Forms of mercury Properties of mercury Exposure routes Mercury concentrations in people Health impacts from mercury exposure Health protection techniques

Figure 4.1
Components in the mercury and health system (referenced in the text in italic type).

People have come in direct contact with *mercury in production processes* as *workers in mercury-related sectors*. Most of this exposure, going back thousands of years, was to inorganic mercury and its compounds, including elemental mercury. Knowledge of how different *forms of mercury* have affected human health has developed over time; health effects resulting from exposure have long been observed in miners and other workers. For example, hat-makers who started to use a mercury-based felting technique in the seventeenth century suffered from repeated and dangerous exposure to mercury. It is often thought that the character known as the Mad Hatter in the nineteenth century books of Lewis Carroll behaved strangely because of mercury poisoning. Others believe instead that Carroll modeled the character on an eccentric furniture dealer and amateur inventor named Theophilus Carter, but the saying "mad as a hatter" nevertheless remains a common colloquial expression in English (Waldron 1983). Additional workers who faced particular risks from mercury use included gilders, potters, tinsmiths, painters, glass workers, mirror makers, and scientists (Goldwater 1972). Industrialization increased the number of workplaces in which workers were exposed to mercury, where *employers in mercury-related sectors* largely determined workplace conditions and thus potential mercury exposure.

Medical professionals have prescribed and applied *mercury-containing medicines and medical treatments* to *medical patients* over a long history. Shen Nong, the mythological father of Chinese medicine who was credited with writing the 40-volume Great Herbal in the twenty-eighth century BCE, included mercury among the many drugs he listed (Hayes and Kruger 2014). Early writings indicate that mercury was also used for medical purposes in ancient India, Greece, and Rome, and later in the Arab world and in medieval Europe (Goldwater 1972). The Greek physician Hippocrates (460–370 BCE), known for the Hippocratic oath that still guides the actions of doctors, was likely to have prescribed mercury compounds as ointments. The use of mercury-containing salves for treating skin lesions continued into the second half of the twentieth century (Goldwater 1972). Calomel (mercurous chloride) and other mercury compounds were used to treat a wide range of illnesses, including syphilis beginning in the fifteenth century. Calomel was also used in laxatives. Using cinnabar to tattoo the surrounding area was a treatment against pruritus ani (intense anal itching) by medical doctors well into the twentieth century (Granet 1940).

Mercury and mercury compounds were also used in other medical areas. Mercuric chloride was applied as an antiseptic to disinfect wounds, including during World War I and World War II (Hylander and Meili 2005). Phenylmercury compounds were used as disinfectants based on their ability to prevent bacterial growth. They were also used as vaginal spermicides beginning in the 1930s (Baker et al. 1939; Löwy 2011). Dental use of mercury likely began in China in the first century CE, and physicians in China and Europe wrote about its use in amalgam in the sixteenth century (Bharti et al. 2010). Mercury-based dental amalgams are still used to restore teeth affected by dental caries. Ethylmercury thiosalicylate (thimerosal, or thiomersal outside the United States) was introduced into vaccines in the 1930s as a preservative to prevent growth of bacterial and fungal contaminants (Baker 2008). Thimerosal is still used when vaccines against diseases including diphtheria, hepatitis B, and influenza are distributed in multiple-use vials where repeated syringe insertions can cause contamination (Clarkson and Magos 2006; World Health Organization 2011b). Thimerosal and other mercury compounds have also been used as preservatives in very small amounts in pharmaceuticals such as ear and eye drops, contact lens solution, nasal spray, and hemorrhoid relief ointment (US Environmental Protection Agency 2020).

People around the world are exposed to methylmercury as *consumers of food*, predominantly seafood, whether harvested from *ecosystems near mercury sources* or *ecosystems far from mercury sources*. The mercury in *fish, shellfish, and marine mammals* sometimes comes from local sources, as was the case in Minamata, but it can also come from long-range transport. The fish that people consume in modern societies originate from many different locations, as *producers of commercial market food* may sell their fish and shellfish worldwide. People are also exposed to mercury by eating seafood caught by *non-commercial harvesters*. For example, consumers of community-harvested marine mammals and whales in places such as the Faroe Islands and the Arctic are exposed to particularly high doses of methylmercury, as are subsistence fishers who catch their food in waters near point sources and contaminated sites. Methylmercury has also been measured in *rice*, where rice plants that are cultivated in submerged or flooded paddies have the ability to take up methylmercury (Meng et al. 2014; Kwon et al. 2018). This can be the dominant source of dietary mercury for some consumers of large quantities of rice where mercury concentrations are elevated.

Understanding the human health impacts of mercury depends on knowledge of the *properties of mercury*, different human *exposure routes* of mercury, and *mercury concentrations in people*. In living people, measuring mercury levels requires taking samples of bodily fluids or tissues. Urine is often used to identify elemental mercury exposure, whereas hair concentrations are a preferred indicator of methylmercury exposure (Clarkson et al. 2007). Both elemental and methylmercury have a half-life in the human body of a few months (Clarkson et al. 2003). Elemental mercury is most dangerous when it is inhaled in its gaseous form: it largely passes through the body when ingested in its liquid form (Ha et al. 2017). In contrast, methylmercury is almost completely absorbed into the bloodstream via the gastrointestinal tract (Mergler et al. 2007). Methylmercury travels throughout the body, including to the brain and to a developing fetus (Clarkson et al. 2007). Mercury is present in all people, but those who live near artisanal and small-scale gold mining (ASGM) activities and indigenous peoples in the Arctic have relatively high methylmercury concentrations, and elevated levels have been measured in seafood consumers in the Asia/Pacific region and the Mediterranean (Sheehan et al. 2014).

Knowledge of the *health impacts from mercury exposure* has developed through time. The term "erethism" was used until the early nineteenth century to describe the impacts of inorganic mercury exposure (Clarkson and Magos 2006). Its principal features were excessive timidity, increasing shyness, a desire to be unobtrusive, and anxiety. H. A. Waldron (1983, 1961) notes that victims of mercury poisoning "had a pathological fear of ridicule and often reacted with an explosive loss of temper when criticized." High exposure, for example at levels experienced by miners, led to fatalities. Symptoms of organic mercury poisoning are different. Starting with numbness of the hands and feet, subsequent symptoms include lack of muscle coordination, loss of vision and hearing, and other signs of rapidly increasing damage to the central nervous system (Clarkson and Magos 2006). High dose exposure to methylmercury can be lethal. Methylmercury at low doses can cause neurological effects on small children and developing fetuses. Other kinds of health effects, for adults exposed to low doses, include cardiovascular problems, impacts on endocrine function, risks of diabetes, and immunological impacts (Eagles-Smith et al. 2018). Knowledge of *health protection techniques*, including occupational safety measures such as increased ventilation and wearing gloves and masks, can help limit exposure to some forms of mercury.

Institutions such as *national and local laws and regulations* include provisions that address mercury use in workplaces and in medicine, and also set labor and food safety standards related to mercury exposure. Governments and non-governmental organizations have issued *dietary recommendations* in the form of advice on the consumption of specific species of fish and types of seafood. Recommendations like these are designed to reduce human exposure and mitigate mercury risks, especially among more vulnerable populations such as pregnant women, small children, and people who regularly consume food containing relatively high levels of methylmercury. Dietary exposure to methylmercury can be influenced by long-standing *cultural norms*, especially for people in indigenous and other communities who hunt and consume seafood and marine mammals as part of their traditions and traditional diets. The *Minamata Convention* acknowledges the important link between mercury and human health, and contains several health-related provisions that are intended to guide its implementation in countries.

Interactions

Mercury use and exposure affects human health in several different ways. Figure 4.2 shows interactions in the mercury and health system. We have selected three interactions in that matrix (the items in bold type in boxes 2-1 and 3-1) through which mercury affects people's health; we then trace the pathways that influence these interactions, which we summarize in figure 4.3 (where the bold boxes correspond to the selected interactions). First, mercury in mining and production processes affects the health of workers (box 2-1), and this is determined by their employment in mercury-related sectors and use of mercury (boxes 1-1 and 1-2). Second, mercury-containing medical treatments affect patients (box 2-1), and this is influenced by the actions of medical professionals (box 1-2). Third, methylmercury-contaminated food affects consumers (box 3-1), as mercury is transported and transformed through ecosystems where human health impacts also depend on biological and socio-economic factors (boxes 2-3, 3-3, 1-3, 1-1, 3-1).

	1. Human	2. Technical	3. Environmental
1. Human	(1-1) Workers are employed in mercury-related sectors; Health impacts are shaped by biological and socio-economic factors	(1-2) Workers use mercury in mining and production processes; Medical professionals prescribe and patients use mercury-containing medical treatments	(1-3) Non-commercial harvesters and commercial food market producers harvest mercury-containing food from ecosystems
2. Technical	**(2-1) Mercury in mining and production processes affects health of workers; Mercury-containing medical treatments affect patients**	(2-2)	(2-3) Industrial uses of mercury contaminate ecosystems near sources
3. Environmental	**(3-1) Seafood and rice lead to methylmercury-related damages to consumers;** Seafood and rice provide health benefits to consumers	(3-2)	(3-3) Ecosystem processes transport mercury and lead to methylmercury production

Above the table: **Knowledge Institutions**

Figure 4.2
Interaction matrix for the mercury and health system.

Occupational Exposure and Health Impacts

Mercury in mining and production processes has affected the health of workers for millennia (box 2-1). As we mentioned in chapter 3, cinnabar may have been mined in Turkey at least as early as 6300 BCE, making miners some of the earliest workers employed in mercury-related sectors (box 1-1). Mercury mining in the Wanshan district in southwestern China goes back two millennia. The mine in Almadén, similarly, is believed to have been in operation for 2,000 years—the first known recorded reference to mining in the area dates back to the fifth century BCE (Coulson 2012)—and may have produced one-third of all mined mercury in the world. Mercury mining in Idrija began in the late fifteenth century. The various operators of the Almadén and Idrija mines used slaves, convicts, and salaried workers at different times to mine for cinnabar under extremely harsh conditions.

a) Occupational exposure and health impacts: Workers in different sectors come into contact with mercury, and different forms of mercury affect their health

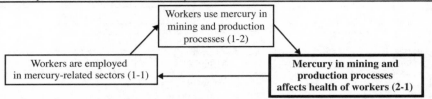

b) Medical use and health impacts: Medical professionals prescribe and patients use mercury-containing medical treatments, which affect patients

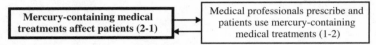

c) Dietary exposure to methylmercury: Mercury-contaminated ecosystems provide food for consumers, affecting their health

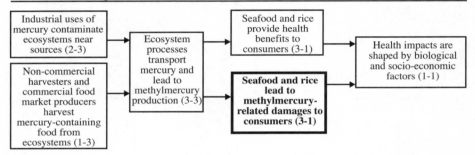

Figure 4.3
Pathways of interactions for the mercury and health system. Bold box indicates focal interaction for each subsection.

One reason for the hazardous working conditions was that the mercury in the mines was not only embedded in cinnabar but also present in the ores as droplets of liquid elemental mercury large enough to be visible to the naked eye. This liquid form of elemental mercury was quick to volatilize into its much more dangerous gaseous form under the very warm conditions in mine shafts, with mercury vapor quickly reaching dangerous levels in the narrow and crowded shafts deep underground (Hamilton 1943).

Much human suffering also took place in the mercury mine in Huancavelica, Peru. Colonial Spanish rulers appropriated this mine in 1563 from the Inca, who had used it as a source of vermilion (Robins 2011). The Spanish introduced a mandatory system called the *mita* to conscript Indian labor into the mercury mine. This devastated indigenous populations through a combination of high mortality rates in mining and the flight of laborers to avoid conscription (Robins 2011). Potential miners considered a two-month stint at Huancavelica a death sentence: mercury exposure combined with accidents, the inhalation of silica dust, and carbon monoxide poisoning gave Huancavelica a reputation as the *mina de la muerte* (the mine of death) (Brown 2001, 468). Mercury mining expanded in the United States starting in the 1800s, including in the mines in New Almaden and New Idrija in California, feeding a growing demand from the gold rush. Mercury from large mercury mines further exposed many additional workers in the mining sector when it was subsequently used in silver and gold mining (box 1-2). We further discuss mercury use in contemporary ASGM in chapter 7.

Adverse human health impacts of mercury exposure were also seen in workers in other sectors, going back at least five hundred years, as workers used mercury in production processes (box 1-2). The doctor Ulrich Ellenbog expressed one of the earliest known concerns in a 1473 pamphlet, writing about the risk to goldsmiths of inhaling mercury vapor in his city of Augsburg, Germany (Goldwater 1972). In the German city of Fürth in 1885, 72 percent of the 7,500-plus sick days taken by mirror makers were reportedly the result of mercury poisoning (Teleky and Kober 1916). Effects of mercury poisoning were also frequently observed in hat makers who used a mercury-based fur-felting technique that was developed in France in the 1700s. This revolutionary technique for making fur hats subsequently spread to other European countries and continents. In the United States, Danbury, Connecticut, became known as "the Hat City of the World" in the mid-1800s (Wajda 2019). The tremors that workers in the many hat factories there experienced from mercury exposure gave rise to the expression "the Danbury Shakes" (Hightower 2009).

Worker exposure to mercury in several manufacturing industries continued, and in some cases increased, during the twentieth century. These industries included hat making as well as the production of mercury lamps and several kinds of manufacturing processes, which we discuss further in chapter 6. Incidents occurred in other workplaces as well. Several people at a seed packing facility in Norwich, England, in the 1930s were poisoned by methylmercury that was used as a disinfectant and preservative to protect seeds against mold (Hunter et al. 1940). Furthermore, members of the Lancashire Constabulary in England who specialized in taking and developing fingerprints using a mixture of mercury and chalk powder were found in 1949 to suffer from the effects of mercury exposure. Their symptoms included tremors in their hands, lips, tongue, and eyelids as well as loose teeth and "irritability and embarrassment which caused the men to blush easily" (Anonymous 1949, 231).

Health problems from industrial mercury use continue in more recent times. In one high-profile case, workers were exposed to mercury in a thermometer factory that was built in 1983 in Kodaikanal in the Indian state of Tamil Nadu. This factory was built with equipment from a decommissioned US plant, and was initially operated by the US company Chesebrough-Pond's. Former workers from the factory show symptoms of mercury vapor poisoning, and this exposure has been linked to premature deaths (Dev 2015). The Tamil Nadu Pollution Board forced manufacturing to shut down in 2001 after public protests. Advocates fought for compensation for exposed workers from the Indian subsidiary of the British-Dutch conglomerate Unilever, which had acquired the factory from Chesebrough-Pond's in the 1990s. An out-of-court settlement was eventually reached in 2016, providing an undisclosed amount of compensation to 591 former workers (Agnihotri 2016).

Alchemists and scientists are another category of workers extensively exposed to mercury. Paracelsus, Jan Baptist van Helmont (1580–1644), Blaise Pascal (1623–1662), and Michael Faraday (1791–1867) are all thought to have suffered from chronic mercury poisoning without realizing what caused their condition (Giese 1940; Goldwater 1972). The difficulty in identifying mercury as the source of illness is influenced by the fact that mercury vapor is colorless and odorless, and symptoms can be caused by many different factors. Another person whose health has been alleged to have been affected by mercury is Isaac Newton (1643–1727), an avid alchemist who used mercury extensively in his experiments. Some scientists have speculated that the psychological illnesses Newton suffered in 1693 were a result of mercury poisoning (Broad 1981). Analysis of Newton's hair centuries after his death revealed levels of mercury high enough to suggest that Newton suffered from chronic mercury poisoning (Johnson and Wolbarsht 1979). Other scientists, however, have questioned not only these measurements but also whether Newton's symptoms were consistent with mercury poisoning or stemmed from other causes (Ditchburn 1980).

The documentation of health effects in scientists from mercury use and exposure in laboratories increased in the twentieth century. In 1926, the

German chemist Alfred Stock wrote a vivid firsthand account of "insidious mercury vapor poisoning" that he and his colleagues suffered from for decades without knowing what caused their illness. Symptoms included memory loss; headaches; nervous restlessness; strong vertigo and visual disturbances; nose, throat, and sinus infections; loosening of teeth; sudden bladder pressure; and isolated bouts of diarrhea (Stock 1926). Scientists nevertheless continued to work with mercury. Clark Goodman at the Massachusetts Institute of Technology noted in 1938: "Mercury is such a common laboratory companion that little consideration is usually given to the possibility of the toxic effects which may result from handling it or from breathing the vapor which is inevitably emanating from the liquid metal" (Goodman 1938, 233). Gradually, scientists became more aware of the health risks. Arthur Giese at Princeton University warned in *Science* in 1940 about the dangers of mercury vapor in air in closed laboratories (Giese 1940).

Much of the human health concern expressed about the laboratory use of and exposure to mercury involved its metallic and inorganic forms, but scientists were also among the first workers who were exposed to organic mercury. When methylmercury compounds were first synthesized in London in the 1860s, two laboratory technicians died from poisoning (Clarkson 2002). In a more contemporary case with a lethal outcome, Karen Wetterhahn, an organometallic chemist and professor at Dartmouth College, died in 1987 from exposure to one or two small drops (less than half a milliliter) of dimethylmercury (Science News Staff 1997). This highly dangerous form of organic mercury penetrated not only the latex gloves she wore in the laboratory but also her skin. Her illness became apparent five months after the accident as it affected her balance, speech, vision, and hearing. Her condition worsened; she fell into a coma and then passed away 10 months after the minuscule spill (Clarkson and Magos 2006).

Medical Use and Health Impacts

Over long periods of time, mercury-containing medical treatments have affected patients (box 2-1). The intentional use of mercury in medicine began more than 2,000 years ago, and persisted through time. Exposure occurs when medical professionals prescribe, and patients use, mercury-containing medical treatments for various ailments (box 1-2). Chinese emperor Qin Shihuang, looking to discover an elixir of immortality, ingested a large quantity of cinnabar (Dubs 1947). This may have contributed to his death at age 49 in 210 BCE (Wright 2001). US president John Adams complained about mercury treatment for smallpox causing his teeth to loosen (US National Archives 2019); some historians have hypothesized that his contemporary, George Washington, may also have lost his teeth in part because of calomel use for smallpox or other ailments (George Washington Foundation 2017). Members of the Lewis and Clark expedition that started in 1804 from St. Louis carried 50 dozen laxative pills containing calomel made by Benjamin Rush, a renowned colonial physician and a signer of the US Declaration of Independence. The effectiveness of these pills earned them the nicknames "Thunderbolts" and "Thunderclappers" (Woodger and Toropov 2004). Two hundred years later, their use helped researchers trace part of the route of the expedition, by identifying latrine pits with elevated mercury levels (Fessenden 2015).

Mercury-containing medicines were extensively used in the past against syphilis. Calomel was administered to syphilis patients beginning in the fifteenth century by rubbing it on the skin, as well as by fumigation or in oral form (Goldwater 1972). Patients were often treated while sitting in a tub, and the children's rhyme "rub-a-dub-dub, three men in a tub" is thought to be a reference to syphilis treatment (Magner and Kim 2017). Another remedy included wearing underpants coated on the inside with mercurial ointment (Parascandola 2009). High-dose applications of mercury for syphilis treatment led to mercury poisoning, but doctors and patients often mistook the symptoms for those of syphilis itself. It was also thought that these highly toxic and agonizingly painful treatment methods for syphilis were "peculiarly appropriate for a disease that was commonly regarded as the wages of sin" (Snowden 2006, 144). Drugs for syphilis were also prescribed for other ailments. Novasurol, for example, containing almost 34 percent mercury, and first introduced by Bayer in the early 1900s to treat syphilis, became widely used as a diuretic starting in the 1920s (Anonymous 1926). It was seen as an effective part of treatment for liver cirrhosis and kidney failure despite reports of fatalities from the 1930s onward. Novasurol may not have caused all of these deaths, however, because the drug was largely given to people with already severe and life-threatening conditions (Goldwater 1972).

Mercury compounds had several other medical uses in the twentieth century. Mercury solutions and ointments were used, for example, by US soldiers beginning in World War I to protect against the spread of venereal diseases such as syphilis and gonorrhea. Soldiers were required to report to army prophylactic stations no later than three hours after sexual contact (Tone 2002). As part of the standardized treatment process, their genitals were washed with a solution of bichloride and mercury, liberally smeared with a calomel ointment, and wrapped in wax paper (Farwell 1999; Shaffer 1920). One treatment for those afflicted with gonorrhea involved injecting a solution containing mercury compounds into the urethra (Young et al. 1919). Starting in the 1930s, phenylmercury compounds were active ingredients in contraceptive suppositories, gels, and foams used by women. The product Volpar, for example, became commercially successful, and contraceptive uses of mercury compounds continued for decades (Löwy 2011). In addition, mercurochrome, the trade name for the phenylmercury compound merbromin, a red-colored disinfectant, was widely applied to prevent infection from minor cuts and scrapes.

In one particularly ill-fated study with human subjects, mercury was administered in an attempt to combat malaria (Snowden 2006). Two Italian researchers—Giacomo Peroni and Onofrio Cirillo—gained Mussolini's personal approval in 1925 to carry out an experiment on workers in Apulia and Tuscany, two areas where malaria was endemic. The effort to eradicate malaria was a central part of the Fascist revolution, and the two researchers were convinced that mercury was a better remedy against malaria than quinine. Against established medical expertise, in a four-year project ending in 1929, a first group of workers received no treatment even though many of them suffered from life-threatening malaria. A second group of workers received mercury through intramuscular injection. In total, 395 people with malaria were administered mercury as an alleged treatment. Peroni and Cirillo claimed that their study was a great success, and wanted to treat the entire Italian army with mercury injections. However, an outside medical expert hired by the Italian authorities forcefully rejected all their findings, and their suggestion of mass vaccination with mercury was never carried out.

The use of mercury in medicine is linked to many additional unintended health problems. This includes the disease acrodynia in children, which was caused by chronic exposure to calomel and other mercury compounds in teething powders, in treatments against intestinal worms, and as diaper disinfectants. Also known as pink disease, acrodynia manifested itself through pink discoloration of hands and feet as well as painfully inflamed nerves and sensitivity to light—"a typical picture is that of a child with head buried in a pillow and continually crying" (Clarkson 2002, 16). Cases of acrodynia were recorded in Europe beginning in the late 1800s, and later also in Australia and North America, with documented fatalities up until at least the 1950s (Dally 1997). It remains unclear exactly why, but only some of the mercury-exposed children developed acrodynia. Symptoms could last for weeks and months, and the disease had an estimated mortality rate of 10 percent in the early twentieth century. It took roughly half a century for researchers to make the connection to calomel, and evidence suggests that at least some adult survivors suffered from chronic chest disease and infertility (Dally 1997; Black 1999).

Acute health effects have also been observed when people have taken mercury-containing medicines without medical prescriptions and recommendations (Goldwater 1972). This included people swallowing mercuric chloride tablets, having mistaken them for other (less harmful) pills. In other cases, people intentionally swallowed mercury-containing tablets for suicidal purposes. One study noted that in 1934, mercury was the fifth most common substance detected in poisoning cases at the Montreal General Hospital, behind lead, morphine, carbon monoxide, and alcohol (Rabinowitch 1934). In addition, pregnant women have taken cinnabar and other mercury compounds in attempts to induce abortions (Goldwater 1972; Severyanov and Anisimova 2013). People have even committed murders using highly toxic mercury salts such as mercuric chloride (Blum 2011). In Germany in 2019, a worker was sentenced to life in prison for using methylmercury and other toxic substances to poison his coworkers' lunch sandwiches. One of his victims was in a coma with serious brain damage for four years before passing away; others suffered lasting kidney damage (Schuetze 2019; Stegemann 2019; Anonymous 2020).

Some mercury-containing medicines and medical treatments, together with the use of mercury-containing medical devices such as thermometers and sphygmomanometers, have had positive impacts on human health (see chapter 6). Mercury use in dental amalgam provided an effective treatment for dental cavities, greatly improving the ability of dentists to repair damaged teeth with many benefits for oral health. The antimicrobial properties of thimerosal and other mercury compounds have made them effective preservatives for different kinds of medical products, including vaccines as well as eye and ear drops. Historically, expanded vaccination has had tremendous health benefits in preventing the spread of serious and sometimes lethal diseases worldwide, and the presence of very small amounts of mercury facilitates the transport to, and use of, vaccines in remote areas. The application of eye and ear drops addressed a wide range of conditions for a large number of patients. All of these largely beneficial uses of mercury continued into the twenty-first century.

Dietary Exposure to Methylmercury

Seafood and rice consumption can lead to methylmercury-related health damages to consumers (box 3-1). As was the case in Minamata, industrial uses of mercury can contaminate ecosystems near sources (box 2-3). People who do not live near a major point source can still be exposed to dangerous levels by consuming seafood containing high levels of methylmercury because ecosystem processes transport mercury and lead to methylmercury production (box 3-3). As a result of such long-range transport, aquatic ecosystems far from point sources can contain much methylmercury. Non-commercial harvesters as well as commercial food market producers harvest mercury-containing food from ecosystems (box 1-3). Whale consumers in places such as the Faroe Islands and indigenous peoples in the Arctic can be particularly highly exposed, because they eat marine mammals that are high on the food chain, and methylmercury biomagnifies and bioaccumulates. People in all regions of the world who purchase and consume seafood from commercial markets are exposed to methylmercury, but in some places like China, exposure from contaminated rice can also be important (Rothenberg et al. 2014).

A recent study from the United States showed that roughly 45 percent of methylmercury exposure to the US population comes from consuming fish from the open ocean, 37 percent from domestic coastal systems, and 18 percent from aquaculture and freshwater fisheries (Sunderland et al. 2018). More than 95 percent of mercury in predatory fish is methylmercury, while the fraction that is methylmercury in shellfish can be lower (Bloom 1992; Sunderland et al. 2018). Methylmercury concentrations are particularly elevated in marine mammals and large fish, such as tuna and swordfish, near the top of food webs. Data from adult American women suggested in the early 2000s that more than 300,000 babies were born annually to women whose mercury levels exceeded the US guidelines intended to protect against adverse neurodevelopmental effects (Mahaffey et al. 2003). More recent studies suggest this number may have declined: although data from

1999 to 2000 showed that 7 percent of women had mercury concentrations above guideline levels, only 2 to 3 percent of women had concentrations that high in subsequent studies up to 2010 (Birch et al. 2014).

Ingesting very high doses of methylmercury through eating contaminated fish leads to symptoms like those seen in the victims of Minamata disease. A Swedish expert group, assessing evidence from Minamata and a similar poisoning case in Niigata, Japan, concluded that clinical symptoms of methylmercury poisoning occurred at blood levels above 200 micrograms mercury per liter, and with concentrations in hair as low as 50 parts per million (Swedish Expert Group 1971). These clinical symptoms include severe neurological symptoms such as numbness and paralysis and permanent damage to the nervous system (Clarkson and Magos 2006). In some other places that were not contaminated by a local point source, people who consumed large amounts of high-mercury fish were reported to have experienced deleterious effects (Korns 1972; Silbernagel et al. 2011). In a well-publicized but unverified case, the actor Jeremy Piven in 2008 stepped down from his role in the Broadway show *Speed-the-Plow* due to fatigue and exhaustion, which he claimed stemmed from high levels of mercury from eating large quantities of sushi and other forms of fish (Itzkoff 2008).

Exposure to methylmercury at lower doses also has health impacts, especially for fetuses and newborn children whose mothers were exposed to methylmercury during pregnancy. Much scientific evidence on the dangers of low-dose exposure comes from epidemiological studies in the Faroe Islands, where people have harvested and consumed large amounts of pilot whale meat and blubber since at least the sixteenth century (Fielding et al. 2015). In the present day, whale meat and blubber contain high levels of methylmercury. Studies that tracked children in the late 1990s linked prenatal exposure to methylmercury to cognitive deficits years later (Grandjean et al. 1997). Two other similar large-scale studies were also conducted around the same time. One of these (in New Zealand) found evidence of neurological impacts from methylmercury exposure while the other (in the Seychelles) did not (US National Research Council 2000). An integrated analysis of data from all three studies found a relationship between increased mercury exposure and decreased IQ (Axelrad et al. 2007). More recent epidemiological data show that methylmercury exposure may lead to neurological impacts at even lower doses than previously thought

(Grandjean 2016). No threshold has been identified below which methyl-mercury exposure is safe (Sunderland et al. 2016).

Developing research since the 2000s shows that low-dose exposure to methylmercury can also harm adult non-pregnant fish consumers. Knowledge of low-dose methylmercury effects in adults is less certain than the effects that have been scientifically documented in newborns and small children in the epidemiological studies in the Faroe Islands and elsewhere. Nevertheless, there is scientific evidence that methylmercury is also damaging to adults (Sunderland et al. 2016). Evidence for methylmercury's cardiovascular impacts continues to grow (Roman et al. 2011; Genchi et al. 2017). Studies have also demonstrated that methylmercury can cause damage to the immune system and the reproductive and endocrine systems, and increase risks for diabetes (Eagles-Smith et al. 2018). In addition, methylmercury has been identified as a possible carcinogen (US National Research Council 2000).

Health impacts from the dietary intake of methylmercury are shaped by biological and socio-economic factors (box 1-1). Scientific knowledge about such relationships is relatively recent, and still emerging (Eagles-Smith et al. 2018). Studies reveal that certain genetic characteristics—variations in a single element of a deoxyribonucleic acid (DNA) sequence—can either enhance or decrease the impact of toxic substances (Basu et al. 2014). This has been shown to be important for mercury, where certain genetic variations are associated with higher and lower mercury levels in exposed populations, and thus with differing adverse health outcomes. Because the timescale of human health impacts from methylmercury can stretch across generations, a growing area of research involves epigenetics, which refers to genetic changes that can be inherited without underlying alterations in DNA (Baccarelli and Bollati 2009). Epigenetic changes as a result of environmental contaminants can have long-lasting, multigenerational impacts (Feil and Fraga 2012). Evidence of epigenetic effects is emerging in the case of mercury (Basu et al. 2014).

A number of studies in different parts of the world have estimated the population-wide impacts of low-dose exposure to mercury in different countries on IQ deficits as well as cardiovascular damages. Some researchers have extended this work to calculate monetary costs as a result of these exposures, or the monetary benefits of action to reduce exposure. One study estimated the burden of mercury pollution for population-wide IQ damages in Europe to be between 8 and 9 billion euro (Bellanger et al. 2013). In the United States, median cumulative benefits estimated in monetary terms to the year 2050 as a result of policies implemented under the Minamata Convention could reach over USD 300 billion (Giang and Selin 2016). For China, the benefits of reducing mercury exposure via emission reduction policies by 2030 were calculated to exceed USD 400 billion (Zhang et al. 2017). These estimates use different methods of analysis, and are not strictly comparable, but taken together they suggest that although the economic burden of mercury exposure in multiple regions of the world is large, it can be reduced. These benefits can far exceed the costs of reducing mercury emissions (Sunderland et al. 2018).

Seafood and rice also provide many health benefits to consumers, as these foods can be critical sources of important nutrients (box 3-1). These nutrients, including n-3 polyunsaturated fatty acids (n-3 PUFAs), have been shown to have positive benefits for brain and visual system development in infants and reduce risks of certain forms of heart disease in adults (Mahaffey et al. 2011). If people stop eating fish or other forms of seafood, the reduction in the consumption of fatty acids, antioxidants, vitamins, and protein must be made up for with other food choices (Fielding 2010). In many indigenous communities, such as in the Arctic and around the Great Lakes on the US-Canadian border, fishing and eating fish and other kinds of seafood are not only important from a dietary perspective, but also economically, socially, and culturally (AMAP 2011; Gagnon 2016). In addition, for many people in the Faroe Islands, the communal hunting of pilot whales (the *grindadráp*) and the consumption of whale meat and blubber is a proud tradition linked to their national identity (Fielding et al. 2015).

Interventions

In varying ways, interveners including industries, governments, and professional organizations have aimed to reduce mercury-related health damages and improve human well-being by targeting different routes of mercury exposure. Figure 4.4 identifies key interveners and their efforts to address the human health impacts of mercury. First, we discuss actions to address exposure to elevated levels of mercury in specific locations (boxes 1-2, 2-1, and 1-3). We next examine efforts to restrict mercury use in medicine

		Knowledge Institutions	
	1. Human	**2. Technical**	**3. Environmental**
1. Human	(1-1)	(1-2) Industries and national and local governments restrict mercury use in mining and production processes; Professional organizations and national governments restrict mercury use in medicine	(1-3) National and local governments ban harvesting seafood from specific waters
2. Technical	(2-1) Industries and national and local governments take measures to protect workers from mercury exposure	(2-2)	(2-3)
3. Environmental	(3-1) National and local governments ban the sale of food containing mercury; National and local govern- ments publish guidelines on dietary consumption	(3-2)	(3-3)
	Interveners		
	Industries; National and local governments; Professional organizations		

Figure 4.4
Intervention matrix for the mercury and health system.

(box 1-2). We then discuss interventions to mitigate the harms of exposure through dietary consumption, including to lower doses from long-range transport (box 3-1).

Addressing Elevated Exposure to Mercury in Specific Places

Industries as well as national and local governments have taken measures to protect workers from mercury exposure (box 2-1). Mining is one major sector of mercury exposure where interventions to reduce health risks have been applied (see chapter 7 on ASGM for a related discussion). Some early human health–related interventions in the mercury mines in Almadén and

Idrija focused on prolonging the ability of miners—many who were slaves and convicts, and thus forced into mining—to labor under brutal conditions. These interventions included actions to limit each miner's exposure to mercury vapor in the underground mining shafts: daily working hours in the mercury mine in Idrija were reduced to six hours in 1665 due to health concerns (Goldwater 1972). Working hours were gradually limited in other mercury mines as well: one system in Almadén counted eight days of four-hour shifts in the mines as equivalent to a month's work (Hamilton 1943). In addition, the royal miners' hospital in Almadén opened in 1752, and the first doctor in Idrija was hired in 1754.

Spanish colonial authorities in the early 1600s debated how to address the dangers to indigenous workers in the mercury mine in Huancavelica, and in the silver mine in Potosí in Bolivia. Arguments made for and against action appealed to conscience, on the one hand drawing attention to the moral obligations of protecting indigenous lives (even as they worked as slaves), and on the other hand weighing the loss of those workers as a labor force relative to the benefits of mercury in the silver mining economy (Brown 2001). Spanish authorities closed underground shafts and returned to open pit mining in Huancavelica in 1604 (Robins 2011). The economic benefits of silver production, however, prompted authorities to reintroduce shaft mining just a few years later, trumping any human health concerns about the risks to the forced workers (Brown 2001). Rotations among jobs in the mine area were introduced to prevent workers' continuous exposure to the most toxic areas underground (Brown 2001). The opening of new tunnels increased the productivity of the mine as well as the ventilation, but only modestly improved the very harsh (and often lethal) conditions for workers, who continued to inhale mercury vapors and other hazardous materials such as silica dust (Robins 2011).

With more workers exposed to mercury and other hazardous substances in manufacturing and industrial production came calls for better worker protections. In the United States, Alice Hamilton played a pioneering role in advocating for workers' rights in the early twentieth century (Sicherman 1984). Her research was among the earliest in the United States to document many occupational risks, including from mercury (Hamilton 1943). Because of her groundbreaking work, Hamilton became in 1919 the first woman hired to the faculty at Harvard University, where she took up a position in industrial health. Hamilton expanded her work at Harvard to address the risks to workers in mercury mining and mercury-using industries (Hamilton 1943). She visited the mercury mines in New Almaden and New Idrija, and traveled to workplaces that produced mercury lamps and felt hats using a mercury-based technique. The US Post Office issued a 55-cent stamp featuring her image in 1995 in recognition of her contributions to society.

Some of the actions Alice Hamilton and other early advocates called for to protect workers in the manufacturing sectors from mercury exposure were taken by governments long after the occupational dangers of mercury use were first known by health professionals and factory owners. Public authorities, consistent with Minamata Convention provisions, today interact with workers in multiple sectors to develop guidelines for the handling of remaining mercury use, consistent with International Labour Organization (ILO) guidance on occupational exposure. Governments also increasingly require firms that use mercury to put procedures in place for the environmentally safe storage and disposal of excess mercury to prevent human exposure as well as environmental discharges. Scientific laboratories and dental practices in many countries are required, as part of worker safety standards and codes, to ensure good ventilation to protect against mercury vapor, to immediately clean up mercury spills, and to provide appropriate protective clothing such as face masks and rubber gloves.

In some cases, industries and national governments have taken action to restrict mercury use in mining and production processes (box 1-2), and we discuss many of these actions focused on the use of mercury in products and processes in chapter 6. For example, the use of mercury in the hat industry was eventually banned in the United States in 1941 (Goldwater 1972). Some actions included addressing the mining of mercury itself. The phaseout of primary mercury mining helped mitigate much damage to workers' health. This process included the closings of the world's two leading mercury mines in Idrija in 1995 and Almadén in 2002, but also mines in Huancavelica in 1974, in Monte Amiata in Italy in 1982, and in the United States and elsewhere during the late twentieth century. The last mercury mine in the US, the McDermitt Mine in Nevada, closed in 1992 (Tepper 2010). Yet many environmental and human health consequences of mercury mining persist. A sign in the municipal museum in Idrija, which is largely dedicated to the city's mining history, notes: "River Idrijca is still flushing away mercury-laden slime, burdening the environment all the way to the Gulf of Trieste." Some of this mercury may be converted into methyl-mercury and continue to cause harm to seafood consumers.

Closed mercury mines are increasingly seen as historical sites, although some mining continues. The two mines at Idrija and Almadén were jointly given World Heritage Site status in 2012 by the United Nations Educational, Scientific, and Cultural Organization (UNESCO), noting the worldwide importance of mercury extraction to gold and silver mining and to industry, finance, and technology. In 2019, local authorities in China applied for World Heritage Site status for the Wanshan mercury mine, which closed in the early 2000s. Formal mercury mining is ongoing in a few places, including in China and the Kyrgyz Republic, and illegal mining occurs in Indonesia and Mexico. This results in continuing exposure to miners. Because the Minamata Convention allows for existing mercury mining up to 15 years after a country becomes a party, China, as mentioned in chapter 3, plans to continue mercury mining until 2032.

Some national and local government interventions included bans on harvesting seafood from specific waters (box 1-3). The detection of high mercury levels in fish in the Great Lakes region caused both Canadian and American authorities to act in 1970 (Anonymous 1970). The Canadian government first placed an embargo on all fish from Lake St. Clair and the St. Clair River, banning commercial and sport fishing (unless it was for catch and release). Authorities in Michigan and Ohio shortly thereafter also stopped all fishing in Lake St. Clair as well as in the Detroit River and parts of Lake Erie. The need for North American governments to ban the harvesting of seafood because of mercury contamination is not just a thing of the past. In 2014, the Department of Marine Resources in the US state of Maine banned the harvesting of lobster and crab from a seven-square-mile area of the Penobscot River due to elevated mercury levels. The size of the ban area was further expanded in 2016 (Overton 2016).

Lack of scientific knowledge and economic and political interests at times affected responses to mercury pollution, as in the case of Minamata. Researchers at Kumamoto University put together in 1957 a list of 64 possible substances (including mercury) that might cause Minamata disease (George 2001). The researchers, unaware of prior work on organic mercury poisoning, did not pay much attention to mercury initially because they assumed that the factory would not waste such a valuable substance. The British neurologist Douglas McAlpine, however, who visited Kumamoto

University in 1958, speculated that the patients' symptoms were similar to those of methylmercury poisoning, also referred to as Hunter-Russell syndrome. These effects had been described in 1940, and the name Hunter-Russell syndrome came from the 1930s study of workers at the seed packing facility in Norwich, England, who were exposed to methylmercury (Hunter et al. 1940; Rice et al. 2014). Kumamoto researchers paid more attention to organic mercury after McAlpine's suggestion because of a possible link between Minamata disease and Hunter-Russell syndrome (Takeuchi et al. 1959).

Laboratory experiments by researchers at Kumamoto University showed that methylmercury and ethylmercury produced Minamata disease–like symptoms in cats. One puzzling factor, however, was that the Chisso factory used inorganic (not organic) mercury as a catalyst in its chemicals production. Not until a few years later did the researchers figure out that the inorganic mercury was converted into methylmercury inside the factory during the manufacturing process and then subsequently discharged with the wastewater, something that people at the factory were aware of but hid from the researchers and the public. The finding that methylmercury was discharged from the factory was published in an English-language article in the *Kumamoto Medical Journal* in 1962, and was picked up by the Japanese media in early 1963 (Irukayama et al. 1962). A former factory manager testified during the trial against Chisso, which started in 1971, that he had already detected a causal connection between the wastewater and the damage to fish around 1954, but that no one took any steps to further investigate the exact cause (George 2001). Instead, Chisso attempted to discredit researchers and resisted regulatory controls.

Phasing Out Medical Uses of Mercury

Professional organizations and national governments have taken measures to restrict mercury use in medicine (box 1-2), yet heated arguments have taken place for centuries over continuing applications. Medical societies in Europe in the sixteenth and seventeenth centuries opposed mercury-based syphilis treatments (Crosland 2004). These treatments were largely carried out by alchemists, barber-surgeons, and "an array of charlatans" rather than by formally trained doctors (Goldwater 1972, 220). Therapeutic uses of mercurial drugs increased in the eighteenth and nineteenth centuries, and this "resulted in the production of a literature of prodigious proportions" (Goldwater 1972, 244). In an article from 1930, three scientists at Boston City Hospital and the Harvard Medical School noted that mercury and its salts ranked "among the most widely used drugs in medicine" and that their therapeutic value was "well established" (Young et al. 1930, 539). Almost all medical doctors and governments agree today that most mercury uses in medicine should be eliminated, but evidence exists that cinnabar is still included in some traditional medicine, in China for example (Liu et al. 2008).

On some occasions, doctors voluntarily phased out certain medical uses of mercury when they believed that there were better and (sometimes) safer alternatives for treating patients. Many doctors began to use arsenic or bismuth instead of mercury for treating syphilis beginning in the 1910s. Penicillin, a much superior treatment option for syphilis and other illnesses, became the common choice in the medical profession by the 1940s (Svidén and Jonsson 2001). Regulatory interventions by governments drove the phaseout of medical uses of mercury in some cases, especially in the second half of the twentieth century. For example, public authorities in the 1950s began to ban the use of mercury compounds in teething powders to address problems with acrodynia in children (Black 1999). In 1980, the US Food and Drug Administration (FDA) identified over-the-counter vaginal contraceptives containing phenylmercuric acetate or any other mercury compound as "not generally recognized as safe" to use (US Food and Drug Administration 1980). These were formally removed from the US market in 1998, together with mercurochrome and other mercury-containing topical antimicrobials and diaper disinfectants (US Food and Drug Administration 1998). Phenylmercury compounds have been phased out in eye drops and other antibacterial applications, but some uses continued into the 2010s (Kaur et al. 2009; Ezenobi and Chinaka 2018). Mercurochrome and other mercury-containing antimicrobials were also still in use in some countries outside North America and Europe in the 2010s (Bell et al. 2014). The Minamata Convention mandated the phase out of mercury in topical antiseptics by 2020.

The continuing use of mercury, both in dental amalgams and in vaccines, remains a high-profile issue. A growing number of dentists began to use mercury-containing amalgams in the early nineteenth century, but some who preferred the older gold-based method of filling cavities resisted the change. This led to the "Amalgam War" in the United States in the mid-1800s, as dentists who used mercury-based amalgams were expelled from the American Association of Dental Surgeons. At that time, if affected teeth were not removed, cavities were often filled with gold foil that was "pounded into cavities, a process which was time-consuming and painful" (Goldwater 1972, 279). Use of mercury-containing amalgams increased around the world in the late 1800s with the development of more reliable amalgams. A second round of amalgam-based disputes started in the 1920s when some chemists who were concerned about mercury vapor, including Alfred Stock, argued that mercury amalgam constituted a serious health threat to patients and dentists (Goldwater 1972). Nevertheless, mercury amalgam remained common because of its durability even as other mercury uses were phased out during the twentieth century. Dental amalgam accounted for one third of all mercury use in Sweden, for example, by the late 1990s (Hylander and Meili 2005).

Current scientific evidence does not support the argument that the typically very small amount of mercury vapor inhaled after being released from mercury amalgams is dangerous to patients' health, except in a very small number of people developing contact allergies (Clarkson and Magos 2006). However, dentists can be highly exposed if they do not take appropriate measures to protect themselves, and mercury waste from dentistry can enter the environment if not properly disposed of. Dental interest groups continue to weigh in on both sides of the debate. Large organizations such as the World Dental Federation and the American Dental Association support the continued use of mercury in some instances. In contrast, smaller groups such as the World Alliance for Mercury Free Dentistry advocate for banning all mercury amalgams (Mackey et al. 2014). A substantial argument against bans is the cost of substitutes, which tend to be more expensive. The World Health Organization (WHO) argues that in low- and middle-income countries, the extra cost could put necessary dental treatment out of reach for people who need it (World Health Organization 2010). Based largely on the WHO's position that mercury-based amalgam still plays an important role in many parts of the world, the Minamata Convention stipulates a phase-down of dental amalgam use, but without setting a deadline (H. Selin 2014). However, discussions continue, with some parties in favor of taking stronger action under the Minamata Convention on dental amalgam.

Governments in some countries and regions including Europe have taken steps to phase out mercury in dental amalgam to protect the environment as well as for precautionary reasons related to human health. In 2009, Sweden introduced a national ban on mercury in products, but kept an exception for the use of dental amalgam for special medical reasons. This exception was removed in 2018, but dentists can continue to apply for dispensation for dental amalgam on a case-by-case basis (Swedish Chemicals Inspectorate 2017). A European Union (EU) regulation prohibits the use of mercury-based amalgams for vulnerable populations (pregnant or breastfeeding women and children under the age of 15). The EU also mandates the use of pre-dosed encapsulated amalgam and amalgam separators in dental clinics to prevent mercury releases into sewage systems and water bodies. After other mercury uses were phased out, dental amalgam was the largest category of mercury use in the EU by 2018 (European Environment Agency 2018). The European Commission is required in 2020 to report to the European Parliament and the member states in the Council of the European Union on the feasibility of ending all dental amalgam use by 2030 (European Commission 2017b).

The contemporary discourse on mercury in vaccines is similar in focus to debates about the use of dental amalgam: how to appropriately balance clear and important health benefits of specific mercury uses in the medical sector with sensible precautionary actions to protect human health. Specifically, this issue concerns the use of very small amounts of thimerosal as a preservative in vaccines to remove the need for refrigeration. The addition of thimerosal, as noted above, makes it much easier to store, transport, and safely use life-saving vaccines, especially in remote and less populated areas. Ethylmercury, the type of mercury in thimerosal, is different from the other mercury forms, such as methylmercury, that are well known for their toxic effects. As a result, it distributes differently in human tissues and metabolizes at a different rate from methylmercury (Clarkson and Magos 2006). The low levels of ethylmercury that US infants were exposed to through vaccines exceeded health guideline levels set for the different form methylmercury (Baker 2008), but no comparable guideline for ethylmercury exists.

Mercury is intertwined in a growing public discourse about vaccine safety. Vocal parents and advocacy groups surrounding children with autism have influenced public controversies about mercury, which are connected to a discredited but publicly influential study in 1998 by the British gastroenterologist Andrew Wakefield. This study, which was retracted in 2010 after it was found to be falsified, linked autism with the measles-mumps-rubella vaccination. It did not specifically focus on mercury, as that vaccine never contained thimerosal. Yet, some US parents who connected the features of autism with those of mercury poisoning published a study alleging a link between thimerosal and autism in the controversial journal *Medical Hypotheses* (Bernard et al. 2001; Kirby 2006). This publication reflected an organized movement against the use of mercury in vaccines. One example of an advocacy organization focused on this issue is the US-based SafeMinds, created in 2000 to argue that toxic substances (including mercury) play a role in autism. American attorney and environmental activist Robert F. Kennedy Jr. has also repeatedly alleged that mercury in vaccines is poisoning American children, a controversial position sharply rebuked by other members of his famous family (Mnookin 2017; Epstein 2019).

Some medical associations and authorities acted on thimerosal starting in the 1990s. The American Academy of Pediatrics and the United States Centers for Disease Control and Prevention proposed in 1999 that despite no conclusive evidence of harm, thimerosal should be removed from vaccines (Baker 2008). This precautionary measure prompted manufacturers to eliminate thimerosal from routine childhood vaccines in the United States by 2001 (Halsey and Goldman 2001), but it also contributed to the continued public debate over the safety of vaccination more generally. Thimerosal is still used in the United States in some common vaccines, including sometimes against influenza. The European Agency for the Evaluation of Medicinal Products in 1999 recommended that although no evidence exists of harms from thimerosal in vaccines, thimerosal-free vaccines should be used for infants and toddlers based on the precautionary principle (EMEA 1999). In Australia, childhood vaccines have been thimerosal-free since 2000, with the exception of a minute amount in vaccinations against hepatitis B (Australia National Centre for Immunisation Research and Surveillance 2009).

The international debate about the use of thimerosal in vaccines continued in the 2010s. During the negotiations of the Minamata Convention, a group called the Coalition for Mercury-free Drugs argued that allowing continued use of thimerosal reflected a double standard where children in developing countries were not protected by the precautionary principle applied to children in developed countries (Sykes et al. 2014). Yet, eliminating all multi-use vials containing mercury could jeopardize critical public health interventions in developing countries, where vaccine distribution faces challenges because cold storage is not readily available in all areas

(Tavernise 2012). In response to arguments about the need for its continued use in global health protection, the use of mercury in vaccines is specifically exempted from requirements on mercury in products under the Minamata Convention. This decision was in large part influenced by the continued support for the use of thimerosal in multi-dose vaccines by the WHO, GAVI (The Vaccine Alliance), and the United Nations Children's Fund (UNICEF), three organizations that collectively argued strongly against a ban (GAVI 2012; Earth Negotiations Bulletin 2013a; Vaccine News Net 2013).

Addressing Exposure through Dietary Consumption

National and local government interventions that addressed dietary mercury exposure included bans on the sale of food containing mercury, including fish (box 3-1). Fish bans began in the 1960s, but food-focused interventions addressing mercury go back to at least 1938, when the US Department of Health, Education and Welfare set a "zero" tolerance for mercurial pesticide residues in food (Goldwater 1972). The Swedish Medical Board in 1967 halted the sale of fish from 40 lakes and rivers due to high levels of methylmercury, and warned that people should moderate their consumption of freshwater fish (Barnes 1967). The US FDA in 1969 set an action level for banning the sale of fish that contained more than 0.5 ppm mercury. In 1970, the FDA removed tuna and later swordfish from the market because of mercury contamination (Mazur 2004; Bell 2019). A lawsuit by swordfish distributors challenged the limit, arguing that mercury was naturally occurring in fish and should not be seen as a contaminant (US Court of Appeals 1980). Partly as a result of industry pressure, the FDA raised the limit to 1 ppm in 1979. The FDA stopped routine testing for mercury in 1998, though many fish on the US market exceeded 1 ppm (Braile 2000). The EU maximum safe limit for mercury is 0.5 ppm for most fish species, but concentrations of 1 ppm are allowed for select species, including swordfish, tuna, and shark (European Union 2006).

National and local governments have recently focused on publishing guidelines on dietary consumption (box 3-1). These guidelines are issued either as a complement to or as a replacement for setting allowable concentrations of mercury in commercial seafood. The WHO develops guidance on balancing the risks and benefits of fish consumption, together with the UN Food and Agriculture Organization (FAO) (FAO/WHO 2011). Domestic dietary guidelines also weigh the risks of methylmercury against the health benefits and cultural importance of seafood consumption. Many national authorities in Europe have issued varying dietary guidelines to minimize methylmercury exposure. In the United Kingdom, pregnant women and children under 16 are advised to avoid eating shark, marlin, and sword-fish, and minimize consumption of tuna to four medium-sized cans or two steaks per week (European Environment Agency 2018). Swedish authorities recommend that pregnant or nursing women avoid eating high-mercury fish more than two or three times a year (Sweden National Food Agency 2019). In France, pregnant and breastfeeding women are cautioned to limit their consumption to one portion (defined as 150 grams) per week of fish likely to contain high levels of methylmercury, including tuna, sea bream, halibut, and monkfish (French Agency for Food 2016).

Authorities set dietary guidelines using a process that combines informa-tion on methylmercury concentrations in fish, fish consumption, and body weight. A person's intake of methylmercury is calculated by multiplying the concentration of mercury in food by the total intake of mercury-containing food, and dividing that number by body weight. A concentration of 1 ppm is equal to 1 microgram of methylmercury per gram of food. If a person who weighs 50 kilograms (110 pounds) eats 100 grams (3.5 ounces) of fish with a concentration of 1 ppm, it would result in a dose of 2 micrograms of methylmercury per kilogram of body weight. This individual methylmercury dose can then be compared to a reference dose: an amount derived by cal-culating the lowest level at which effects are observed in epidemiological studies and then reducing that level for safety to account for variability among exposed individuals. The US National Research Council in 2000, using epidemiological data from the Faroe Islands study, set a reference dose at 0.1 microgram of methylmercury per kilogram body weight per day (US National Research Council 2000). This equals 0.7 micrograms per week. The meal described above would exceed this guideline. Some researchers argue that this reference dose should be halved based on more recent data on the dangers of methylmercury (Grandjean and Budtz-Jørgensen 2007; Grandjean 2016). The existing US level, however, is lower than some inter-national guidelines. The Joint FAO/WHO Expert Committee on Food Addi-tives in 2003 set its estimate of a provisional tolerable weekly intake of methylmercury to 1.6 micrograms per kilogram for childbearing women.

The US FDA and Environmental Protection Agency (EPA) in 2017 agreed on coordinated advice for pregnant women and children; the advice highlights which fish are "best choices," "good choices," or "choices to avoid" based on the differences in methylmercury levels between species. Some US states also formulate their own recommendations on the consumption of fish from local waters that contain elevated levels of methylmercury. The 2017 guidelines were the latest revisions of federal-level voluntary guidelines; previous versions met much opposition from the private sector. In the mid-2000s, the tuna industry lobbied the FDA to exclude canned tuna from dietary guidelines, and funded a $25 million campaign to make the case for the benefits of eating fish. The industry's efforts included placing newspaper ads targeting the public, and funding scientific work at the Harvard Center for Risk Analysis that questioned the relative importance of mercury risks. In a related effort echoing others that have questioned risk-based government policies, the Center for Government Freedom, a nonprofit organization founded by the tobacco industry, created a website to argue that standards for mercury in food were unwarranted (Mencimer 2008).

Well-intended efforts to implement sound dietary advice can have unintended consequences. A study in the United States found that a 2001 federal advisory on mercury content in fish resulted in an overall decline in total fish consumption by pregnant women (Oken et al. 2003), but consuming fish during pregnancy can have important nutritional benefits. To minimize potential unintended consequences, today's communication on the risks of methylmercury consumption for pregnant women and children emphasizes balancing the benefits and risks of fish consumption. People are encouraged to eat fish low in mercury to maximize positive benefits and minimize negative impacts. A point made by Pál Weihe, a local physician and researcher who worked on many of the Faroe Islands studies, sums up the importance of dietary choice: "For every portion of whale you could eat 100 portions of cod" (Fielding 2010, 436). More recent dietary advice also acknowledges the complexity and variability of risk, including approaches to protect sensitive populations. Emerging scientific knowledge about the genetic markers that influence a person's sensitivity to mercury may be used in the future to better identify susceptible subpopulations, and thus guide intervention efforts (Basu et al. 2014).

Insights

The Minamata pollution story that opened this chapter captures the dangers to human health that mercury and mercury compounds pose, as well as the societal struggles to effectively address them. In this section, we draw insights from the multiple ways that mercury has affected human health. First, the dangers that mercury poses are determined by its inherent properties as a hazardous substance as well as its interactions with technology and society. Second, understanding and implementing transitions toward greater human well-being relate to how harms from mercury to human health are valued, and transitions focused on enhancing human health protection have largely been incremental to date. Third, governance to protect human health from mercury involves integrating risk reduction strategies at levels from local to global, using various types of interventions.

Systems Analysis for Sustainability

Understanding how people are harmed by mercury requires examining the ways in which people's exposure to differing forms and amounts of mercury is influenced by technologies, knowledge, institutions, and environmental processes. Interactions between workers and employers, for instance, led to changes in working conditions that affected mercury use and exposure. In some cases, technological change, such as increased ventilation in mines and workplaces and the use of protective equipment, lowered people's exposure to mercury. In other cases, the use of scientific and technical knowledge about mercury's properties to develop new mercury-based products and manufacturing processes increased the number of workers who came in contact with mercury. But as knowledge of mercury's toxic properties and techniques for health protection improved, local exposure to mercury in laboratories and other workplaces decreased. The dispersal of mercury through the environment, however, connects sources of mercury to health impacts in remote locations. Environmental processes transform mercury into methylmercury, which contaminates fish; dietary choices to eat or not eat certain species of fish influence people's intake of methylmercury; and dietary recommendations from governments or agencies can influence these consumption choices.

The mercury health system has shown a resistance to change over time, even though continuing mercury use has largely resulted in harms to human well-being. Mercury mining, and uses of mercury in silver and gold mining, other workplaces, and in medicine, continued for centuries despite much evidence of health dangers. The environment responds slowly to changes in discharges, and people (and other organisms) are not able to biologically adapt to higher exposures to mercury. Some resistance to change, however, can be seen as beneficial to human health. The WHO and many dentists and doctors oppose calls for immediate bans on the use of mercury, both in dentistry and vaccines, on the grounds that such measures would end up harming human well-being. However, some adaptations, like dietary changes, can have short-term health benefits by reducing methylmercury exposure. All such changes are more supportive of human health when implemented carefully, as less-informed changes can be detrimental. This was the case when pregnant women ate less fish because of methylmercury advisories, thereby forgoing nutritional benefits from eating fish.

Mercury is a global-scale pollutant, but its health impacts are largely local. Furthermore, the severity of local health impacts differs greatly. Mercury is an inherently dangerous substance, but different forms of mercury have highly varying levels of toxicity; as a result, the "dose makes the poison." Vapor from elemental mercury in ambient air does not constitute a direct threat to human health, but concentrations can reach acutely toxic levels in unventilated workplaces and underground mine shafts. The same amount of mercury deposition in distinct geographical areas can cause different degrees of damage to local populations, either because the combination of chemical and biological processes in the environment can convert more or less of the mercury into methylmercury, or because people do not make the same choices about seafood consumption. Individual genetics furthermore influence the ability of mercury exposure to cause health damages. Although some health-based guidelines for maximum tolerable intake such as those issued by the WHO provide global guidance, there is variability in both their uptake by and applicability to local communities and individual people across the world.

Sustainability Definitions and Transitions
Defining sustainability focused on human well-being, now and for future generations, requires assessing health impacts and valuing them appropriately.

The mercury health system illustrates that both the knowledge of mercury's adverse health effects and the perceived importance of those harms has changed over time. Mercury's toxic effects were long documented in miners, and were known to harm goldsmiths in the 1400s. Medical associations in Europe raised concerns about mercury use in medicine in the 1500s. Nevertheless, much mercury use and exposure continued in workplaces and in medicine into the late twentieth century, not only harming people but also adding to the mercury in the environment that converts into methylmercury. Dangers of high dose dietary exposure to methylmercury became well known in the 1950s, effects of low dose methylmercury exposure were discovered late in the twentieth century, and knowledge about effects at even lower doses continues to grow. The value placed on mercury relative to the health damages it causes is influenced by the context of power dynamics. The effects of mercury on miners and other workers were known for centuries, but their lives were often not highly valued by those who had the ability to control exposure. More recently, lobbyists for the seafood industry argued that mercury standards in food to protect consumers were unwarranted or too strict.

Improvements in human health from reduced mercury exposure have mostly occurred as a result of incremental change. As we discussed earlier in this chapter, these include the application of worker protection measures and the implementation of dietary guidance. Many interventions targeted toward human health by governments, including under the Minamata Convention, emphasize efforts to protect populations from exposure through dietary guidelines, stressing changes that people can make in their own lives by choosing fish lower in mercury. While these efforts do not eliminate exposure, they can have substantial health benefits, especially for pregnant women and small children. However, historical efforts to protect human health from the dangers of mercury did not follow a clear linear path of improvement toward a steady strengthening of restrictions on the mining and occupational handling of mercury. Throughout most of the history in which mercury has affected human health, those few measures that were taken in occupational settings were dwarfed by the increased use of mercury, especially since the beginning of the Industrial Revolution. Similarly, the use of mercury in medicine continued to grow centuries after the first warnings from medical societies.

Transitions in the mercury health system were driven by a variety of factors, with innovation and regulations playing important and interacting roles. Many early efforts to mitigate human mercury exposure in work-places were largely ad hoc. It was not until the mid-twentieth century that governments started to take on a more active and coordinating role in regulating mercury uses to protect human health. New scientific knowledge, growing societal concerns about mercury's impact on human health, and technological development drove more recent initiatives to protect workers from mercury exposure. Some substitutions and bans in medicine were triggered by mercury-specific concerns, such as the switch away from using calomel in teething powders as the result of its observed negative impact on small children. In other instances, mercury was phased out because of the availability of new and better treatment options. For example, the penicillin doctors began to use for syphilis instead of mercury was introduced as a superior treatment for a range of serious illnesses, not primarily to reduce a patient's exposure to mercury compounds.

Sustainability Governance

The Minamata Convention is the first multilateral environmental agreement to include a separate article on health (World Health Assembly 2014). The treaty links the WHO and ILO guidelines on dietary intake and occupational exposure, respectively, with on-the-ground implementation that is targeted to specific local conditions. This approach is intended to address risks in ways that correspond with the multi-scale nature of the mercury health system. People eat seafood that is sourced from international and national commercial markets as well as from local waterways, and their consumption patterns are affected by personal preferences and culture. Local workplace conditions, as well as international markets for goods made by using mercury, influence occupational exposure to mercury. Targeting mercury's health impacts in ways that correspond to the multi-scale nature of the mercury health system requires governments and other possible interveners in all parts of the world to have both the knowledge and capacity to act. To that end, information sharing and capacity building are important to better protect human health from mercury.

Governance strategies need to consider interventions at different points along the causal pathways of mercury use and exposure. Interventions that prevent human mercury exposure directly can have substantial benefits. Phasing out mercury mining and mercury use in manufacturing, and using protective clothing in workplaces where mercury is still used, are two

methods of protecting workers from harmful health effects. Carefully implemented dietary advice can dramatically reduce methylmercury exposure to
vulnerable populations. At the same time, interventions at earlier points
on pathways of mercury use and environmental discharges can prevent
longer-term harm before it occurs. Controlling the amount of mercury that
enters ecosystems from major point sources and that converts into methylmercury is necessary in the longer term to protect the health of future generations. Some such efforts can address multiple pathways simultaneously.
Reducing or eliminating industrial uses of mercury not only helps prevent
negative health impacts to workers, but also stops mercury discharges from
entering the environment and ultimately affecting the health of populations both nearby and far away.

Many risk-reduction measures that target the dangers from mercury
exposure to human health are more effective if the government agencies
or industries that develop and implement them consider the influence
of socio-economic factors. Reducing mercury use in dental amalgam and
childhood vaccines can be appropriate where there are safe alternatives,
but such options are not readily available in all parts of the world. Precautionary measures must also be carried out carefully to avoid feeding mis-
information about the safety of life-saving vaccines. Scientific evidence of
health damages from low-dose exposure to methylmercury supports the
necessity to set standards, not only when formulating maximum allowable
content in seafood but also when targeting dietary advice to particularly
vulnerable populations, including pregnant women and small children. At
the same time, because benefits for cardiovascular health and neurological
development in children come from the important nutrients provided by
fish, the dietary guidelines focusing on methylmercury exposure need to be
communicated carefully to ensure they do not inadvertently cause harm by
changing people's diets in a negative way.

*The history of mercury's impacts on human health shows repeated efforts to dress
up a dangerous trickster as a beneficent deity, similar to the mistaken use of the
caduceus as a symbol for the health profession. Employers disregarded the health
of their workers who came into contact with mercury, doctors prescribed damaging
mercury-containing treatments to their patients, and governments only recently
began to respond to the human health dangers of mercury exposure. Many (but not
all) mercury uses in workplaces and medicines have been phased out. Remaining*

mercury uses in dental amalgam and vaccines still have positive consequences for human well-being, but methylmercury in food remains a health challenge. All people on Earth carry the signals of industrialization in the form of concentrations of mercury and other toxic substances in their bodies. For people who are highly exposed or particularly vulnerable to methylmercury, the persistence of mercury in the environment and its continuing conversion into methylmercury will have severe and long-lasting consequences for many generations to come.

5 Energy, Industry, and Pollution: Mercury, the Winged Messenger

The large expansion in the use of fossil fuels in the industrial era increased access to energy and supported economic growth. Yet, the burning of fossil fuels, especially coal, emits the hazardous substances they contain into the atmosphere. These substances include mercury, other heavy metals, sulfur, and carbon dioxide. Hazardous byproducts of combustion that travel via air currents harm populations both nearby and far away. Past regulatory efforts to mitigate air pollution reduced emissions of mercury and other harmful substances in some places. These efforts had measurable benefits for the environment and human health, and are sometimes hailed as local and regional regulatory successes. Challenges remain in other places, however, and for other pollutants such as carbon dioxide. Efforts to control mercury emissions and deposition illustrate that there can be tradeoffs between incremental actions with local and regional benefits and efforts that seek to address the fundamental activities that ultimately produce pollution. However, interventions aimed toward incremental changes are not always in conflict with those that aim to fundamentally restructure underlying processes. In some cases, these two approaches can be synergistic.

The US Environmental Protection Agency (EPA) began measuring mercury in Steubenville, Ohio, in 2002. Data there showed elevated levels of mercury in rainfall, and nearly 80 percent of this mercury was estimated to come from nearby coal-fired power plants (Keeler et al. 2006). Researchers and pollution-control advocates have referred to places like Steubenville, where local sources cause elevated mercury deposition, as "hot spots" (Beusse et al. 2006). This was not the first time that Steubenville became known for its severe pollution. A group of Harvard University researchers began to study the effects of air pollution in Steubenville and five other locations in the United States in 1974. These sites were the so-called six cities in a groundbreaking study that sought to quantify the mortality risks of inhaled fine particulate matter. The researchers found that risk of death in Steubenville, the city among the six with the worst air quality, was 26 percent higher than in the least polluted city in the study (Dockery et al. 1993). Steubenville was called a "test lab" for dirty air and its impact on human health and the environment (Barringer 2006).

Air quality has improved in Steubenville and many other US cities since the 1970s, but air pollutants remain a problem. Researchers analyzed changes in air pollution in the six cities in a follow-up study in the early 2000s, and found that concentrations of fine particulate matter in Steubenville had nearly halved (Laden et al. 2006). Current pollution levels, including mercury deposition, are nevertheless still elevated. Steubenville's air quality improvement is also embedded in a larger socio-economic context. The city's predominant industries in the 1970s were coal mining and steel production; economic stagnation and declines in industrial production beginning in the 1980s deeply affected its residents. In 2006, one resident wondered aloud to a *New York Times* reporter whether cleaner air was worth the trade-offs (Barringer 2006). As mercury emissions from US coal burning have declined, decreasing deposition from domestic sources has in part been offset by increases in mercury coming from elsewhere. In the early 2000s, scientists at the top of Mount Bachelor in Oregon measured mercury in air transported across the Pacific Ocean that had chemical signals characteristic of industrial air pollution from China (Weiss-Penzias et al. 2006).

The story of air pollution in Steubenville is not unique to Ohio or the United States. As we highlight in the chapter title with a reference to the god Mercury as an airborne messenger, atmospheric mercury emissions from industrial point sources in many regions of the world connect geographically distant places across the globe. Like Steubenville, many cities in other industrialized countries have suffered pollution problems and undergone difficult economic transitions as once-dominating and prosperous industries closed down. Contemporary atmospheric mercury emissions from large stationary point sources mostly occur in developing countries, and a large fraction of these emissions results from burning coal to produce energy and industrial goods. Industrialization and rapid economic development in many developing countries over the last few decades, in particular in urban areas in China and India, led to increased emissions of mercury and other air pollutants. Fossil fuel–based energy production is at the same time coming under growing regulatory scrutiny worldwide due to its effect on local air quality as well as its influence on global climate change.

Contributions from local and distant emission sources to mercury deposition in any given region have varied in the past, and will continue to change. These variations result from the adoption of pollution prevention laws, the application of pollution control technology, transnational economic forces, and changes in weather patterns and climate conditions. The relative contribution of nearby and faraway sources is also a matter of definition. The statistic that 80 percent of mercury deposition in Steubenville in the early 2000s came from nearby sources is contingent on how "nearby" is defined, whether on scales of tens or hundreds or thousands of kilometers. Regardless of the definition, it is clear that long-range atmospheric transport is important to mercury deposition. For example, for the region of the United States that includes Ohio, Illinois, Indiana, Michigan, and Wisconsin, one study from the early 2010s estimated that 25 percent of mercury deposition came from sources beyond those states, located either in other US states or in other countries of the world (Lin et al. 2012).

We focus in this chapter on atmospheric mercury emissions from coal burning and other major point sources, and on the development of strategies to mitigate these emissions. In the section on system components, we discuss the human, technical, and environmental components that influence emissions from point sources to the atmosphere, and the institutions and knowledge that affect the amount of mercury emitted to air. We trace how industrial activities lead to pollution, examine how pollution controls influence mercury emissions, and address the transport and fate of mercury in the atmosphere in the section on interactions. Our discussion of interventions explains how governments have mandated, and the private sector has applied, new pollution abatement technology for mercury without, in most cases, making fundamental changes to the energy and production processes that result in mercury emissions. In the final section on insights, we discuss how atmospheric transport and socio-economic factors connect mercury from local to global scales, how reductions in mercury emissions have come largely from incremental changes, and why cross-scale action is important when further addressing mercury emissions to air.

System Components

Much of the mercury that has been historically emitted to the atmosphere by human activities has come from industrial point sources. Industrial energy production based on fossil fuels contributed greatly to human development by providing an expanded supply of energy to billions of people. This had many societal benefits, but also led to sharply reduced air quality and ecological damages, including from mercury emissions. Other industrial activities producing a range of different products similarly provided material benefits to individuals and modern societies while causing much harmful pollution. Mercury pollution from contemporary industrial sources continues to be a global problem, but with important regional variations in emission levels. Figure 5.1 shows the human, technical, environmental, institutional, and knowledge components in the atmospheric system for mercury.

People have mined *coal in geological storage* and used this *extracted coal* as an energy source for thousands of years. During most of this time, coal burning was small in scale. This could cause major indoor air quality problems, but contributed relatively little to atmospheric pollution worldwide. The Industrial Revolution led to a massive increase in coal use and associated mercury emissions—enough to increase mercury emissions from coal burning by 100-fold over the past 150 years (Streets et al. 2018). Present-day coal burning and mercury emissions to the air in many countries all over

Human components	Technical components	Environmental components
Producers and consumers of energy and industrial goods People living near industrial point sources People living far from industrial point sources	Extracted coal Energy and industrial goods Industrial point sources of mercury emissions Air pollution control devices	Coal in geological storage Atmosphere Ecosystems near emission sources Ecosystems far from emission sources
Institutional components		**Knowledge components**
Markets for energy and goods National and local laws and regulations International air pollution agreements Global Mercury Partnership Minamata Convention		Mercury concentrations in the atmosphere Long-range transport Mercury deposition Techniques for air pollution control

Figure 5.1
Components in the atmospheric system for mercury (referenced in the text in italic type).

the world are deeply embedded in a long-standing fossil fuel–based economy: coal burning has been essential to the production of *energy and industrial goods* during the past three centuries, and people play important roles as *producers and consumers of energy and industrial goods*. Often, *markets for energy and goods* cross multiple spatial scales, and are influenced by forces of supply and demand. Many markets for energy are domestic or regional, and shape the prevalence of coal and other energy sources in the energy mix. Markets for goods have become more global over time given increased trade across international borders and continents.

Present-day anthropogenic sources make up roughly a third of total mercury emissions to the *atmosphere*, as noted in chapter 3, with the remainder coming from natural sources and reemission of historically mobilized mercury (legacy mercury). Human activities emitted approximately 2,500 tonnes of mercury to air globally in 2015, according to the most recent global mercury assessment (UNEP 2019). Table 5.1 summarizes total mercury emissions to air by region, the fraction contributed by fuel combustion and industry in each region, and each region's contribution to global emissions. Artisanal and small-scale gold mining (ASGM), which we discuss further in chapter 7, is thought to be the largest source globally, at 38 percent of total emissions. Stationary coal combustion contributed 21 percent of total emissions. Other types of *industrial point sources of mercury emissions* are also important, including production of non-ferrous metals such as copper, zinc, and lead (15 percent), cement production (11 percent), and production of ferrous metals (iron and steel) (2 percent). Emissions from mercury-added product waste contributed 7 percent, with other sources such as biomass and large-scale gold production contributing the remaining 6 percent. Emissions from some smaller sectors were not quantifiable.

One study estimates that 38,000 tonnes of mercury has been mobilized over all time through coal combustion (Streets et al. 2018). The vast majority of that mercury has entered the environment after 1850, with 70 percent being emitted to the atmosphere. In the early 1970s, a back-of-the-envelope calculation—made by multiplying the average value of mercury in coal by the total amount of coal burned—estimated that some 3,000 tonnes of mercury were being emitted globally to the atmosphere by coal burning each year (Joensuu 1971). This estimate was too large by an order of magnitude because of errors in the average value of mercury in coal: different kinds of coal contain varying trace amounts of mercury (although

Table 5.1
Mercury Emissions to Air by Region for 2015

Region	Mercury emissions from fuel combustion and industry, tonnes	Total emissions, tonnes (uncertainty range)	Percentage of regional mercury emissions from fuel combustion and industry	Regional contribution to total global emissions from fuel combustion and industry, percent	Regional contribution to total global anthropogenic mercury emissions for all sectors, percent
East and Southeast Asia	536	859 (685–1430)	62%	47%	38.6%
South Asia	184	225 (190–296)	82%	16%	10.1%
Russia, former Soviet Union states, & other Europe	91	124 (105–170)	73%	8%	5.6%
Sub-Saharan Africa	91	360 (276–445)	25%	8%	16.2%
European Union	69	77.2 (67.2–107)	89%	6%	3.5%
South America	56	409 (308–522)	14%	5%	18.4%
Middle East	40	52.8 (40.7–93.8)	77%	4%	2.4%
North America	35	40.4 (33.8–59.6)	86%	3%	1.8%
Central America	25	45.8 (37.2–61.4)	54%	2%	2.1%
North Africa	14	20.9 (13.5–45.8)	67%	1%	1%
Australia, New Zealand, & Oceania	8	8.79 (6.93–13.7)	87%	<1%	<1%

Source: UNEP (2019)

this variability does not generally track with particular types of coal or regional origin). More recent estimates suggest that the amount of mercury emitted to the atmosphere from coal combustion in 1970 was between 300 and 400 tonnes (Streets et al. 2018). Estimates still vary about the amount of mercury emitted from coal for the present day—an estimate for 2010 quoted an uncertainty range of 221 to 1,473 tonnes (Streets et al. 2018).

Atmospheric emissions of mercury and other air pollutants affect humans and the environment both near and far away from sources. Mercury in air from nearby industrial point sources does not reach levels that typically cause health impacts from inhalation, but other air pollutants from the same sources can and do. Some forms of mercury remain in the atmosphere only a short time, and deposit to *ecosystems near emission sources*; other forms travel long distances through the air before undergoing chemical reactions and depositing to *ecosystems far from emission sources*. A fraction of the mercury deposited both locally and far from point sources is converted into methylmercury and accumulates in aquatic food webs. People's health is affected by eating fish with elevated methylmercury concentrations, including *people living near industrial point sources* and *people living far from industrial point sources*. *Air pollution control devices* prevent varying amounts of mercury and other pollutants from point sources from entering into air. Many of these are end-of-pipe technologies that capture mercury emissions right before they would be emitted (Mukherjee et al. 2008).

Scientific knowledge about *mercury concentrations in the atmosphere*, the chemical behavior of different mercury compounds, and the propensity for *long-range transport* of different mercury compounds has developed over time. Some of the first measurements of atmospheric concentrations of mercury date back to the 1960s, when researchers detected high concentrations of mercury in the air in urban areas, such as around San Francisco (Williston 1968). These elevated mercury concentrations often coincided with high levels of other air pollutants. The first routine atmospheric mercury measurements began at a station in 1980 in Rörvik, Sweden (Iverfeldt 1991). Researchers at other sites began to measure atmospheric mercury concentrations in the 1990s, including at Wank Mountain in Germany in 1990, Cape Point in South Africa and Alert in Canada in 1995, and Mace Head in Ireland in 1996 (Slemr et al. 2003). Increasingly accurate measurements from these stations helped to document and determine regional and global atmospheric transport patterns of mercury and trends over time.

Today, atmospheric mercury monitoring stations exist in all regions of the world (Sprovieri et al. 2016). However, measurement data remain sparse in several regions, especially in the tropics and in the Southern Hemisphere (Obrist et al. 2018). Analytical techniques could be improved, specifically for measuring different forms of mercury in the atmosphere (Jaffe et al. 2014). The exact chemical composition of some of these forms of atmospheric mercury also remains unknown. Several measurement networks generate knowledge about *mercury deposition*, with most sites in the United States and Europe (Lindberg et al. 2007; Prestbo and Gay 2009). Historical records of mercury levels in environmental archives such as lake sediment cores are used to understand atmospheric deposition in the past (Biester et al. 2007). Measurements of mercury isotopes (atoms of mercury with different numbers of neutrons) are increasing in accuracy, and can help identify different sources of natural and anthropogenic mercury emissions as well as environmental processes such as chemical transformations in the atmosphere (Blum et al. 2014).

Institutions such as *national and local laws and regulations* address mercury emissions from various point sources. *International air pollution agreements* started to tackle mercury emissions in the 1980s. The Convention on Long-Range Transboundary Air Pollution (CLRTAP), originally developed in the 1970s to reduce pollution from sulfur dioxide and associated acid rain issues, began to consider mercury in the late 1980s (Selin and Selin 2006). Efforts under the *Global Mercury Partnership* have not only prompted increased awareness of mercury emissions, but also generated detailed technological information on control options for power generation and other major atmospheric sources. International expert groups, working under the Global Mercury Partnership and the *Minamata Convention*, develop and update technical guidelines for pollution controls. These guidelines communicate knowledge about *techniques for air pollution control,* including options for mercury controls and associated costs, and technical conditions and economic feasibility influence their application in different countries and situations (UNEP 2016).

Interactions

Mercury emissions from stationary sources to air are distributed widely in the environment. Figure 5.2 shows interactions in the atmospheric system

	1. Human	2. Technical	3. Environmental
1. Human	(1-1) **Producers and consumers interact in socio-economic systems**	(1-2) Producers and consumers influence production of energy and industrial goods	(1-3)
2. Technical	(2-1) Energy and industrial goods provide benefits to producers and consumers; Air pollution control devices incur costs to producers and consumers	(2-2) **Air pollution control devices capture mercury and other air pollutants**; Byproducts of combustion containing captured mercury reused in other applications	(2-3) **Industrial point sources emit mercury and other pollutants**; Air pollution control devices reduce mercury emissions
3. Environmental	(3-1) Atmospheric pollution and methylmercury in ecosystems harm people living near and far from industrial point sources	(3-2) Fossil fuels enable operation of industrial point sources and production of energy and industrial goods	(3-3) The atmosphere transports mercury and other air pollutants to nearby and remote ecosystems; Ecosystem processes lead to production of methylmercury

Figure 5.2
Interaction matrix for the atmospheric system for mercury.

for mercury: we have selected three interactions in the matrix (the items in bold type in boxes 1-1, 2-3, and 2-2) to focus on in this section; we then trace the pathways that influence these interactions, which we summarize in figure 5.3 (where the bold boxes correspond to the selected interactions). First, producers and consumers interact in socio-economic systems (box 1-1), producing energy and industrial goods that benefit society, and at the same time lead to emissions of mercury and other air pollutants (boxes 1-2, 2-1, 3-2, and 2-3). Second, industrial point sources emit mercury and other pollutants (box 2-3), and this mercury travels through the atmosphere and deposits in both nearby and remote ecosystems, where some is transformed into methylmercury that harms people (boxes 3-3 and 3-1). Third, air pollution control devices capture mercury and other air pollutants (box 2-2), affecting both emissions and costs to producers and consumers, while some

a) Industrial production and air pollution: Production of energy and industrial goods benefits society and leads to emissions of air pollutants, including mercury

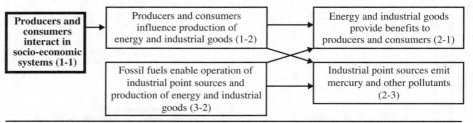

b) Atmospheric transport of air pollutants: Point sources affect air pollution transport and mercury distribution in ecosystems

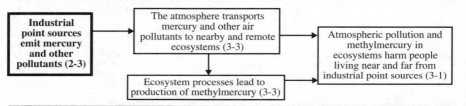

c) Pollution control and mercury emissions: Air pollution controls reduce mercury emissions, but incur economic costs to producers and consumers

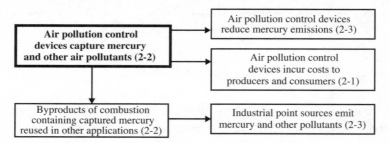

Figure 5.3

Pathways of interactions in the atmospheric system for mercury. Bold box indicates focal interaction for each subsection.

of the captured mercury can be reused in other applications where it may be emitted to the atmosphere (boxes 2-3, 2-2, and 2-1).

Industrial Production and Air Pollution

Industrial production and air pollution originate with producers and consumers interacting in socio-economic systems (box 1-1). The industrial era marked the beginning of a dramatic expansion of production, consumption, and trade worldwide, which accelerated further after 1950 (Steffen et al. 2007). These interactions, through supply and demand, influence the production of energy and industrial goods (box 1-2). Between 1850 and 2017, global energy production increased by almost a factor of 20 (Ritchie and Roser 2019). During the same period, world gross domestic product (GDP)—the monetary value of goods and services in the economy—increased by a factor of more than 50 (Roser 2019). Energy and industrial goods provide benefits to producers and consumers (box 2-1). Increased production raised material standards of living and enabled dramatic improvements in health and educational attainment. But the beneficial impacts of these changes were not evenly distributed globally: the Global North industrialized earlier and faster than did much of the Global South.

The availability and extensive use of fossil fuels, especially relatively cheap coal, enabled the operation of industrial point sources and production of energy and industrial goods (box 3-2). Changes in production in different regions over time altered the location and quantity of coal use (Streets et al. 2018). Coal use in Western Europe peaked in the 1950s, and declined thereafter. In the United States, coal use increased from the 1970s to the 1990s, but declined in the 2000s. Most of the growth in coal consumption since the 1960s was in developing countries, with dramatic increases particularly in China. These trends are related to patterns of industrialization and a shift in energy-intensive production away from North America and Europe to other rapidly industrializing regions, mainly Asia. Global coal-fired power capacity doubled between 2000 and 2018, to 2,000 gigawatts (Carbon Brief 2018). After declining from 2014 to 2016, coal production increased again in 2017 (International Energy Agency 2018). Coal's share of the world's energy mix was 27 percent in 2016 (International Energy Agency 2018). There are an estimated 1.1 trillion tonnes of identified coal reserves worldwide, which translates into another 150 years of coal use at current production rates (World Coal Association 2019).

Industrial point sources emit mercury and other pollutants (box 2-3). Because of an early expansion of coal-based manufacturing and energy generation, point sources in Europe and North America were the dominant contributors of mercury emissions to air from 1850 to the late 1900s (Streets et al. 2018). By 2015, however, the European Union (EU) and North America contributed only 6 percent and 2 percent, respectively, to global mercury emissions from these sources (see table 5.1). Instead, most mercury emissions from fuel combustion and other industrial point sources came from East, Southeast, and South Asia (62 percent), where China and India, by far, are the largest emitters. North and Sub-Saharan Africa were responsible for 9 percent, and Russia and other countries formerly part of the Soviet Union contributed 8 percent. Some mercury emissions are related to the production of exported goods. One estimate is that 33 percent of China's mercury emissions and releases for the year 2010 were a result of production for export (Hui et al. 2016). Thus, the decline in North American and EU mercury emissions has been offset by increases elsewhere that are in part driven by consumption in these two regions.

Atmospheric Transport of Air Pollutants

Industrial point sources emit mercury and other air pollutants (box 2-3). The atmosphere transports air pollutants to nearby and remote ecosystems (box 3-3), where they harm people living near and far from industrial point sources (box 3-1). The influence of local air pollution on human health has long been recognized. For example, coal use in London and other British cities in the nineteenth century led to frequent episodes of dense "fog"— essentially coal smoke that turned the sky ash grey (Fouquet 2011). Extreme pollution events in cities such as London and Donora, Pennsylvania, in the 1940s and 1950s led to cardiovascular and respiratory problems and mortalities (Davis 2002; Jacobs et al. 2018). Awareness of atmospheric long-range transport of pollutants expanded with the acid rain issue in the 1960s and the recognition that substances like sulfur can travel up to thousands of kilometers, crossing national borders (Odén 1967; Odén 1976). People can be exposed to air pollutants like fine particulate matter at dangerously high levels when they breathe air in urban areas, but, as noted above, mercury is not present in ambient air in concentrations that cause health concerns for exposure via inhalation, even downwind of industrial point sources.

As we explained in chapter 3, mercury from industrial point sources is emitted in different forms, and the properties of these different forms determine how far they can travel through the atmosphere. Coal-fired power plants emit both elemental and oxidized (more soluble) forms of mercury. Some oxidized mercury is attached to atmospheric particulate matter. Mercury in its elemental form can travel worldwide, but mercury in oxidized form, including the fraction on atmospheric particles, largely deposits near to sources. Experts are still uncertain about the proportion of these different kinds of mercury emitted from different sources: the fraction of mercury emitted in each form from an industrial point source depends on the characteristics of the fuel, operational procedures, and the control technologies applied (Muntean et al. 2018). Emitted mercury, depending on its form, transports via the atmosphere to nearby or remote ecosystems (box 3-3), where it is deposited. Ecosystem processes then lead to production of methylmercury (box 3-3). Methylmercury in ecosystems, in turn, harms people (box 3-1).

Mercury deposition to ecosystems originates from a combination of local and far away sources. But it was not possible for a long time to attribute measured mercury concentrations to source regions because sophisticated mercury transport models and data on emissions, transport, and deposition were unavailable. Researchers had advanced computer modeling capabilities and conducted enough measurement studies on mercury transport by the early 2000s to be able to get a clearer scientific picture of mercury atmospheric transportation and deposition patterns. Several of the initial studies attempted to connect emission sources to remote concentrations for the first time; some used atmospheric models to calculate the percentage of deposition resulting from domestic sources, particularly in the United States (Seigneur et al. 2004; Selin and Jacob 2008). Other studies calculated intercontinental and interregional transport for different regions, including between Asia and North America (Travnikov 2005; Corbitt et al. 2011). Researchers also used the fraction of different mercury isotopes in environmental measurements to differentiate some of the mercury coming from coal combustion versus other sources (Sherman et al. 2015).

The relative contribution of industrial sources in any given region to mercury deposition in the same region is both uncertain and variable. Nearly all mercury deposition in the Arctic comes from beyond the region, but there are also a few sources within the Arctic that contribute (UNEP and AMAP 2015). In contrast, some models suggest that over half of the mercury deposited in Europe came from within the region, although estimates differ (Kwon and Selin 2016). Estimates for Asia are that 23 to 55 percent of mercury deposition comes from sources within the region (UNEP and AMAP 2015). One study calculated that 80 percent of nationwide deposition of mercury in the United States originated from sources in other parts of the world (Selin and Jacob 2008). Variation exists within regions as well: the percentage of domestically emitted mercury depositing to areas of the northeastern United States could reach as high as 60 to 80 percent, though many other US areas predominantly receive deposition that originates from outside North America (Seigneur et al. 2004; Selin and Jacob 2008). Steubenville, the city whose story begins this chapter, is one of the US locations that receive a large amount of deposition from nearby sources.

Pollution Control and Mercury Emissions

The operation of air pollution control devices captures mercury and other pollutants before they are emitted to the atmosphere (box 2-2). These control devices on industrial point sources often exist in the form of end-of-pipe technologies. One of the simplest examples of an end-of-pipe pollution control technology dates back to at least ancient Egypt and uses fabric, such as a piece of cloth or a bag, to filter out particles from an effluent stream (Billings and Wilder 1970). Starting in the nineteenth century, another action to mitigate the impacts of air pollution from industrial sources in urban areas was to build taller and taller smokestacks that were high enough to route the emission out of industrializing and expanding cities (Fenger 2009). Although higher smokestacks mitigated some of the pollution's local effects on air quality and human health, they simply shifted the impacts of pollutants to other geographical areas.

Air pollution control technologies have become more efficient and effective over time. Many of these are still end-of-pipe technologies. More modern fabric filters made of advanced materials designed to optimize airflow were applied in the twentieth century to control emissions of sulfur as well as lead and arsenic (Billings and Wilder 1970). Electrostatic precipitator technology, in which an electrical current is used to attract and capture particles, emerged in the late 1800s, but its first larger-scale applications on point sources were in the early twentieth century (Parker 1997). Other technologies chemically treat or "scrub" the flue gas produced by the burning of fossil fuels to reduce harmful emissions, such as emissions of sulfur dioxide or nitrogen oxides. Flue gas desulfurization, during which sulfur dioxide undergoes a chemical reaction to remove it from the flue gas before being emitted into the atmosphere, began to be applied at large scale in the early 1970s (Lefohn et al. 1999). Other air pollution control techniques that are not installed at the end-of-pipe include processes to wash the coal before it is burned (UNEP 2016). This is often done to reduce sulfur dioxide emissions or to increase combustion efficiency, but the washing can reduce mercury content as well. More recent technologies that control mercury specifically include activated carbon injection, a process that can capture 90 to 98 percent of mercury emissions (Srivastava et al. 2006).

Different air pollution control devices capture different forms of mercury, and this has implications for how far emitted mercury can travel in the atmosphere. Some control technologies, especially those applied earlier on for controlling other air pollutants, preferentially capture more soluble (oxidized) or particulate forms of mercury that tend to travel regionally. For example, control equipment that captures particulate matter such as fabric filters and electrostatic precipitators also reduces emissions of particle-associated mercury. Similarly, flue gas desulfurization captures mercury in its oxidized form. This was the case historically in North America, and is also the case in contemporary application of mercury controls in rapidly industrializing economies such as China and India (Giang et al. 2015). More recent controls such as activated carbon injection can control elemental mercury emissions, which travel globally. The performance of modern air pollution control devices in capturing mercury depends on several factors, including the type of coal, operating conditions such as temperature and air-to-fuel ratio, and other components affecting the chemical reactions that mercury undergoes in the burning process (UNEP 2016).

The application of air pollution control devices reduces mercury emissions (box 2-3). Atmospheric mercury emissions in the United States decreased by nearly 80 percent from 1990 to 2011 because of the implementation of air pollution control technologies (US Environmental Protection Agency 2015). In the EU, mercury emissions in 2016 were 71 percent lower than they were in 1990 (European Environment Agency 2018). Mercury emissions have not always risen and fallen proportionally with coal use over time. David G. Streets and colleagues (2018, 133) note that the level of mercury releases "at any given time is determined by competition between production growth and technology improvement." For example, coal use increased in the United States in the 1970s and 1980s, but its mercury emissions stayed constant. Increases in the number and output of new point sources in recent years have nevertheless outpaced the installation of air pollution control devices: global anthropogenic mercury emissions rose 20 percent from 2010 to 2015 (UNEP 2019).

End-of-pipe controls prevent mercury emissions to the atmosphere, but they do not change the total amount of mercury removed from fossil fuel reserves. The fraction of mercury emitted to the air globally out of the total amount of mercury in burned coal (including power generation and industrial production driven by coal) dropped from 77 percent in 1950 to 55 percent in 2010 (Streets et al. 2018). Because the amount of emitted mercury went down, the amount of captured mercury increased. Byproducts of combustion that contain captured mercury are sometimes reused in other industrial applications (box 2-2). Mercury-containing fly ash from a coal-fired power plant, for example, can be later reused. When fly ash is heated to high temperatures as a component in making cement, mercury can be emitted to the atmosphere unless appropriate control technology is present in the cement factory (Wang et al. 2014). Thus, if byproducts are not properly disposed of or used in environmentally safe ways, at least some of the previously captured mercury can be emitted from these industrial point sources (box 2-3).

The installation of air pollution control devices incurs costs to producers and consumers (box 2-1). For producers, those include the fixed capital costs of the equipment itself as well as variable operating and maintenance costs once the equipment is installed. At least some of these increased production costs are passed on to consumers in the form of higher energy prices and more expensive goods. Government regulations can help drive down pollutant control costs by providing incentives for technological innovation. This was, for instance, the case for the US federal cap-and-trade system for sulfur dioxide (Schmalensee and Stavins 2013), which led to lower costs for air pollution controls that also reduced mercury emissions. Some people, like the Steubenville resident who wondered whether cleaning up the air was worth the cost to jobs, may view costs of pollution control as a tradeoff against their environmental and human health benefits. Several studies that have attempted to monetize the benefits of mercury controls, as mentioned in chapter 4, find those benefits to be substantial (Trasande et al. 2005; Bellanger et al. 2013; Giang and Selin 2016; Zhang et al. 2017). In the US, such benefits have been found to be greater than economic costs of mercury reductions (Sunderland et al. 2018).

Interventions

Interveners such as national and local governments, industries, and international bodies have taken a number of intentional actions to address atmospheric emissions of mercury at varying scales. Figure 5.4 highlights interventions in the atmospheric system for mercury. First, we discuss

Knowledge
Institutions

	1. Human	2. Technical	3. Environmental
1. Human	(1-1)	(1-2)	(1-3)
2. Technical	(2-1)	(2-2) National and local governments mandate use of air pollution control devices; Industries deploy air pollution control devices; International bodies develop guidelines for use of air pollution control devices	(2-3) National and local governments set air emission standards
3. Environmental	(3-1) National and local governments set ambient air and environmental quality standards	(3-2) International bodies, national and local governments, and industries influence fossil fuel use	(3-3)
Interveners			
National and local governments; Industries; International bodies			

Figure 5.4
Intervention matrix for the atmospheric system for mercury.

standards that regulate mercury emissions through technological and other approaches targeting pollution controls and mercury emissions; emission-control approaches include developing, mandating, and optimizing relevant technologies (boxes 2-2, 2-3, and 3-1). Second, we examine mercury controls involving industrial production, focusing on efforts to reduce dependence on fossil fuels (box 3-2).

Laws and Regulations Addressing Mercury

Laws address atmospheric mercury emissions in different ways. Some of the earliest mercury-specific regulations set by national and local governments were ambient air and environmental quality standards (box 3-1). Ambient air quality standards identify a maximum permitted concentration of mercury in air, and environmental quality standards introduce a maximum allowable mercury concentration in other environmental media such as water, soil, or fish. But these regulatory efforts on industrial point sources were not the first ones that led to the reduction of mercury emissions. Initial efforts that reduced mercury emissions in many countries instead took the form of legislation and regulations that were primarily developed for the purpose of controlling other air pollutants. Although air pollution controls implemented during the 1800s and much of the 1900s aimed to reduce atmospheric emissions of sulfur dioxide and particulate matter from industrial sources, they also captured mercury, even though the capture was largely unintended. Governments began specific interventions to address atmospheric mercury emissions in the 1970s.

National and local governments have set air emission standards to regulate the amount of mercury emitted from a particular source (box 2-3). Regulatory approaches can take several different forms. Emission limit values (ELVs) set the maximum allowable amount of mercury (or another pollutant) that can be emitted from an individual point source. National and local governments have also mandated the use of air pollution control devices on specific sources (box 2-2). These control devices are then deployed by industries (box 2-2). One specific type of technology-based approach is setting best available techniques (BAT). BAT is often defined as the techniques most effective to prevent or reduce emissions and releases, taking into account both economic and technical considerations. The use of best environmental practices (BEP) is defined as the most appropriate combination of environmental control strategies and measures for reducing pollution. Applying BEP, for example, can help ensure the appropriate management of mercury captured as a byproduct of coal combustion so that it is not reemitted elsewhere.

Policies and regulatory standards that both indirectly and directly reduce mercury emissions can be, and have been, strengthened over time. ELVs can be lowered, and the suite of technologies that constitute BAT can be made more stringent. Correspondingly, and often prompted by regulatory actions, engineers and industry have developed and deployed new and more effective devices and techniques for emission controls. Combinations of these techniques applied together can lead to different rates of mercury capture, and the performance of emission controls for reducing mercury emissions can be enhanced (UNEP 2016). Multiple pollutants including mercury can also be captured simultaneously as part of a broader approach to controlling air pollution. Studies point out that there is not one single approach or best solution for mercury control from stationary sources such as coal-fired power plants; the choice depends in part on plant-specific characteristics (Sloss 2008).

The United States adopted some of the earliest policies that specifically addressed mercury emissions. Regulations under the 1970 US Clean Air Act named mercury, along with asbestos and beryllium, to an initial list of regulated hazardous air pollutants (National Emission Standards for Hazardous Air Pollutants 1973). The US EPA established emission limits in 1973 for two types of mercury emission sources—mercury ore processing facilities and mercury cell chlor-alkali plants—but it did not address emissions from coal burning. The formulation and implementation of regulatory standards at the time were hampered by lack of scientific knowledge on emission sources and levels. The 1973 regulation listing mercury as a hazardous air pollutant notes: "Current data on the environmental transport of mercury do not permit a clear assessment of the effect of mercury emissions into the atmosphere on the mercury content in the aquatic and terrestrial environments" (National Emission Standards for Hazardous Air Pollutants 1973, 8825). Few actions on atmospheric mercury emissions in the United States were driven by concerns about deposition until more scientific data emerged two decades later.

The United States strengthened its mercury emission controls in the 1990s and into the 2000s. The US Clean Air Act Amendments of 1990 mandated controls on mercury emissions from waste incinerators. However, language on regulating mercury emissions from power generation in an earlier draft of the amendments was dropped in favor of a clause that merely called for the EPA to conduct further study before issuing regulations (Lee 1991). The amendments require a Maximum Achievable Control Technology approach (a type of BAT standard) to hazardous air pollutants whereby the EPA determines a technology-based standard by averaging the best-performing 12 percent of facilities. Yet, the EPA initiated no regulatory action on mercury emissions from the power sector at that time. In contrast, states in New England took further action on mercury emissions, also working together with eastern Canadian provinces on an action plan on mercury dating back to 1997 (Selin and VanDeveer 2005). This plan set aggressive goals that went beyond national commitments in both the United States and Canada at the time, aiming for a 50 percent reduction in regional mercury emissions by 2003 and a 75 percent reduction by 2010, which were also achieved (Smith and Trip 2005; New England Governors and Eastern Canadian Premiers 2011).

Environmental groups used legal means in the 1990s to pressure the US federal government to adopt regulations on mercury emissions from coal-fired power plants. This included litigation by the Natural Resources Defense Council (NRDC) and other advocacy groups to force the EPA to control such emissions under the Clean Air Act (Walke 2011a). This litigation led the EPA, following a protracted process and a settlement agreement, to determine in 2000 that regulation of mercury emissions from power generation was "appropriate and necessary" (US Environmental Protection Agency 2000). The Clean Air Mercury Rule, introduced in 2005, was an effort by the George W. Bush administration to meet the requirement to control mercury emissions from coal-fired power plants. Rather than mandating technology-based control standards, the administration proposed a cap-and-trade scheme similar to that used to reduce emissions of sulfur and nitrogen oxides. The administration argued that this would result in a more economically efficient outcome, and its relatively weak requirements were crafted with strong influence by the power generation industry (Walke 2011b).

Many US states and environmental groups argued that differential applications of control technologies to power plants under the Clean Air Mercury Rule would create mercury "hot spots" in areas such as those around Steubenville, resulting in harm to people who lived near power plants. Shorter-lived forms of emitted mercury travel a distance comparable to sulfur dioxide, for which cap-and-trade approaches were successfully implemented, but the fact that methylmercury is a toxic substance that bioaccumulates in ecosystems made the "hot spot" issue an effective argument. The US Court of Appeals for the District of Columbia Circuit in 2008 vacated the Clean Air Mercury Rule, finding that it did not adequately control mercury emissions based on requirements under the Clean Air Act. Under President Barack Obama, the EPA in 2013 finalized new Mercury and Air Toxics Standards that set technology-based limits on coal-fired power plants. These standards survived repeated court challenges by industry groups, and utilities invested in pollution control measures to comply with them by 2015. In late 2018, the EPA under the administration of President Donald Trump proposed to roll back these standards. The EPA reversed the "appropriate and necessary" finding in spring 2020 (Friedman and Davenport 2020).

Canada introduced federal standards related to mercury emissions in 1996 with the adoption of national guidelines on the use of wastes, including mercury-containing wastes, as supplementary fuels in cement kilns (Government of Canada n.d.). Two years later, the federal government set mercury emission guidelines for cement kilns. The federal government in 2000 formulated technology-based control standards for mercury emissions from smelting and roasting processes in non-ferrous metal production, and ELVs for mercury emissions from waste incineration (Canadian Council of Ministers of the Environment 2000; Canadian Council of Ministers of the Environment 2019). The province of Manitoba had adopted its own standards on waste incinerators two years earlier. The federal government set nationwide limits and provincial caps on mercury emissions from coal-fired power plants in 2006, consisting of overall mercury emissions goals as well as requirements for technology-based controls on new sources (Canadian Council of Ministers of the Environment 2006). Canada took these measures shortly after the United States had introduced the Clean Air Mercury Rule in 2005.

The EU adopted its first regulation to address atmospheric mercury emissions in the late 1980s: a 1989 directive on new waste incineration plants set ELVs for mercury. This directive was superseded by stricter regulations in the 2000 Waste Incineration Directive. Like the United States, the EU did not take action on mercury emissions from power generation in the 1990s, aside from capturing mercury as a side effect of controls that were mandated for other pollutants. The EU's Industrial Emissions Directive from 2010 introduced technology-based standards to control mercury emissions from coal-fired power plants together with emission limits for waste incineration plants and cement kilns. Germany was the only EU member state that had specific mercury emission limits for power generation sources prior to the 2010 directive, which it originally set in 2004 as part of its large combustion plant ordinance. These standards were further revised in 2013 (Mayer et al. 2014). The revised German standard, however, is still roughly five times less stringent than the current US standard (Lin et al. 2017).

Some countries outside of North America and Europe also began to set mercury emission limits starting in the 1990s. For example, in a 1999 air pollution control law, the Philippines set mercury emission limits from industrial sources (Republic of the Philippines 1999). China, by far the world's largest user of coal, established a technology-based standard for mercury emissions from coal-fired power plants in 2011 that was set at the same level as the German standard (Chen et al. 2017; Lin et al. 2017). This emission standard was easily met by most Chinese power plants with basic emission controls, and more advanced air pollution controls for sulfur dioxide and nitrogen oxides capture an even greater percentage of mercury in coal (Sloss 2012). Indonesia introduced mercury-specific emission standards for coal-fired power plants starting in 2020, but these were set at a modest level. Benefits from these control measures will be dwarfed by an increased use of coal, however, if coal use increases as expected over the next decade.

Some countries recently strengthened their legislation on mercury emissions, while others continue to rely on capturing mercury as a result of controls on other pollutants. South Korea strengthened its national mercury emission limits in 2010 (Sloss 2012). These limits, however, are less stringent than both the German and US standards (Sloss 2012). Japan amended its air pollution control act in 2015 to include a mercury emission control system, including technology-based standards (Government of Japan n.d.). India, the world's second largest user of coal, is increasing its coal capacity without mercury-specific emission controls. Similarly, South Africa relies on technology-based controls for particulate matter, sulfur dioxide, and nitrogen oxide to reduce mercury emissions. This may result in a capture of 6 to 13 percent of mercury emissions by 2026 compared to a 2011 to 2015 emissions baseline (Garnham and Langerman 2016). However, an increase in coal use could result in a net increase in mercury emissions in South Africa. Russia is another country that lacks mercury-specific controls on coal-fired power plants, but has standards for non-ferrous metals production.

International bodies have developed guidelines for the use of air pollution control devices (box 2-2). Some mercury air pollution controls are included in regional agreements covering North America and Europe. Canada, the United States, and Mexico, acting under the Commission for Environmental Cooperation that was set up as part of the implementation of the North American Free Trade Agreement (NAFTA), adopted a joint action plan on mercury in 1997 with general commitments to address pollution (Commission for Environmental Cooperation 1997). A second phase of the action plan, launched in 2000, set an aspirational goal to reduce national mercury emissions by 50 percent from 1990 levels by 2006 (Commission for

Environmental Cooperation 2000). The US and Canada both exceeded this goal. The parties to CLRTAP adopted a protocol on heavy metals in 1998 that entered into force in 2003 (Selin and Selin 2006). Parties to the protocol, which was made more stringent in 2012, include the United States, Canada, the EU, and a few other European countries. The protocol targets mercury emissions from large stationary sources and mandates that parties reduce their mercury emissions below their levels in 1990 (or an alternative year between 1985 and 1995). It sets ELVs and stipulates that parties apply technology-based controls in the form of BAT.

The existence of national and regional regulations and the availability of effective end-of-pipe control technology facilitated the decision to include industrial atmospheric emission sources in the Minamata Convention. The Global Mercury Partnership built on this information to produce a summary of available control technology for power plants, informing treaty negotiations (Krishnakumar et al. 2012). The Minamata Convention mandates the application of controls on new and existing emission sources in five different categories: coal-fired power plants, coal-fired industrial boilers, non-ferrous metals production, waste incineration facilities, and cement production. The regulatory approach is similar to the one under the CLRTAP heavy metals protocol. Parties must apply BAT, BEP, or ELVs to new point sources to control emissions, and where feasible reduce them, no later than five years after they join. Parties must also control, and where feasible reduce, emissions from existing point sources through BAT, BEP, or ELVs, or a multi-pollutant control strategy no later than 10 years after the treaty becomes legally binding for them.

The Minamata Convention largely leaves it up to each party to define BAT, BEP, and ELVs for stationary sources based their own socio-economic and technical context, but the treaty's Conference of the Parties (COP) is tasked to develop technical guidance documents to assist this process. This decision was based on the recognition that some parties had already set their own (and sometimes diverging) standards. It was also a way to allow developing countries to initially adopt standards that may not be as stringent as in industrialized countries. As a result, different "best available" domestic standards will capture varying amounts of mercury. Countries may also formulate standards using different metrics. For example, the United States currently uses performance metrics (lb/Btu), while others including China and the EU apply concentration limits ($\mu g/m^3$). In addition, the language in the Minamata Convention regarding "control, and where feasible reduce" mercury emissions from stationary sources represents a compromise that allowed for increases in coal use—and may, even with stringent application of BAT, result in increasing mercury emissions globally. Increased waste incineration, cement production, and mining may also result in higher mercury emissions despite the introduction of stricter emission controls.

Industry resistance to mercury emission controls is strong in some countries, and lobbying by industry can influence government positions, including on joining the Minamata Convention. Australia's decision so far to not ratify the Minamata Convention is attributed to opposition from the fossil fuel sector to stricter controls on mercury emissions (Bramwell et al. 2018). Australia lacks a federal standard on mercury emissions. Some Australian states have set no mercury-specific standards on coal-fired power plants while other states have only lax ones—some plants in New South Wales are allowed to emit 33 times more mercury than they would under the German and Chinese standards and 666 times more mercury than if they operated under the US standard (Environmental Justice Australia 2017). New Zealand, which similarly is not a party to the Minamata Convention, also has not set any mercury-specific emission standards. New Zealand's largest energy retailer, Genesis Energy, announced in 2015 that it would close the country's last two coal-fired power plants in 2018, but this decision was later revisited, and coal burning is expected to continue until at least 2025 and maybe longer (Coughlan 2018).

Scientific data show that stringent emission controls on industrial sources can reduce the environmental and human health burden of mercury emissions. Domestic and regional controls through the 1990s and 2000s reduced atmospheric mercury emissions, concentrations, and deposition near regulated point sources in North America. A study of mercury air concentrations from two sites in Vermont and New York found declines of a few percent per year from 1992 to 2014, with faster declines in the beginning and then slowing down or leveling off (Zhou et al. 2017). The study attributed these declines to reduced mercury emissions from regional sources, and correlations with declines in sulfur dioxide indicate that reductions in emissions from coal-fired power plants contributed to these reductions in atmospheric mercury concentrations. Declining emissions are also linked to reductions in mercury concentrations in fish (Hutcheson et al. 2014).

Another study reported that mercury concentrations in rainfall declined overall between 1998 and 2005 in the northeastern United States (Butler et al. 2008). Much of this decline was attributed to the domestic controls on municipal waste incineration that were implemented in the 1990s.

Measurement studies carried out outside of North America also find declines in atmospheric mercury emissions, concentrations, and deposition in some regions starting in the 1980s, concurrently with the introduction of domestic and regional emission controls. Records from a few long-term stations in Europe show that atmospheric mercury concentrations declined substantially from 1980 to 1993 together with regional mercury emission reductions (Tørseth et al. 2012). Atmospheric concentrations at the monitoring site in Sweden with the longest time series of data declined about 60 percent between 1980 and the early 1990s (Tørseth et al. 2012). Another study showed a 7 percent yearly decline in atmospheric concentrations at Wank Mountain in southern Germany between 1990 and 1996 (Slemr and Scheel 1998). In addition, scientists have attributed measured decreases in atmospheric mercury concentration in the early 2010s in parts of China to declining mercury emissions from nearby domestic sources (Tang et al. 2018). Declines in mercury emissions also have longer-term benefits, as the amount of mercury available to remobilize and cycle later on is reduced (see chapter 3).

Trends in mercury concentrations and deposition at other sites, including those farther away from point sources, are less consistent. Measured mercury deposition in rainfall has increased at some sites—even in the United States where national emissions have declined—complicating interpretation of overall trends (Butler et al. 2008; Weiss-Penzias et al. 2016). Variations in weather patterns also affect trends in mercury deposition (Giang et al. 2018). Scientists debate whether observed atmospheric concentration trends in past decades at remote monitoring sites in North America and Europe—as opposed to those directly downwind of controlled sources—reflect regional or global emission declines (Zhang et al. 2016; Obrist et al. 2018). Global mercury emissions were relatively stable between 1980 and 2000, and increased slightly after that, but their regional distribution changed (Streets et al. 2017). The timescales of these changes are also affected by the cycling of historical emissions in the oceans (Soerensen et al. 2012). Some evidence suggests that concentrations in the Southern Hemisphere may have begun increasing again after 2007 (Martin et al.

2017). The more recently reported 20 percent increase in global emissions between 2010 and 2015, driven by emissions in Asia, may also affect future trends in atmospheric concentration and deposition.

Efforts to Reduce Dependence on Fossil Fuels

International bodies, national and local governments, and industries influence fossil fuel use in different ways (box 3-2). These types of interventions are often associated with efforts to address carbon dioxide emissions, such as those under the Paris Agreement. Substituting the underlying activity leading to pollution differs from efforts to reduce pollution that increase the efficiency of coal-fired power plants and other stationary sources or that apply end-of-pipe controls to capture emissions. Moving the world's energy system away from fossil fuels—necessary from a climate change perspective—would prevent mercury mobilization from this source entirely, including air emissions and releases to land and water. It would also eliminate the need to properly dispose of captured mercury. Future mercury emissions from combustion and industrial production will largely be determined by how much coal is used, in Asia in particular, and the implementation of pollution controls in countries with the highest coal use (Streets et al. 2009). The benefits of climate policy for mercury pollution are thus likely to be largest in places that rely heavily on coal (Rafaj et al. 2014).

Few policy-makers so far have shown an appetite for linking mercury reduction efforts with the more contentious politics of fossil fuel reductions. China and India stressed the importance of coal burning for economic development during the Minamata Convention negotiations, and India linked coal use to its push for greater electrification. Negotiators from the United States and the EU who were pushing for action on atmospheric emissions also noted the continuing need for coal burning in some places. Moving away from coal completely would eliminate its associated mobilization of mercury, but state-of-the-art, end-of-pipe technologies can reduce atmospheric mercury emissions by 90 percent or more. China, India, and Indonesia are projected to significantly increase their use of coal from today's levels despite their climate change commitments under the Paris Agreement (Edenhofer et al. 2018). In these and other countries, the control technologies that address mercury can be implemented far more cheaply than large-scale changes to a fossil fuel–free energy supply. But provisions under the Paris Agreement and the Minamata Convention can be synergistic, as they can result in reductions of mercury from different sectors (Mulvaney et al. 2020).

Mandates to apply end-of-pipe technologies can either extend or shorten the lifetime of polluting sources. With less environmental burden due to the application of pollution control technology, more modern coal-fired power plants may remain in the energy system for longer than they otherwise would. In China, for example, new coal-fired power plants are being built that accommodate state-of-the-art, end-of-pipe pollutant control technology to capture mercury and other air pollutants (but not carbon dioxide), reducing reasons to close them down. In contrast, technologies can shorten the lifetime of some, especially older, power plants. In the run-up to the Mercury and Air Toxics Standards taking effect in 2015 in the United States, anticipation of future regulatory actions reportedly prompted the retirement of several older coal plants, whose operators cited the cost of compliance with federal regulations as a reason for their premature closure (Storrow 2017). Mercury standards are also thought to have been a factor in the closure of coal-fired power plants in Canada, or their conversion to other fuels such as natural gas or wood (Sloss 2012).

Insights

The story of air pollution in Steubenville at the beginning of this chapter illustrates both the ability and difficulty of addressing mercury emissions to the atmosphere from the large point sources that are integral to industrial and energy production in contemporary societies. In this section, we draw insights from the interactions and interventions that characterize the atmospheric system for mercury. First, atmospheric mercury connects places across the world not only through its environmental transport but also through connections facilitated by socio-economic factors and institutions. Second, actions to address mercury have largely involved incremental transitions, but these have had substantial benefits for human well-being in particular places. Finally, efforts to govern mercury emissions should consider their multi-scale, complex nature.

Systems Analysis for Sustainability

Understanding how mercury emissions to air disperse in the environment and impact human well-being requires paying attention to local as well as long-range dynamics. Mercury from industrial point sources is emitted in different forms, which can deposit either nearby or far away. Mercury emissions that deposit regionally in the near term can be remobilized and cycle further through the environment, later affecting other populations far away in time and space. End-of-pipe control technologies change both the total amount of mercury emitted and the relative fraction of emissions in shorter- and longer-lived forms. Mercury captured in air pollution control technology creates mercury storage needs, and this mercury can be further distributed when using coal-combustion byproducts in industrial processes. The introduction of stricter mercury-specific pollution controls primarily in North America and the EU beginning in the late 1900s resulted in emission reductions. In contrast, in Asia, mercury emissions from coal burning and industrial production continue to rise.

A systems perspective on atmospheric mercury emissions from industrial point sources should take into account how the locations and levels of such emissions have changed over time, as the provision of energy and consumer goods has increased worldwide. Interventions to reduce mercury emissions in North America and Europe have accomplished their goals through setting technology-based emission controls from large point sources. This allowed the production of energy and consumer goods to continue providing benefits to human well-being, without altering underlying processes of coal burning and industrial manufacturing. Governments in other regions of the world to date have intervened less to target mercury emissions from industrial point sources. This allows mercury emissions to continue in these regions, and also shifts the relative importance of different regions for global emissions. Mercury emissions between 2010 and 2015 increased in all regions outside North America and Europe, but Asian mercury emissions are particularly high. China has introduced mercury-specific emission controls, but India has yet to move in this direction.

Mercury travels globally, but global average concentrations alone do not reflect the harms posed by mercury to the environment and human well-being. This is because the different forms of mercury emitted to the atmosphere from large point sources can either disperse globally or deposit regionally and locally. Ecosystems downwind from coal-fired power plants and other industrial sources may experience high levels of mercury deposition, leading to exposure to nearby populations. These "hot spots" of mercury emissions and deposition are a political issue of importance to human health and the environment in many regions of the world. Trends in environmental concentration and deposition of mercury track national and global emission trends in some places, but not in other locations. Even in the United States, where national mercury emissions declined dramatically over the past three decades, there remain places where measured mercury deposition increased during the 2000s. As a result, studies of environmental and human health impacts of mercury emissions from large point sources must consider local, national, regional, and global trends in emissions and deposition simultaneously.

Sustainability Definitions and Transitions

The underlying production and consumption processes resulting in atmospheric mercury emissions lead both to benefits and harms to human well-being that have been valued differently in different places over time. This complicates efforts to define what constitutes progress toward greater sustainability. The production of fossil fuel–based energy and industrial goods has had substantial material and health benefits to people for over 250 years. At the same time, mercury that was emitted into the atmosphere through these processes has damaged human health and the environment, and created a legacy of long-lasting environmental pollution. The burning of fossil fuels and industrial manufacturing, and associated mercury emissions, continue decades after mercury was identified as a major air pollutant. Some governments and private sector actors have attempted to reduce the amount of mercury that is emitted into the atmosphere together with other air pollutants, but governments in other countries do not view mercury emissions to be dangerous enough to warrant the economic costs of mandating pollution controls. As a result of this and other factors, the negative impacts of mercury emissions are unevenly distributed across the planet and influence future generations.

Most national and regional action on mercury emissions from large point sources have relied on an incremental introduction of stricter pollution control technologies rather than on seeking shifts in underlying activities of coal burning and industrial production. Such incremental transitions toward reducing mercury emissions facilitated the continued burning of fossil fuel and industrial manufacturing with fewer environmental and public health burdens. These actions had substantial benefits in regions that took action, but mercury remains a global problem. The

Global Mercury Partnership and the Minamata Convention largely codify this vision of a technology-based approach to addressing atmospheric mercury emissions at the global level. Neither the Global Mercury Partnership nor the Minamata Convention has encouraged the deeper systemic change of rapidly phasing out the use of fossil fuels. Pollution prevention has reduced the quantity of mercury that would otherwise have entered the atmosphere, with both near-term and long-term benefits, but the limits of a technology-based approach to controlling mercury emissions are evident by the fact that global mercury emissions still grew in the early 2010s.

The atmospheric mercury system shows that incremental, technology-based change and long-term systemic change can be both conflicting and synergistic, sometimes simultaneously. Technology-based approaches can, under some conditions, lead to more fundamental change. Mercury standards in the United States and Canada accelerated the closing of at least some coal-fired power plants that were too old or unprofitable to warrant the addition of new end-of-pipe technology. This illustrates how dynamics that lead to gradual progress toward sustainability and to more disruptive change can be simultaneously present, and potentially reinforcing. In the near-term, however, there is a potential for tension between policies that support incremental change and those that call for quicker and more radical change. The introduction of emission standards and control technology on industrial sources can mitigate mercury emissions and other local pollution problems, but the resultant longer operability of modern coal-fired power plants can leave the longer-range effects and broader implications of carbon dioxide emissions on climate change largely unaddressed.

Sustainability Governance

Efforts to reduce atmospheric mercury emissions rely on creating and adapting institutions that fit the multi-scale nature of the problem. Addressing only national and regional mercury emissions—illustrated by government-led actions in North America and Europe in the 1990s and early 2000s—can lead to emission reductions, but both regions still have problems with mercury deposition from the long-range transport of mercury emissions from sources located elsewhere. This shows the importance of a global governance approach to controlling mercury emissions. The Minamata Convention covers mercury emissions globally, but also attempts to balance its global scope with regional and national flexibility in formulating air pollution laws and setting mercury emission standards. It requires all parties to control, and where feasible, reduce their mercury emissions from certain categories of point sources, but it is largely left to the individual parties to define their own standards and approaches to mercury and multi-pollutant controls. International cooperation helps diffuse knowledge about regulatory practices and different air pollution control techniques.

A range of regulatory strategies can reduce mercury emissions. Multi-pollutant control technology can increase the efficiency of pollutant capture and reduce implementation costs. Some end-of-pipe technologies allow countries to control mercury emissions using strategies that are primarily designed to address other air pollutants such as fine particulate matter. Taking advantage of material synergies can be politically attractive in some cases, but not in others. Reductions in mercury emissions from industrial point sources during the twentieth century came as an unintended side effect of efforts to reduce emissions of other pollutants. But links between addressing mercury emissions from coal-fired power plants and the more politically contentious issue of climate change mitigation create governance challenges. Reducing mercury emissions from industrial sources is both technically feasible and politically attractive in many countries across the world, but phasing out coal is less so. Countries decided during the negotiations of the Minamata Convention that it was preferable to have a treaty that merely mandates end-of-pipe measures for large stationary sources without any further restrictions on coal burning than to have no mercury treaty at all.

It is difficult to assess the ultimate effectiveness of governance efforts to protect the environment and human health from mercury emissions from large point sources. Mercury emissions from these sources can decrease due to factors other than regulatory controls, including market-driven changes in coal burning and industrial manufacturing. Variations in weather patterns and ecosystem processes furthermore make interpreting mercury concentration trends in the environment difficult. This affects the ability to monitor the impacts of implementation and compliance with pollution standards. Mercury captured by pollution control technology can also later be discharged into the environment through reuse in other industrial manufacturing processes unless there are proper standards in place. In the absence of a large societal shift away from fossil fuels—whether driven by economic factors in energy markets, by concerns about air pollution and climate change, or by a combination of those factors and concerns—the greater application of technology-based controls addressing mercury emissions as mandated by the Minamata Convention remains critical to protecting human well-being.

Mercury, the winged messenger god, conveys a mixed message in terms of efforts to control mercury emissions and deposition. Industrialization resulted in emissions of mercury to the atmosphere from coal burning and other point sources. Some early efforts to address other pollutants from these sources also reduced mercury emissions, but governments have taken specific actions targeting mercury emissions only over the past few decades. Regulatory efforts to control emissions of mercury and other harmful pollutants to date have focused on technology-based approaches that have both near-term and long-term environmental and human health benefits. Mercury emissions have declined in North America and Europe, but emissions have grown elsewhere, particularly in Asia. The geographical origin of emissions and their local impacts have thus shifted, but atmospheric transport carries much mercury across borders. Future policy efforts that focus more aggressively on eliminating fossil fuel use by restructuring energy systems could ultimately prevent industrial mercury emissions at their source.

6   Assets and Liabilities: Mercury, God of Commerce

Those who study environmental science and governance are familiar with a historical evolution common to many substances: they are at first highly valued by society, then progressively seen as liabilities through their societal and environmental distribution, as their dangers to human and ecosystem health become increasingly appreciated. Useful properties of such substances, including mercury, supported scientific and technological advances and provided many other social and economic benefits to people and societies. However, some substances like mercury also caused severe human health damages in occupational settings and through their local and global dispersion in the environment. Changes in perceptions of hazardous substances, and societal responses to their dangers, occurred at different times and in different ways in places across the globe. Understanding the benefits and harms that stem from the intentional uses of hazardous substances is important for better managing societal uses of materials toward greater sustainability.

Daniel Gabriel Fahrenheit constructed the world's first mercury thermometer in 1714 in Amsterdam. This groundbreaking invention was a significant improvement over existing, much less accurate, temperature-measuring devices. More than a century earlier, in the 1590s, Galileo Galilei had developed a thermoscope, or air thermometer, as a crude way to measure changes in ambient temperatures (Camuffo and Bertolin 2012). The thermoscope consisted of a glass tube that was submerged in water at one end; the rise and fall of the water line indicated relative atmospheric temperature differences. This construction of the thermoscope built on the fact that air expands when it is heated, a scientific discovery made by Philo of Byzantium in the third century BCE. Galileo was not able to measure temperatures against a fixed degree scale, however, and his rudimentary thermoscope was largely unreliable because its measurements varied with atmospheric pressure (Rasmussen 2012). The invention of the first modern thermometer is often credited to Santorio Santorio, who applied a simple measurement scale to Galileo's thermoscope (Middleton 1966).

The introduction of fully sealed liquid-based thermometers using alcohol—developed beginning in the 1640s by several people including Evangelista Torricelli and the Grand Duke of Tuscany Ferdinand II—represented another important technological step forward (Wright and Mackowiak 2016). But these alcohol-based thermometers did not solve the fundamental problem: a lack of measurement precision and comparability. This all changed with the invention of the mercury-in-glass thermometer, which allowed Fahrenheit to conduct and record measurements with much more accuracy and detail. With his new thermometer, Fahrenheit was also able in 1724 to devise the first standardized scale to measure temperatures: the Fahrenheit scale where water freezes at +32 degrees and boils at +212 degrees. Using the new mercury thermometer, Anders Celsius in 1742 invented an alternative scale that reversed the hot and cold ends and set the water boiling point at 0 degrees and the water freezing point at +100 degrees. Proposals to invert this scale came a year later, resulting in the contemporary Celsius scale where water freezes at 0 degrees and boils at +100 degrees (Bolton 1900).

The story of the invention of the mercury thermometer is one example of how mercury has been an important component in a broad range of consumer products and production processes. Electronic thermometers are increasingly replacing the basic mercury thermometer, but the mercury thermometer was the dominant technology for nearly 300 years because of its simplicity, precision, and reliability. It was successfully used to advance scientific knowledge in meteorology and climatology, to develop new manufacturing techniques that required the ability to read accurate temperature measurements, and to diagnose and treat illnesses in patients all over the world. Many other commercial products and production methods also relied on the unique properties of mercury to provide a wide range of social and economic benefits. But as the chapter title, referring to Mercury, the god of commerce, indicates, societal attitudes about mercury fundamentally shifted over time, from seeing the element as a valuable asset that can be beneficially used, to viewing it as a liability whose intentional uses should be regulated and eventually phased out altogether.

Many governments have introduced increasingly strict controls on the use of mercury, both in consumer products and production processes. Their efforts to phase out mercury uses mirror other attempts to regulate the commercial use of toxic substances. Some of these regulated substances are naturally occurring elements like mercury, whereas others are synthesized in laboratories. Well-known examples include other heavy metals such as lead and cadmium and organic chemicals like DDT, PCBs, and CFCs. Plastics—a building block of modern consumer societies and now ubiquitous in landfills and oceans—are in the process of a similar transition. Other substances that societies currently place high value on—such as rare earth elements—might become the next generation of liabilities. Continuing uses of mercury and other hazardous substances may be warranted in some applications, at least in the short term, where decision-makers judge that societal benefits outweigh risks and liabilities. Ultimately, these decisions reflect broader issues of how societies identify and govern risks from materials, which vary in their utility as well as their hazard.

In this chapter, we examine intentional uses of mercury in products—excluding medicine, which we discussed in chapter 4—as well as in production processes, and societal efforts to reduce and eliminate such use. In the section on system components, we outline the human, technical, and environmental components that influence the use of mercury in products and processes, and the institutions and knowledge that surround these uses. We cover in the interactions section how the development of technologies for mercury-added products and processes led to social and economic benefits as well as negative environmental and human health consequences. We discuss efforts to reduce and eliminate mercury uses in products and processes, and address as well the negative consequences of related mercury discharges, in the section on interventions. In the final section on insights, we discuss how mercury's benefits and costs to different populations were affected by system interactions over time, how largely incremental transitions reduced or eliminated mercury in products and processes, and how policy efforts can more effectively govern products and processes across scales.

System Components

Mercury that has been used in consumer products and production processes originated from two main sources: primary mercury mining and the reuse of mercury already in commerce. Of the 1 million tonnes of elemental mercury estimated to have been extracted from cinnabar and other ores since the year 1500, roughly half was used in a multitude of products and processes, while the other half was used in gold and silver mining (Hylander and Meili 2003). Much commercial mercury has been recycled and reused, sometimes multiple times, in other products and production processes. This means that some of the mercury in products and processes has cycled through society for long periods. The present-day main human, technical, environmental, institutional, and knowledge components for the products and processes system for mercury are summarized in figure 6.1.

Humans have been *producers and consumers of mercury-added products and other goods made using mercury* for millennia. Globally, annual use of *mercury in commerce* peaked in 1970, associated with a peak in annual mercury mining of 10,000 tonnes (Hylander and Meili 2003). Commercial mercury demand has declined since then, but mercury is still used in a variety of *mercury-added products* as well as *other goods made using mercury*. An estimated 2,040 to 3,600 tonnes of mercury were consumed in mercury-added products and production processes in 2015 (UNEP 2017). This estimate

Human components	Technical components	Environmental components
Producers and consumers of mercury-added products and other goods made using mercury Workers and employers in mercury-related sectors People living near mercury discharges or contaminated sites People living far from mercury discharges	Mercury in commerce Mercury-added products Other goods made using mercury Mercury in production processes Extracted fossil fuels Mercury in stockpiles and landfills Industrial point sources of mercury discharges	Geological reservoirs Ecosystems near mercury uses Contaminated sites Ecosystems far from mercury uses

Institutional components	Knowledge components
Markets for mercury-added products and goods made using mercury National and local laws and regulations Mercury markets Global Mercury Partnership Minamata Convention	Properties of mercury Mercury-based product development and production techniques Individual and societal beliefs Environmental impacts from mercury discharges Health impacts from mercury exposure Health protection techniques Mercury-free product development and production techniques

Figure 6.1
Components in the products and processes system for mercury (referenced in the text in italic type).

excludes mercury use in artisanal and small-scale gold mining (ASGM), which can also be considered a process-based use (we address that use separately in chapter 7). The term "mercury consumption" refers to the mercury content of all mercury-added products used, as well as the gross mercury input into all industrial processes during a given year. Production processes consumed 46 percent of this non-ASGM mercury; 44 percent was used in a variety of products, including batteries, measuring and control devices, lamps, and electrical and electronic devices; and the remaining 10 percent was used for dental amalgam.

Human uses of mercury in products and processes relied on the availability of mercury in *geological reservoirs* as well as knowledge about the *properties of mercury* and *mercury-based product development and production techniques*. Some of the earliest uses of cinnabar, connected to *individual and societal beliefs*, had religious and mystical connections in Asia, Africa, and Latin America (Mahdihassan 1985). Vermillion, a pigment made from grinding cinnabar powder, was applied to preserve human bones in graves from the Neolithic period in Spain in 3000 BCE (Martín-Gil et al. 1995). Cinnabar, which has a deep red hue, was also used as a coloring agent. Asian uses of "Chinese Red," a cinnabar-based color for painting vases made from lacquer, go back at least to the first millennium BCE (Gettens et al. 1972). The Roman author Pliny the Elder (23–79 CE) noted that "the ancients" used to paint with cinnabar that was "adulterated by the agency of goats' blood, or of bruised sorb-apples" (Pliny the Elder n.d., chapter 32: Quicksilver). Romans used mercury as an ingredient in makeup and other beauty products. For example, women used cinnabar as a rouge to heighten the color of their cheeks (Stewart 2014). People today use skin-lightening creams that contain mercury.

Alchemy, an ancient branch of natural philosophy and the forerunner of modern chemistry, focused on ways to improve, purify, or otherwise transform certain materials; mercury was central to alchemical concepts and work, and alchemists were the ones who gave the name quicksilver to mercury (Goldwater 1972). Attempts to unify the physical world linked the sun, the moon, and the five nearer planets to seven metals, including mercury, more than 2,000 years ago in Europe (Crosland 2004). Alchemy included a long-standing search for the philosopher's stone—an elusive but potentially very valuable substance capable of turning base metals such as mercury into much more valuable gold (Goldwater 1972). Alchemical efforts continued at major universities well into the 1900s. As late as 1924, two chemists at Berlin Technical University claimed to have turned mercury into gold (Anonymous 1924). Alas, the search for the philosophers' stone and efforts to change other metals into gold have proved unsuccessful.

Workers and employers in mercury-related sectors have used mercury and mercury compounds extensively. Mercury use in the silver-plating industry dates back at least to Roman times (Pliny the Elder n.d.). The use of *mercury in production processes* expanded beginning in the fourteenth century, first in Europe and then elsewhere (Svidén and Jonsson 2001). These early mercury uses included the production of mirrors and hats. Mercury use also grew in laboratories as science progressed. Industrial applications of mercury increased in the nineteenth century. Mercury fulminate was used in percussion and blasting caps starting in the 1800s. Mercury was a component of lamps beginning in the 1800s, and of batteries in the second half of the twentieth century. Mercury compounds were used in agricultural pesticides beginning in the 1910s (Novick 1969). In the 1900s, the pulp and paper industry added phenylmercury acetate to prevent the growth of fungi in the pulp during manufacturing, and to protect wood pulp during storage and shipment (Löfroth and Duffy 1969; Novick 1969). Phenylmercury was also applied as a biocide in indoor and outdoor paint (including in antifouling paint for ships).

The chemicals industry developed several different production processes that involved mercury to transform *extracted fossil fuels* such as coal and oil to make societally important high-volume chemicals. These included vinyl chloride monomer (VCM, a building block of plastic polyvinyl chloride, or PVC), acetaldehyde (an intermediate in the production of chemicals and plastics), and polyurethane (polymers used in many products). The chemicals industry also used mercury in processes that produced chlorine and alkali (caustic soda and sodium hydroxide used to make soap, textiles, and chemicals) from brine (sodium chloride solution). In the mid-twentieth century, mercury was used to make enriched lithium-6 for the development of thermonuclear weapons. In the nuclear power sector, lithium-7—a byproduct of mercury-based lithium-6 production—is a key component in the fluoride cooling systems of molten salt nuclear reactors developed in the 1950s.

Mercury that is no longer in commercial use either exists as *mercury in stockpiles and landfills* or has been discharged into the environment. *Industrial point sources of mercury discharges* have historically been a large source of releases of mercury to land and water as well as emissions to air. Some of the mercury releases first entered *ecosystems near mercury uses*. Mercury from industry, agriculture, and leaks and wastes from mercury-added products have combined to create a large number of *contaminated sites*. It is estimated that there are currently over 3,000 mercury-contaminated sites in different parts of the world (Kocman et al. 2013). Releases of mercury have created health risks to some *people living near mercury discharges or contaminated sites*. In addition, a portion of the discharged mercury that was used in products and processes has entered the atmosphere, where it can travel long distances before depositing in *ecosystems far from mercury uses*. As a result, *people living far from mercury discharges* are also exposed to mercury in the environment that originates from its use in products and processes.

Mercury's use in a growing number of products and processes depended heavily on the development of knowledge of its properties. As *markets for mercury-added products and goods made using mercury* diffused these products and goods throughout societies, knowledge began over time to include information about negative *environmental impacts from mercury discharges* as well as *health impacts from mercury exposure*. The formulation of labor standards was facilitated by knowledge of *health protection techniques* for workers. Governments also adopted *national and local laws and regulations* on mercury uses and discharges. Uses of and human exposure to mercury were reduced through new knowledge of *mercury-free product development and production techniques*, especially since the 1970s. The development and commercialization of alternative products and processes that did not rely on the use of mercury have had a large impact on *mercury markets* by significantly reducing the demand for mercury. At the international level, the *Global Mercury Partnership* supports the further development and introduction of mercury-free alternatives, while the *Minamata Convention* mandates reductions and phaseouts of mercury use in products and processes.

Interactions

Intentional uses of mercury in consumer products and production processes have had both positive and negative influences on human well-being. Figure 6.2 shows interactions in the products and processes system for mercury: we have selected two interactions (the items in bold type in boxes

Knowledge

Institutions

	1. Human	2. Technical	3. Environmental
1. Human	(1-1) Producers and consumers interact in socio-economic systems	(1-2) Producers and consumers sell and buy mercury-added products and other goods made using mercury	(1-3) Producers and consumers intentionally discharge mercury-added products into ecosystems
2. Technical	(2-1) **Mercury-added products and other goods made using mercury provide benefits to and harm people**	(2-2) Mercury in products and production processes influences quantities of mercury in commerce and stockpiles	(2-3) **Mercury-added products and industrial point sources discharge mercury into ecosystems**
3. Environmental	(3-1) Methylmercury in ecosystems and contaminated sites affects people	(3-2) Cinnabar ores provide mercury for use in products and processes	(3-3) Ecosystem processes transport mercury and lead to production of methylmercury

Figure 6.2
Interaction matrix for the products and processes system for mercury.

2-1 and 2-3) to focus on in this section; we then trace the pathways that influence them, which we summarize in figure 6.3 (where the bold boxes correspond to the selected interactions). First, mercury-added products and other goods using mercury provide benefits for producers and consumers but also harm them (box 2-1), and mercury uses affect and are affected by market dynamics and resource availability (boxes 3-2, 2-2, 1-1, and 1-2). Second, mercury-added products and industrial point sources discharge mercury into ecosystems (box 2-3), where it is transported and transformed into methylmercury and subsequently affects people (boxes 3-3, 3-1, and 1-3).

Commercial Mercury Benefits and Harms

Mercury-added products and other goods made using mercury have provided benefits to as well as harmed people (box 2-1). Primary mining of

a) Commercial mercury benefits and harms: Economic and social conditions and geological reservoirs containing mercury prompt the development of consumer goods that provide human benefits but also cause harms

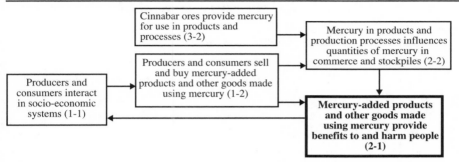

b) Commercial mercury and the environment: Commercial mercury enters ecosystems where it affects human health

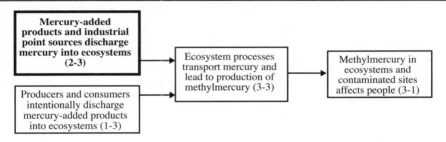

Figure 6.3
Pathways of interactions in the products and processes system for mercury. Bold boxes indicate focal interaction for each subsection.

cinnabar ores provided much of the mercury for early uses in products and processes (box 3-2). Mercury has been mined in all regions of the world, but much extracted mercury from cinnabar during preindustrial times came from a smaller number of mercury mines (see chapter 3). As the commercial demand for mercury in the chemicals industry and other sectors grew in industrial times, recycling of mercury from its original use became an increasingly important source of supply for both domestic use and trade. Mercury in production processes influences quantities of mercury in commerce and the amount of mercury stored in stockpiles (box 2-2). The gradual closing of previously large mercury mines coupled with the US and European Union (EU) elemental mercury export bans that were adopted in the 2000s made recycling and reuse a more important source of commercial mercury. At the same time, the amount of excess industrial mercury grew as industries switched to mercury-free manufacturing alternatives, which increased the amount of mercury in stockpiles in many countries.

Changes in the supply and demand for commercial mercury have been influenced by producers and consumers interacting in socio-economic systems (box 1-1). Producers and consumers for thousands of years have sold and bought mercury-added products and other goods made by using mercury (box 1-2). In addition to mercury's use in traditional medicine (see chapter 4), ancient texts suggest that mercury was used in magico-religious rituals in China and India going back millennia. Mercury was important in Hinduism, including as a representative of the seed of Lord Shiva, and it was also discovered in Egyptian tombs (Masur 2011). Aristotle wrote in the fourth century BCE about the use of mercury in religious ceremonies (Nriagu 1979). In Caribbean religions with African roots such as Santeria, Palo, and Vodou, sprinkling mercury inside homes or mixing it into perfumes, lotions, or soap and water for ritualistic purposes is thought to bring good luck and ward off evil spirits (Wendroff 2005; Newby et al. 2006; Wexler 2016). Some of these practices continue in the Caribbean and in other countries where followers have relocated, including the United States (Yehle 2011).

Many early commercial mercury uses were associated with social status and affluence. Cinnabar imported from Spain and India was used in ancient Rome, where cinnabar-containing products were scarce and associated with prosperity (Stewart 2014). The expanded use of mercury in manufacturing that began in the fourteenth century not only increased the amount of mercury in commerce, but also solidified the metal's links to prominence and wealth. Processes involving mercury were used to produce relatively expensive goods such as large mirrors and fur hats, whose ownership conferred social status, as only people with financial means could afford them. Mercury-based manufacturing techniques also provided economic advantages for producers. The technique of using mercuric nitrate in the fur-felting process in hat making, developed in France, was initially kept a trade secret because it helped French hat makers corner the market for a superior and highly desirable product. As the knowledge of how to use mercuric nitrate spread, many hat makers in other countries switched over to the new mercury-based production method (Svidén and Jonsson 2001).

Mercury and mercury-added products were important to the development of new scientific knowledge and capabilities that had societal benefits, particularly beginning in the seventeenth century. As Leonard J. Goldwater argues: "Without mercury some of the most significant advances in chemistry would have been delayed for years and possibly for centuries" (Goldwater 1972, 98). Experiments using mercury and mercury-added instruments facilitated the discovery of more than 20 chemical elements. In perhaps the most important of these discoveries, Joseph Priestly and Carl Wilhelm Scheele, in the mid-1770s, independently used mercury to identify oxygen—also referred to as "fire air"—which represented a major turning point in the history of modern science. Mercury-containing instruments also facilitated the development of dialysis and osmosis. The mercury thermometer, described at the beginning of this chapter, advanced both meteorology and medicine. The mercury barometer, invented by Evangelista Torricelli in 1643, greatly enhanced the ability to accurately measure atmospheric pressure (Middleton 1963; West 2013). In 1896, Scipione Riva-Rocci revolutionized medicine when he introduced the mercury-based sphygmomanometer for measuring blood pressure by wrapping an inflatable cuff around the upper arm (Roguin 2005).

Mercury continued to provide benefits to more modern societies. Mercury fulminate in blasting caps, including those invented by Alfred Nobel in the 1860s, helped build roads, bridges, and dams. Mercury was also a key component in the development of lighting technology. E. H. Jackson took out the first patent for a mercury-containing lamp in London in 1852, building on scientific discoveries that mercury could produce light when vaporized in an electric arc (Perkin 1911). Commercialization of this technology began in 1901 with a design by the engineer Peter Cooper Hewitt, whose company was acquired later by the General Electric Company (Cleveland and Morris 2014). Higher-intensity mercury lamps were introduced in 1934, and their efficiency and long life made the mercury-vapor lamp useful in public lighting, including for streets and highways (Freeman 1940). Mercury lamps were also used to light the large workspaces required for airplane factories and hangars (Anonymous 1945). Mercury was a key component in the first commercially viable fluorescent lamp, patented in 1926 (Cleveland and Morris 2014), and continues to be used in fluorescent bulbs and compact fluorescent lamps (CFLs). In addition, mercury was added to neon tubes to produce a blue light.

Mercury-containing batteries were first used in the 1940s. Samuel Ruben, the cofounder of Duracell, submitted a patent in 1945 for a new battery containing mercury (Ruben 1945). In zinc–mercuric oxide batteries, mercury is part of the electrode reaction (Naylor 2002). Using mercury allowed for the production of a dry cell battery that had a high capacity-to-volume ratio and could be stored under tropical conditions, criteria that were important to the US military during World War II (Ruben 1947). Mercury was also added to other kinds of batteries to enhance their performance. In small button-cell batteries, including alkaline manganese oxide batteries as well as zinc air and silver oxide batteries, mercury prevented the buildup of hydrogen gas from zinc corrosion (Northeast Waste Management Officials Association 2010).

The development of mercury-based production processes in the chemicals industry during the 1900s helped firms to produce more profitable products that protected human health and improved living standards. One example of such a process is in the production of chlorine and caustic soda (chlor-alkali production). It involves applying an electrical current to sodium chloride to separate positively charged sodium from negatively charged chlorine to make alkali. Initial technologies set up two reservoirs with opposite electrical charges separated by a permeable membrane— the so-called diaphragm-cell process (Crook and Mousavi 2016). In 1895, Hamilton Castner and Karl Kellner commercialized an alternative—the mercury-cell process—whereby sodium forms an amalgam with the mercury while the chlorine gas volatilizes (O'Brien et al. 2005). Mercury-based chlor-alkali technologies were more energy efficient than other techniques, and were widely adopted in Europe and Japan (Crook and Mousavi 2016). The use of chlorine benefited public health. Chlorination of London's water supply in the early 1900s, for example, protected against water-borne diseases (MacKenzie 1945). The mercury-cell process also made it possible to produce a higher quality caustic soda, which beginning in the 1940s was used to make new popular kinds of synthetic fibers such as rayon (O'Brien et al. 2005).

The mercury-cell technology for chlor-alkali production gained ground in the United States as well, but the older diaphragm-cell process remained in use in the majority of US plants throughout the twentieth century. Masaru Yarime (2007) cites two main reasons for this. First, there were abundant brine wells in the United States that provided the raw material for chlorine and alkali in the liquid form used by the diaphragm-cell process;

the mercury-cell process in contrast requires beginning with solid salt. Second, because energy was cheaper in the US compared with other regions, it was less advantageous for manufacturers to switch to the more energy-efficient mercury-cell technique. The diaphragm-cell process, however, relies on asbestos, another highly hazardous substance. The mercury-cell process was also used in a few other places, but in countries where chlorine production was developed later, it was more common to rely on a newer membrane-cell technology that needed neither mercury nor asbestos. There were roughly 75 plants in 40 countries that still used the mercury-cell process in 2015, accounting for 8 percent of the global chlorine production capacity of 60 million tonnes (UNEP 2017).

Mercury was applied in other chemical production processes as well. In VCM manufacturing, mercury is a catalyst for a reaction between acetylene and hydrochloric acid that produces the VCM, which is in turn used to make PVC. China is currently the world's largest manufacturer of PVC, and its domestic use of PVC is increasing rapidly. A reported 85 percent of all VCM production in China as of 2014 still used a mercury-based production technique (UNEP 2017). Mercury-based technology was also adopted widely in the chemicals industry for the production of acetaldehyde, a technique that began in Germany in 1912 (Eckert et al. 2006). In this process, a sulfuric acid/mercury sulfate solution is used as a catalyst. The Chisso factory in Minamata used a slightly different method that was developed locally starting in 1932 (George 2001). The mercury-based technique that produced acetaldehyde from acetylene was the dominating production technique for acetaldehyde manufacturing until the early 1960s (Eckert et al. 2006). By the time of the negotiations of the Minamata Convention, there was no longer any known mercury-based manufacturing of acetaldehyde.

While some uses of mercury were beneficial, others caused harm. Many workers suffered, and the use of mercury-added products in some instances also harmed consumers through mercury exposure. In Iraq in 1956, 1960, and 1972, people used imported seeds treated with organic mercury compounds, which had been intended for planting, to make bread; thousands of people fell ill and hundreds died (Jalili and Abbasi 1961; Bakir et al. 1973; Rustam and Hamdi 1974). Other known fatalities from people eating mercury-treated seeds occurred in Pakistan in 1961 and 1969, in Guatemala from 1963 to 1965, and in Ghana in 1967 (Haq 1963; Bakir et al. 1973; Derban 1974). In the United States, three members of a New Mexico family were poisoned in 1969 after eating meat from hogs that had been fed methylmercury-treated seeds (Waldron 1970).

The use of mercury also had mixed societal implications in other ways. Many of the percussion and blasting caps containing mercury fulminate were put toward violent ends. Mercury was also used in the production of thermonuclear weapons. To make enriched lithium-6 for such weapons in the United States, Soviet Union, and China, mercury was run against a solution of lithium hydroxide in a column exchange process whereby lithium-6 accumulated in the mercury phase. Between 1950 and 1963, an estimated 11,000 tonnes of mercury—an amount more than four times greater than today's global annual anthropogenic emissions—was used in Tennessee at the Oak Ridge National Laboratory in this process (Brooks and Southworth 2011). The United States stopped making lithium-6 in 1963, and the major mercury-based production of lithium-7 is currently in China and Russia (Mazur et al. 2014). Recent reports that North Korea acquired large quantities of mercury fueled speculation about the country's capacity to produce thermonuclear weapons (Albright et al. 2017). In addition, chlorine made from mercury-based processes was used in chemical weapons production.

Commercial Mercury and the Environment

Many mercury-added products and industrial point sources have discharged mercury into ecosystems (box 2-3). Roughly three quarters of the mined mercury that was used in products and processes is estimated to have been discharged into the atmosphere, land, and water, or has ended up in landfills (Horowitz et al. 2014). But there are many global and local data uncertainties regarding mercury discharges into the environment from individual sources. The fate of many mercury-added products discarded for centuries all over the world is unknown, including their location in landfills. There are also large uncertainties about mercury discharges from individual point sources. One primary example of this was seen in Minamata. It is unclear how much methylmercury Chisso released into Minamata Bay and neighboring waters by the manufacturing of acetaldehyde and vinyl chloride, but the company claimed in 1972 that it lost roughly 82 tonnes of mercury between 1932 and 1971. One year later, the Japanese Ministry of International Trade and Industry calculated that it was more likely over 224 tonnes, while other estimates go as high as 600 tonnes (George 2001).

Ecosystem processes transport mercury and lead to the production of methylmercury (box 3-3). In turn, methylmercury in ecosystems and contaminated sites affects people (box 3-1). The industrial discharge by Chisso and its related damages to the environment and human health was not an isolated event. A similar case, also stemming from the use of mercury in acetaldehyde production, happened in Niigata, Japan, where the chemical company Showa Denko discharged methylmercury into the Agano River. Mercury poisoning in Niigata was detected in the fall of 1964. Fewer people were affected by Minamata disease in Niigata than in Minamata, but the discharges of methylmercury that began in 1936 and lasted until 1965 still caused much damage. The leadership at Showa Denko initially denied that it had caused the problem and deflected blame by coming up with alternative (false) explanations, including pesticides released from a 1964 earthquake (George 2001). In 1967, 13 victims of Minamata disease filed a suit against Showa Denko in Niigata District Court. In 1971, two years before the legal ruling against Chisso, the court found Showa Denko responsible for causing the outbreak of Minamata disease in Niigata (Funabashi 2006). By 1999, 690 cases of Minamata disease were recorded in Niigata (Eto 2000).

Industrial releases of mercury also caused damages in Ontario, Canada. The Dryden Chemicals company released an estimated 9 to 11 tonnes of mercury into the Wabigoon-English River system between 1962 and 1970 (Takeuchi et al. 1977). This mercury came from a chlor-alkali plant that used mercury cells to produce sodium hydroxide and chlorine for bleaching pulp at the nearby Dryden Paper Company. Chlor-alkali plants and paper mills were sometimes co-located for practical purposes. These releases caused an outbreak of Minamata disease in downstream communities of First Nations peoples in the Grassy Narrows and White Dog Reserves, where researchers found many symptoms in cats and humans that were similar to those previously documented in Minamata (Takeuchi et al. 1977; Harada et al. 2005; Takaoka et al. 2014). The provincial government in Ontario ordered Dryden to stop the mercury discharges in 1970, but it did not acknowledge the presence of Minamata disease (Mosa and Duffin 2017). A decision by the government in 1970 to close down all fishing significantly affected people for generations; similar to the fishers in Minamata, local indigenous groups depended on fish for both food and income (Jago 2018).

Producers and consumers have also intentionally discharged mercury-added products into ecosystems (box 1-3). Panogen and other liquid preparations of organic mercury compounds for use in agriculture against fungi were introduced in the 1940s. By 1950, the use of Panogen in Sweden was "as routine in farming as plowing" (Löfroth and Duffy 1969, 10). Swedish conservationists and ornithologists were among the first to notice fatalities in birds that had eaten treated seeds (Egan 2013). Panogen was also widely used in many other countries. A 1964 advertisement in the US publication *The National Future Farmer*, targeted toward young people in agricultural regions, highlighted the Morton Chemical Company's efforts to educate farmers about the benefits of treating seeds with Panogen to prevent mold and encourage root and foliage development. It noted that farmers all over the world had been using Panogen for 25 years (Anonymous 1964). Direct release of mercury to ecosystems in these applications was both widespread and extensive. It was estimated that a total of 2,100 tonnes of mercury worldwide were used in agriculture in 1965 (Smart 1968).

Interventions

Different interveners have attempted to address problems stemming from mercury use in products and processes, including national and local governments, industries, experts, and international bodies. Figure 6.4 identifies the central interveners and interventions in the products and processes system for mercury. Two main categories of interventions have addressed commercial mercury use. First, some interventions have focused on reducing or eliminating the uses themselves (boxes 2-2 and 1-2). Second, other interventions have addressed mercury from an environmental perspective, by managing stockpiles, setting limits on discharges, and cleaning up contaminated sites (boxes 2-2, 2-3, and 3-3).

Reducing or Eliminating Commercial Mercury Uses

Many interventions that reduced or eliminated commercial mercury uses originated in the private sector, as industries and experts designed mercury-free products (box 2-2). Economic considerations, rather than environmental and health concerns, sometimes drove initiatives to either reduce the amount of mercury used or move to mercury-free alternatives in products. For example, making large mirrors using silver nitrate, a technique developed in the middle of the nineteenth century, was quicker and easier than using mercury to make them. It was not until around 1900, however, when

	Knowledge Institutions		
	1. Human	**2. Technical**	**3. Environmental**
1. Human	(1-1)	(1-2) National and local governments adopt laws and regulations that control mercury use in products and processes	(1-3)
2. Technical	(2-1)	(2-2) Industries and experts design mercury-free products and production processes; National and local governments implement stockpile and landfill standards; International bodies develop guidelines for managing stockpiles and landfills	(2-3) National and local governments ban and restrict discharges of mercury from point sources; International bodies develop mercury waste guidelines; National and local governments implement standards for waste management
3. Environmental	(3-1)	(3-2)	(3-3) National and local governments clean up mercury-contaminated sites
	Interveners		
	National and local governments; Experts; Industries; International bodies		

Figure 6.4
Intervention matrix for the products and processes system for mercury.

technological advances made silver-backed mirrors more durable than mercury-containing mirrors, that they became competitive on the market (Hadsund 1993).

National and local governments have adopted laws and regulations that control mercury use in products (box 1-2). One of the earliest examples of controls on mercury-added products occurred when Swedish authorities banned Panogen and restricted the use of other mercury-containing pesticides in 1966. Their application quickly declined together with concentrations of mercury in food and the environment. Some use of less toxic mercury compounds nevertheless continued in Sweden, including Panogen

Metox (methoxyethylmercuric acetate), which was not banned until 1988. In the United States, the Department of Agriculture in 1970 ordered that methylmercury seed treatments be taken off the market after the 1969 incident in which the New Mexico family was poisoned with Panogen was reported on television (Waldron 1970). After industry challenged this decision in court, an appellate court ruled that scientific evidence connecting Panogen to health impacts was insufficient, and that farmers had no economically viable substitute to mercury-treated seed. The US Environmental Protection Agency (EPA), however, issued a notice in 1972 that it intended to cancel registration for all mercury-treated pesticides, and it announced a ban on almost all such pesticides in 1976 (United Press International 1976).

Phaseouts of phenylmercury in paints involved both conflict and cooperation between governments and industry. The US EPA was forced to withdraw a notification to ban the use of phenylmercury in paints in 1976 (when mercury use in pesticides was prohibited) after industry successfully challenged the decision in court on the grounds that there were no available substitutes (Meier 1990). In 1990, the EPA reached a voluntary agreement with the National Paint and Coatings Association to cease manufacturing of mercury-containing paints. This decision followed a case in 1989 where a four-year-old boy in Michigan was hospitalized for several months with severe symptoms of acrodynia after the inside of his house was painted with mercury-containing paint (Agocs et al. 1990). This case gained much public attention and led authorities in Michigan and, shortly thereafter, the EPA to act. Japan banned the use of mercury compounds in paints in 1980 (UNEP/FAO 1996). Phenylmercury was used in paint production in Europe at least into the late 2000s (European Chemicals Agency 2011), but a 2012 regulation banned the sale of all products in the EU containing phenylmercury compounds in more than trace amounts starting in 2017 (European Commission 2012). Phaseouts of phenylmercury in paints have been slow in other parts of the world as well.

Countries have also initiated several regulatory measures on other mercury-added product categories, including electronics. The EU member states, together with the European Parliament, took on a leadership role in regulating the content of hazardous substances in electronics both regionally and globally in the early 2000s (Selin and VanDeveer 2006). A 2002 directive on the restriction of the use of certain hazardous substances in electrical and electronic equipment (RoHS), which was updated in 2011, limits the use of mercury in electrical and electronic equipment. The EU RoHS directive spurred similar regulatory measures in other countries including China, Japan, Taiwan, and South Korea. In the United States, California acted along the same lines (Wright and Elcock 2006). Following additional regulatory measures, a 2017 EU directive banned the manufacturing, export, and import of all mercury-added products covered by the Minamata Convention by either the end of 2018 or the end of 2020 (European Union 2017).

Actions taken by both industry and government influenced trends in mercury use in batteries, with the total amount of mercury used declining with time. Municipalities in Japan started battery recycling in the 1980s (Pollack 1984). US states around the Great Lakes and in the Northeast took action to restrict mercury use in batteries in the early 1990s (Cain et al. 2011). New Jersey restricted the sale and disposal of many mercury-containing batteries in 1992, and, along with Arkansas and Minnesota, banned mercuric-oxide batteries in 1993 (Sznopek and Goonan 2000). Federal legislation in the US phasing out mercury use in batteries through the Mercury-Containing and Rechargeable Battery Management Act followed in 1996. Much state and federal legislation exempted mercury use in small button-cell batteries because of a lack of alternatives, but mercury-added button-cell batteries were banned by some US states in the 2010s (Zero Mercury Working Group 2012). Major battery manufacturers advertise that all of their batteries are mercury free; Duracell, for example, has been making button-cell hearing aid batteries without mercury since 2011 (Panasonic 2018; Duracell 2019).

Countries in other regions implemented bans and restrictions on mercury in batteries later than in the United States. A 2006 EU directive prohibited batteries with mercury content greater than 0.0005 percent by weight, but exempted button-cell batteries with mercury content less than 2 percent (similar to earlier US legislation). Another 2015 directive updated the prohibition to include mercury-containing button-cell batteries. China banned mercuric oxide batteries in 1999 and set regulatory limits on mercury content for other types of batteries starting in the early 2000s (Cheng and Hu 2011). Statistics on trends in mercury use for battery production in China are conflicting for the late 1990s and early 2000s, but overall use was more than an order of magnitude greater than in the United States (Feng 2005).

Mercury use in Chinese batteries was reported to drop after the year 2000 (China Council for International Cooperation on Environment and Development 2011). However, in China, which is the main producer of alkaline button cells (used, for example, in toys and remote controls), only 10 percent of battery production was reported to be mercury free in 2012 (Zero Mercury Working Group 2012).

Actions on mercury-added measuring devices involved interplay between governments and industrial innovation, including the introduction of digital thermometers. US states took the lead in controlling the sale of mercury-added measuring devices in the 2000s. New Hampshire passed a law in 2000 that prohibited the sale of mercury thermometers and other measuring devices (Health Care Without Harm 2018). This was followed by similar legislative measures in 13 other states over the following three years, shrinking the US market for such products. The US EPA largely relied on such state action to encourage firms to voluntarily cease production and sale (US Environmental Protection Agency 2014). In addition, the US National Institute of Standards and Technology in collaboration with the EPA in 2011 declared that it would no longer provide calibration services for producers and users of mercury thermometers, something it had been doing since 1901. This put further pressure on the private sector to end the production of mercury thermometers. By the early 2010s, there was only one manufacturer of mercury thermometers left in the United States (Roylance 2011).

Other jurisdictions have also acted on restricting the availability of mercury-added measuring devices. The EU in a 2007 directive banned the sale of mercury-added measuring devices including thermometers and barometers starting in 2009, although some member states had acted earlier. For example, Sweden banned the manufacturing and sale of medical mercury thermometers in 1998, and subsequently implemented additional controls on other measuring devices (Swedish Chemicals Inspectorate 2004). Other countries began to introduce controls on mercury-added measuring devices in the early 2010s, spurred on by the negotiations of the Minamata Convention. The Canadian federal government adopted a ban on the manufacturing and import of mercury-added measuring devices in 2014 (Government of Canada 2014). Other countries have also taken legislative steps, but with extended time lags before the controls enter into force. China, a main producer of mercury-added measuring devices, announced in 2017 that it would prohibit the production and export of mercury thermometers as of 2026 (Chemical Watch 2017).

Governmental efforts to address mercury in lamps and light bulbs have focused on a combination of bans and controls on maximum allowable mercury content, creating incentives for innovation and product development. For example, the US banned mercury vapor lamps as of 2008 (Energy Policy Act 2005). In contrast, most controls on CFLs have focused on limiting their mercury content rather than on phasing them out. Early traditional fluorescent lamps, containing mercury, were more efficient and provided more light per unit of energy than the older incandescent lamps (Bright and Maclaurin 1943), and these tubular-shaped lamps were adopted for certain lighting applications. CFLs using mercury were invented in 1976, but were not marketed widely until the 1990s, when they were hailed as an environmentally friendly substitute for the more energy-inefficient light bulb–shaped incandescents (Smithsonian National Museum of American History 2017). Although CFLs are still widely sold all over the world, governments have taken steps to phase out the early versions, which have relatively high mercury content, while still allowing (and in many cases encouraging) the continued use of those that contain less mercury.

Some government policies in other areas have, at least in the short-term, expanded markets for CFLs. The US Energy Independence and Security Act of 2007 set energy efficiency standards requiring that new light bulbs after 2012 use at least 28 percent less energy than the incandescent light bulbs. Several different kinds of light bulbs on the market met this requirement, including CFLs that use 75 percent less energy than a comparable incandescent light bulb (Natural Resources Defense Council n.d.). This may appear to be an increase in mercury use with no mercury-related benefits, but the situation is complex when considering both mercury use and discharges. The reduction of energy use can prevent a larger quantity of mercury from entering the environment than results from its use in the bulb itself, depending on recycling rates and levels of mercury emissions from the energy sector. In countries with high levels of mercury emissions from coal-fired power generation, switching to a CFL from an incandescent bulb results in reduced mercury emissions (Eckelman et al. 2008). Light-emitting diode (LED) bulbs, an even more efficient alternative, contain no mercury. In late 2019, however, the Trump administration blocked further implementation of the phaseout of incandescent bulbs.

   Government interventions to control mercury use in products have influenced, as well as been influenced by, international actions. The 1998 heavy metals protocol under the Convention on Long-Range Transboundary Air Pollution (CLRTAP) limits the use of mercury in batteries, and urges parties to manage the risks of mercury for other uses. The Minamata Convention controls mercury uses in nine product categories and requires parties to discourage the manufacture and distribution of new mercury-added products. The nine categories are: pesticides, batteries, switches and relays, CFLs and linear fluorescent lights, high-pressure mercury vapor lamps, cold cathode fluorescent lamps and external electrode fluorescent lamps (CCFL and EEFL) for electronic display, cosmetics, and several kinds of measuring devices (including thermometers, barometers, and sphygmomanometers). The deadline for actions on these products is 2020, but parties can apply for up to two five-year extensions. For some products such as pesticides and non-electronic thermometers and barometers, parties must phase out all uses. For other product categories, the Minamata Convention merely sets maximum mercury amounts allowed. For example, it allows mercury in certain button-cell batteries up to 2 percent, consistent with earlier legislation in the United States and the EU.

   Some Minamata Convention controls on mercury-added products are more difficult to enforce than others, especially if they concern mercury use in relatively low-tech and easily produced products. For example, mercury-containing skin-lightening soaps, creams, and powders are in continuing production and use, mainly in Africa and Asia, but also in some immigrant communities outside of these regions, including the United States (Zero Mercury Working Group 2010; World Health Organization 2011a; Minnesota Department of Health 2019). The sale of these products, which often takes place in market stalls and small shops, is difficult to address because the deep cultural roots of their application are associated with a desire for lighter skin, especially by women. The Minamata Convention also does not address mercury-containing products used in traditional and religious practices. China and Sri Lanka argued for exempting traditional medicines during the treaty negotiations (Earth Negotiations Bulletin 2012). Similarly, Nepal called for excluding religious uses of mercury from the Minamata Convention (Earth Negotiations Bulletin 2013a). In addition, the Minamata Convention explicitly exempts all military uses of mercury—the global scope of such uses is unknown.

Similar to their actions on products, industries and experts also designed mercury-free production processes (box 2-2). Private sector technological and economic considerations drove a gradual phaseout of mercury use in many sectors. This included the switch to mercury-free alternatives in acetaldehyde and VCM production in Europe and North America. The choice between mercury-based and mercury-free manufacturing techniques depended in large part on whether coal or oil was used as a raw material. Acetaldehyde can be produced from acetylene from coal in the presence of a mercury catalyst, or from oil-derived ethylene, a process that does not require mercury (Othmer et al. 1956). The coal-based acetylene process was the common technique in many regions for more than the first half of the twentieth century, but industries in most countries switched to the ethylene-based Hoechst-Wacker process in the late 1950s because the oil-based feedstock was cheaper to produce and easier to handle (Trotuş et al. 2013). A similar choice between coal and oil affected choices in VCM production. The dominant technology for VCM production today relies on a cheaper and more efficient mercury-free process that begins with ethylene, which is produced from oil. It is also possible to make VCM from acetylene produced from coal; this is the reason that countries with large coal reserves, such as China, continue to use the mercury-based process (UNEP 2017).

National and local governments in some instances also played a substantial role in controlling mercury use in production processes by adopting laws and regulations (box 1-2). Government actions often occurred in tandem with technological improvements spurred by the private sector. In some cases, government signals that an area of mercury use would become subject to legislative action pushed industry to look for alternatives. Refinement of the membrane-cell technology in chlor-alkali production, for example, was prompted in part because of growing environmental concerns about mercury in the 1960s that foreshadowed potential controls (O'Brien et al. 2005). In some of these instances, government action pushed industry to switch to a new production method that they later abandoned, despite having made large investments in it. For example, Japan was the first country to phase out the mercury-cell manufacturing process for chlor-alkali production in the early 1970s, in response to concerns about mercury that stemmed from the Minamata disease experience (Ministry of the Environment Japan 2013). This caused Japanese firms to invest in a less optimal technology, the diaphragm-cell process, which resulted in future inefficiencies when those plants were prematurely retired in favor of an improved ion-exchange process that was also mercury-free (Yarime 2007).

Governmental efforts to reduce mercury use in production processes sometimes had unintended effects. For example, the Chinese government attempted to reduce mercury use in the VCM sector in the early 2010s (J. Liu, pers. comm. February 28, 2019). From 2011 to 2015, plants producing VCM were required to switch from using a high-mercury catalyst (10.5–12.5 percent mercury content) to a low-mercury catalyst (4–6.5 percent mercury content), aiming for a 50 percent reduction in mercury use. Some plants failed to make necessary technical changes to adjust to the lower-mercury catalyst, and needed to use more catalyst per production unit or to increase manufacturing capacity to maintain production levels. This offset some of the intended reductions, and also increased levels of catalyst waste. International cooperation has supported the development of mercury-free technologies for coal-based VCM production in China (Zhang et al. 2011). Developing and implementing such technologies will be key, as China plans to continue to use coal as a feedstock in the domestic production of VCM.

The Minamata Convention bans or restricts all major remaining mercury uses in the chemicals industry. Parties were required to stop using mercury in acetaldehyde production in 2018 and to phase out mercury use in chlor-alkali production no later than 2025. Although there was no evidence of any remaining mercury-based acetaldehyde production anywhere in the world when the Minamata Convention negotiations began in 2010, the ban was included in the treaty at the request of Japan for symbolic reasons because of its ties to the Minamata pollution tragedy. Parties can apply for up to two five-year extensions to the phaseout date for chlor-alkali production. The Minamata Convention imposes restrictions on mercury use in three other processes: VCM production (reduce mercury use per unit of production by 50 percent between 2010 and 2020); sodium or potassium methylate or ethylate production (phase out mercury use as fast as possible, within 10 years of entry into force, and reduce emissions per unit of production by 50 percent between 2010 and 2020); and polyurethane production (phase out mercury use as fast as possible, within 10 years of entry into force).

Substitution may not always be unequivocally positive; alternative technologies used to replace mercury in production processes may also be harmful to human well-being. For example, in the chlor-alkali industry, modern plants use membrane technology that consists of per- and polyfluoroalkyl substances (PFAS). PFAS are highly toxic substances that are sometimes referred to as "forever chemicals" because of their inability to break down in the environment (Johnson 2018). The uses for two of the most well-known examples of this larger class of substances—perfluorooctanesulfonic acid (PFOS) and perfluorooctanoic acid (PFOA)—are regulated by several countries and controlled under the Stockholm Convention (Wang et al. 2017). Yet, the use of PFAS membranes for chlor-alkali production is common across the world where chlor-alkali production does not rely on mercury and/or asbestos. PFAS can be released during the production processes involving these membranes (Strynar et al. 2015). The human health impacts of the use of PFAS in chemicals production, however, remain unknown (Cousins et al. 2019). Newer membrane-free technologies are being developed as further alternatives (Hou et al. 2018).

The existence of transnational institutions, such as the Global Mercury Partnership, has driven much of the more recent international technical work to identify and diffuse substitutes for mercury in products and processes. These partnerships link stakeholders and their societal interactions to new technologies. In doing so, they enhance capacity transnationally to develop further technical know-how by raising awareness and building knowledge and support. There are specific partnership areas on mercury reduction in products and in the chlor-alkali sector. Stakeholders involved include different government ministries, private sector actors (producers, importers, and sellers), and civil society organizations. The United Nations Environment Programme (UNEP) and other international organizations, including through their involvement in the partnerships that predated the Minamata Convention, have helped raise awareness, mainly in developing countries, since the early 2000s. Identifying substitutes for mercury uses can be complex, but some actors have argued for broader visions for phase-outs of mercury. The EU has established a goal of a mercury-free economy, arguing that the dangers and costs of mercury to the environment and human health are too extreme to allow any ongoing uses (European Commission 2017a; European Commission 2018).

Preventing Environmental Releases and Promoting Cleanup
Many national and local governments have restricted and banned discharges of mercury from point sources (box 2-3). Such controls on mercury

discharges from industrial sources date back to the 1970s. The US EPA set mercury emission limits on mercury-cell chlor-alkali plants in 1973 (see chapter 5). Regulations under the Convention for the Prevention of Marine Pollution from Land-Based Sources, which set mercury limit values for releases from chlor-alkali production in the early 1980s, encouraged European producers to increase production efficiency (Yarime 2007). The share of chlorine production in Europe based on mercury-cell technology fell from 63 percent to 26 percent between 1997 and 2012 due to a combination of plant retirements and growing environmental concerns about mercury discharges (European Commission 2014). Reduced demand for chlorine because of regulations on ozone-depleting substances and other chlorine-containing products influenced shutdowns of facilities that used mercury-based production techniques in the United States (Snyder et al. 2003). Canadian laws on maximum allowable releases of mercury from chlor-alkali plants led most to convert their underlying technology to mercury-free alternatives (Commission for Environmental Cooperation 1997).

Another area of government controls involves removing mercury from waste streams, where leakages from mercury-added products in landfills can cause water contamination and other problems. A 2002 EU directive on waste electrical and electronic equipment (WEEE) targeted mercury together with other hazardous substances (Selin and VanDeveer 2006). This directive, which was revised in 2012, is designed to operate alongside the RoHS directive on hazardous substances in products that was initially adopted around the same time. The WEEE directive regulates 10 common categories of electrical and electronic equipment that have often contained mercury, including household appliances, information technology and telecommunications equipment, lighting equipment, electrical and electronic tools, medical devices, and monitoring and control instruments. Under the directive, producers take on a greater responsibility for recycling, reprocessing, and safe disposal. This form of extended producer responsibility for dealing with waste is also intended to provide incentives for industry to design more environmentally friendly products that are easier and cheaper to manage safely once they have been returned.

National and local government interventions driven by concerns about environmental discharges have implemented standards for managing mercury in stockpiles and landfills (box 2-2). Sometimes this has involved collaboration with international bodies, which have developed guidelines for managing such stockpiles and landfills (box 2-2). These guidelines aim for the environmentally sound storage of mercury. International bodies have also developed guidelines for managing mercury wastes (box 2-3). National and local governments in turn implement standards for waste management (box 2-3). Mercury has posed storage and waste disposal problems for centuries, but it is only during the last 50 years or so that countries have developed and implemented legislation around the management of hazardous wastes. One Global Mercury Partnership area focuses on mercury supply and storage and another centers on waste management. The increasing phaseout of mercury uses in products and processes creates needs for the institutionalization of environmentally safe storage and disposal mechanisms. If mercury-use bans are combined with export bans, such mechanisms must be set up domestically. It is estimated that globally somewhere between 30,000 and 50,000 tonnes of excess mercury will become available by 2050 (UNEP 2015).

The Minamata Convention mandates that parties shall store and dispose of discarded mercury and mercury waste in an environmentally sound manner. In this area, the Minamata Convention connects with other international agreements including the 1989 Basel Convention on the Control of Transboundary Movements of Hazardous Wastes and Their Disposal (Selin 2010). The Minamata Convention, consistent with the Basel Convention, identifies three categories of mercury wastes: (1) waste mercury or mercury compounds; (2) wastes that contain mercury or mercury compounds (for example, thermometers and CFLs); and (3) wastes that are contaminated with mercury or mercury compounds (for example, residues from mining and industrial processes). Significant variations exist across countries when it comes to developing national inventories of mercury and mercury wastes. To ensure sound management, it will be important to further establish effective mechanisms for both identifying and managing waste that contains or is contaminated with mercury. An important part of the work of the Minamata Convention Conference of the Parties (COP), in collaboration with the Basel Convention, is to further develop technical guidelines for environmentally sound storage and disposal.

High levels of excess mercury in society from phaseouts of commercial uses increase the demand for environmentally safe management of waste mercury. One study estimates that roughly 11,000 tonnes of metallic mercury from the chlor-alkali industry, non-ferrous metal production, and other processes need to be disposed of in the EU alone before midcentury

(Hagemann et al. 2014). EU law mandates that liquid elemental mercury must undergo appropriate conversion before it is transferred to permanent storage because it is so hazardous (Science for Environment Policy 2017). Treatment options for liquid mercury include forming an amalgam with solid metals, such as zinc or copper, or converting it into mercuric sulfide. Long-term storage can be either above or below ground. Above-ground storage requires access to specialty warehouse-style storage facilities. Options for underground storage in Europe include depositing mercury in salt mines or hard rock formations (Science for Environment Policy 2017). The development and implementation of similar kinds of waste management strategies and policies are important to many countries, as waste mercury is a challenge worldwide.

National and local governments have cleaned up mercury-contaminated sites (box 3-3). Sites that are contaminated with mercury (sometimes together with other hazardous substances) pose significant management problems. Many of these sites are located close to factories that used mercury in products and manufacturing processes. The Minamata Convention requires that parties endeavor to develop strategies for identifying and assessing mercury-contaminated sites. One option to address contaminated terrestrial sites is the washing of mercury-contaminated soils, where mercury binds to other particles and is later separated for further processing. Phytoremediation and phytostabilization can use plants and plant roots to immobilize mercury and other contaminants from soil and water before these plants are recovered and stored (Science for Environment Policy 2017). In addition, bacterial remediation is used to remove mercury from water; it can also work for soil (Mahbub et al. 2017).

Addressing contaminated sites can be contentious. The clean-up of Minamata Bay involved a politically sensitive and extensive reclamation project. Sediments were dredged from Minamata Bay in the 1980s and were placed underneath the eco-park in Minamata. Some residents worry that untreated methylmercury in the sediments may leak back into the water, as Minamata is located in an earthquake prone region. Researchers believe this risk is minimal, however, arguing that the methylmercury has transformed into the more stable form of mercuric sulfide (Sokol 2017). More recently, local groups in Kodaikanal and authorities in the state of Tamil Nadu have been in protracted debates with the Indian subsidiary of Unilever about how to address soils contaminated with mercury from the thermometer factory that was forced to close in 2001 (see chapter 4). One area of disagreement has involved how much soil should be cleaned up: the factory has argued that only the most contaminated soil needs to be removed, while local activists believe it is necessary to treat a much larger quantity of soil that has lower mercury concentrations (Dev 2015).

Cleanup of mercury in contaminated sites represents a significant economic cost to society. Based on the well-established principle that a polluter should pay the costs of pollution, at least some of these costs should be borne by the polluting factory. Yet a large portion of the cleanup costs often falls on the public sector (and, by extension, taxpayers). In one of the most expensive mercury cleanups to date, the cost of removing mercury that was released from lithium enrichment for nuclear weapons production at Oak Ridge is estimated at USD 3 billion. During the period from 2012 to 2015, cleanup costs for reported mercury spills by the US EPA ranged from tens to hundreds of thousands of dollars per incident (Wozniak et al. 2017). While North America and Europe struggle to address old contaminated sites, the number and severity of mercury-contaminated sites continue to increase in Asia and other parts of the world (Li et al. 2009; Kocman et al. 2013). The global economic costs for dealing with all of these sites are unknown, but given the fact that many sites in North America and Europe remain untreated, and that the problem continues to increase in many other parts of the world, costs will only grow with time.

Insights

The story of the mercury-based thermometer at the beginning of this chapter shows that even a single product is part of a system in which mercury connects with issues of human well-being, prosperity, industrial production, technological change, and legacy contamination. In this section, we examine insights from the products and processes system for mercury. First, the use and presence of mercury in products and processes has varied over space and time, with different drivers and consequences. Second, mercury has both benefited and detracted from human well-being, and technological change as well as government intervention have driven gradual transitions toward mercury-free alternatives. Third, cross-scale actions by governments and the private sector occurred simultaneously, showing the complex nature of designing effective interventions.

Systems Analysis for Sustainability

The unique physical properties of mercury intersected with technology development, economic factors, and concerns about environmental and human health damages to drive changes in the use of mercury in products and processes. Mercury use at times decreased in some sectors while it increased in others. For example, mercury use was phased out in mirror and hat making around the same time it began and expanded in battery and light bulb production. The continuing availability of mercury mined from cinnabar facilitated its growing use in products and processes, and industrialization dramatically increased its use as a combined result of technological innovations and growing consumer demand for new mercury-containing products and goods made using mercury. Much of this mercury was eventually discharged into the environment, and the application of mercury-treated pesticides added to this environmental burden. Local and national governments increasingly adopted standards and laws controlling mercury uses based on growing concerns about the environmental and human health impacts of mercury in the late 1900s. Advances in scientific and technical knowledge made it possible to switch to mercury-free alternatives for major products and processes.

The products and processes system for mercury has been able to continue to provide societal benefits from goods and industrial manufacturing while gradually reducing its reliance on mercury-based technology. Development of new techniques allowed producers to adapt to mercury phaseouts, some of which were prompted by government mandates. Mercury-free mirrors, pesticides, thermometers, batteries, and light bulbs have continued to provide important societal benefits, and the chemicals industry has found ways to keep manufacturing high-volume chemicals by using mercury-free production processes. In contrast, many environmental components have much more limited capacity to change once mercury has entered ecosystems. Contaminated sites and landfills that contain high levels of mercury can continue to cause local harm to wildlife and people for decades to centuries. Effective cleanup of such sites is both very expensive and technically complicated. Some discharged mercury from products and processes adds to the global environmental cycling of mercury; this cycling continues for a long time and is difficult to alter.

Mercury has been ubiquitous in global commerce, and its use, environmental presence, and behavior in particular places have influenced its effects on ecosystems and human well-being. Relatively small amounts of discharges, compared to the global use of mercury, can have severe negative impacts on the environment and human health in particular places. Industrial discharges of mercury in Minamata, Niigata, and Grassy Narrows led to many human deaths, permanent neurological damages, and highly contaminated aquatic ecosystems. Accidental consumption of mercury-treated seeds in several countries resulted in additional human fatalities, and the application of mercury-containing pesticides caused harm to birds and other local animal populations. The long-range transport of some of the mercury that has been discharged into the environment from products and processes also transfers its environmental and human health damage far and wide, by adding to the amount of mercury globally that can be converted into methylmercury in faraway ecosystems. These distant effects are often not as visible as the more local consequences, but they can also be substantial and long lasting.

Sustainability Definitions and Transitions

Present-day efforts toward sustainability emphasize the reduction and ultimate elimination of mercury use in products and processes. Yet historically, mercury uses have both contributed to and detracted from human well-being in complex ways. Many known societal uses of mercury, going back to ancient China and the Roman Empire, were closely linked with prosperity and high social status; coveted items such as cosmetics, mirrors, and fur hats were made using mercury. Thus, mercury itself had a commercial value, both as an independent commodity and as a contributor to the value of manufactured goods. The positive value of mercury was not just economic. Stocks of mercury contributed to human well-being, for example, through their uses in developing scientific knowledge, in producing scientific instruments, and in allowing for the production of chlorine used in water disinfection. At the same time, much mercury has been dispersed into the environment, adding to contaminated sites and damaging human health. The gradual phaseouts of mercury in products and processes reduced risks of local contamination as well as the amount of mercury going into global environmental cycling.

Gradual transitions away from mercury use in products and processes, shaped by a combination of private-sector innovation and government action, dramatically reduced the global amount of mercury use. In some cases, mercury-specific interventions drove change. In others, the use of mercury played an incidental role in technological development and other environmental concerns. Economic drivers prompted some switches away from mercury-using technologies. One example is the replacement of mercury-using acetylene-based production processes with mercury-free ethylene-based production processes, a change driven largely by the advantages of using oil instead of coal as a raw material. Another example is light bulbs, which first involved increases and then decreases in mercury use, as the switch to compact fluorescent bulbs was driven by concerns about energy efficiency. These mercury-containing bulbs are now being replaced by newer non-mercury technology in the form of LEDs. Some changes in mercury use in products and processes took hold in specific places and then spread internationally. This was seen, for example, in the cases of batteries and chlor-alkali production processes.

Some transitions to mercury-free products or processes occurred relatively rapidly, but many were preceded by a longer history of related actions. It can be difficult to define how quickly change occurred without identifying a clear starting point. For example, is it when the use of mercury started, when such mercury use was discovered to be dangerous, or when the first mercury-free alternative became commercially available? Japan banned the use of mercury in chlor-alkali manufacturing leading to plant closings, a rapid and discontinuous change. In the use of mercury in mirror making, in contrast, slower dynamics of technology improvements occurred over time that ultimately led to silver nitrate mirrors outcompeting those made using mercury. In the case of batteries, innovation and subnational action emerged over time before national regulatory bans took effect. Some dynamics were slowed by the influence of powerful actors who resisted change, such as when chemical companies in Minamata and Niigata denied responsibility for mercury discharges, or when industry fought pesticide bans in US courts. In other cases, such as for batteries, industries accelerated action through technology innovation.

Sustainability Governance

The use of mercury in products and processes is governed by a combination of domestic controls in the context of global goals. Many countries developed their own policies and strategies for addressing mercury use over the past century, but the Minamata Convention now sets deadlines for phaseouts and limits on most of the remaining commercial uses of mercury in products and processes. The Minamata Convention introduces harmonized top-down controls for major current commercial mercury-added products and processes. Many of these controls are of particular importance to developing countries where relatively few regulations on mercury use in products and processes had previously been adopted. The existence of mercury-free alternatives for banned mercury-added products and processes, as well as products that meet the requirements for maximum allowable mercury content, greatly facilitates implementation of the Minamata Convention controls. In addition, domestic and international markets for mercury-added products had already shrunk significantly before the Minamata Convention was adopted. However, the Minamata Convention does not control military uses of mercury or mercury used in traditional medicines and religious practices.

Governance efforts to address mercury in products and processes have affected how economic benefits and mercury impacts have varied across space and time. For instance, the EU emission standard for chlor-alkali production reduced mercury use but delayed the phaseout of mercury to give the industry a longer time to reconfigure its production system. This benefited producers in the short term, but allowed more mercury to circulate in society and the environment in the long term. In addition, had mercury use in chlor-alkali production been phased out before the EU's export ban, it is possible that this mercury might have been resold and potentially used in ASGM (see chapter 3). In contrast, the approach taken by Japan resulted in more near-term adjustment costs to the domestic chemicals industry, but an earlier stop to mercury-containing technology eliminated risks of mercury exposure to workers and also lessened the impact of environmental discharges of mercury to ecosystems and people, both locally and in faraway locations.

Efforts to manage the benefits and harms of mercury have varied in their effectiveness. Different interventions over time had differential impacts, and have sometimes involved trade-offs. Conflicts between public health interests and industry interests are well illustrated by the stories of Minamata, Niigata, Grassy Narrows, and Kodaikanal, which follow an all too common pattern of behavior of entrenched economic interests denying responsibility and fighting the local communities harmed by their activities. This type of clash often involves citizens banding together to take on powerful economic actors, including cases that involve contaminated sites and waters. Many of these are the result of industrial discharges and illegally stored wastes, which add large amounts of mercury to environmental components. This situation is echoed in other communities worldwide. Often, community members may perceive a trade-off between their economic security—where their jobs and livelihoods might be dependent on the polluting industry—and the health and environmental impacts of mercury.

Mercury, the god of commerce, has witnessed a significant change in societal perceptions of the benefits and liabilities of mercury uses, particularly over the past half century. Mercury used in products and manufacturing processes for millennia provided many benefits to society, but also caused a wide range of harms to human health and the environment. Many early substitutions within the private sector, which shifted toward mercury-free products and industrial manufacturing processes, occurred primarily for economic reasons, but more recent changes and government controls have been introduced because of environmental and human health concerns about mercury. Societies use less mercury now than five decades ago, as a result of bottom-up innovation, top-down regulation, and public pressures. Some manufacturers phased out mercury use relatively quickly, while others tried to reduce use over longer periods of time. Large-scale industrial uses of mercury are mostly in the past, but the negative environmental and human health impacts and the economic costs of mercury discharges from products and processes, and from contaminated sites, will persist long into the future.

7 Mining and Sustainable Livelihoods: Mercury, God of Finance

People have long mined and used minerals from the Earth's crust. The extraction of nonrenewable resources continues to be essential to much production and consumption in contemporary societies. Although large companies run many mining operations, artisanal and small-scale mining provides an important source of income for tens of millions of miners and their family members, predominantly in developing countries. Resource extraction also leads to a number of social and environmental harms, including through the use of mercury in artisanal and small-scale gold mining (ASGM). This raises fundamental questions about whether nonrenewable resource extraction can be reconciled with efforts toward greater sustainability both locally and globally. Intervening to improve human well-being in artisanal and small-scale mining communities requires attention not only to local conditions but also to the transnational drivers of production and consumption of minerals. The strong link between mining and poverty in developing regions also raises important issues of equity in promoting human well-being.

Peru's environment minister Manuel Pulgar-Vidal proclaimed a 60-day state of emergency for the country's remote Amazonian region of Madre de Dios on May 23, 2016. The emergency declaration was prompted by concerns about high levels of mercury in people and the environment from widespread ASGM. Government authorities banned catching and consuming a local species of catfish because of elevated mercury levels (Tegel 2016). This decision was announced a little more than four years after the government had launched a violent crackdown using the military and the police against illegal (or "wildcat") miners who did not possess formal mining permits, resulting in arrests and fatalities (Anonymous 2012; Redacción Gestión

2019). At that time, it was estimated that ASGM in Madre de Dios was a billion-dollar industry that produced 19 tonnes of gold annually, involving 40,000 people who largely entered the often-dangerous ASGM sector because there was little or no alternative employment. Despite the government's actions, unlicensed ASGM had surpassed coca and cocaine production by 2016 as the most profitable unlawful activity in Peru (US Agency for International Development 2018).

Unlicensed ASGM in the informal sector continues to expand in Madre de Dios. It is an important source of income for many miners, who have few other options to support their children and other family members. ASGM also exposes people in mining communities to multiple environmental and social risks. It has led to the destruction of over 60,000 hectares (nearly 150,000 acres) of old-growth rainforest (US Agency for International Development 2018). This has major impacts on biodiversity and the carbon cycle. Inhabitants of Madre de Dios have many misconceptions about the dangers of mercury and how to effectively protect themselves from exposure. Some miners believe, for example, that eating large amounts of cilantro can help the body get rid of mercury (Goldstein forthcoming). The mercury used in Madre de Dios is smuggled into the region in large amounts. Some of this mercury likely originates from illegal mercury mines in Mexico. Edwin Vásquez, the coordinator of an umbrella group for indigenous peoples founded in 1984 in Lima, referred to the human toll of ongoing ASGM when he forcefully stated: "We need an alternative to hell, and hell is mining" (Rochabrún 2018).

The story of ASGM in Madre de Dios is only one example of a much larger phenomenon. ASGM occurs in roughly 70 developing countries throughout Africa, Asia, and Latin America, and "has become an indispensable part of the socioeconomic fabric in the developing world" (G. M. Hilson 2002, 866). As we highlight in the chapter title, much use of mercury has had a close relationship with the extraction of gold, which serves as a monetary instrument in many economies. Roughly half of the 1 million tonnes of elemental mercury that were mined during the past 500 years were used to extract gold and silver (Hylander and Meili 2003). Although major gold mining firms have stopped using mercury, large-scale gold mining still results in some mercury emissions and releases due to the extraction and processing of rocks that contain trace amounts of mercury. But total mercury discharges from ASGM are much greater than those from large-scale gold mining. The use of mercury in ASGM causes local problems, and adds to the amount of mercury that is transported long distances in commerce and through the atmosphere (IGF 2017).

The current global boom in ASGM goes back at least to the start of the gold rush in Serra Pelada in Brazil in 1980 (Mallas and Benedicto 1986). ASGM is frequently a low-skill activity that requires little capital investment. Much ASGM worldwide takes place in the informal sector, where vulnerable miners who are often looking to escape poverty can face strong opposition, whether from large mining firms or local and national governments. As a result, ASGM is an area where issues of power relations are integral to both mercury use and abatement efforts. When Pope Francis visited Madre de Dios in January 2018, he specifically highlighted the threat to indigenous peoples and the Amazonian region stemming from environmental problems of gold mining, and from other kinds of damaging natural resource extraction and use (Chauvin 2018). The pope's comments built on his 2015 Encyclical Letter *Laudato Si*, where he noted the existence of an important "ecological debt" between the Global North and the Global South, as raw materials are exported from developing countries to satisfy consumers in industrialized countries (Francis 2015, para. 51).

In this chapter, we examine the extensive use of mercury in the ASGM sector, and the efforts to reduce and ultimately eliminate such use. In the section on system components, we identify the human, technical, and environmental components associated with the use of mercury in ASGM and resulting environmental discharges and human exposure, and related institutions and knowledge. We discuss factors that drive people into the ASGM sector, and those that influence the continued use of mercury in mining processes and its impacts on human health and well-being, in the section on interactions. We detail in the section on interventions the ways in which different actors have attempted to address mercury-related problems in ASGM, including efforts trying to balance the socio-economic benefits with the environmental and human health problems that stem from the use of mercury in ASGM. In the section on insights, we examine local to global factors and dynamics that shape ASGM and related mercury use and discharges, the relationship between mining, mercury use, and sustainability in short-term and long-term focused interventions, and the need for addressing ASGM with locally specific governance approaches that recognize the importance of justice and power.

System Components

Gold has played many important roles in both ancient and contemporary societies, and it has been mined on every continent except Antarctica. It was extracted in what is now Saudi Arabia going back 3,000 years (Kirkemo et al. 2001; Eisler 2003). It is estimated that over 190,000 tonnes of gold have been mined throughout human history, and roughly two-thirds of that amount has been extracted since 1950 (World Gold Council 2019c). Mercury has been used in gold mining since at least Roman times, and its use continues in contemporary ASGM. The ASGM sector and associated mercury demand have grown substantially since the 1980s. Many more people may enter the ASGM sector in the future, as it remains economically attractive to potential miners and others who seek to make a living in mining communities. The main present-day human, technical, environmental, institutional, and knowledge components for the mercury and ASGM system are summarized in figure 7.1.

The ability to mine gold depends on knowledge of different *gold extraction techniques*. Much early gold mining relied on mercury-based amalgamation methods, but major mining companies currently extract gold using a cyanide leaching technique that was first introduced in the late 1800s. In contrast, most *ASGM miners* continue to use mercury-based amalgamation

Human components	Technical components	Environmental components
ASGM miners Gold supply chain participants Other ASGM community members Mercury supply chain participants Gold processors People living far from ASGM sites	Mercury used in ASGM Mercury in commerce Transportation and communication infrastructure Mining and amalgamation equipment ASGM mercury capture devices	Ore at mining sites Ecosystems near ASGM sites Ecosystems far from ASGM sites

Institutional components	Knowledge components
Mercury markets National and local laws and regulations Gold markets International certification schemes Standards set by international organizations Global Mercury Partnership Minamata Convention	Gold extraction techniques Health impacts from mercury exposure Health protection techniques Mercury concentrations in the environment Environmental impacts from mercury discharges

Figure 7.1
Components in the mercury and ASGM system (referenced in the text in italic type).

techniques. Small-scale mining is used as a general label for largely non-mechanized and labor-intensive mining operations when compared with the operations that are run by large mining companies (United Nations 1972; Siegel and Veiga 2010; Spiegel and Veiga 2010). Artisanal mining is the subset of the small-scale operations that use rudimentary techniques to extract minerals from *ore at mining sites*. The Minamata Convention defines ASGM as "gold mining conducted by individual miners or small enterprises with limited capital investment and production" (Article 2). There have been efforts to specify further quantitative criteria including mine output, labor productivity, and levels of technology use and organization, but definitions vary across international organizations and countries (G. M. Hilson 2002).

Roughly 75 percent of the annual gold supply comes from mining, with the rest coming from the recycling of jewelry and technology (World Gold Council 2019a). Approximately 54,000 tonnes of gold remain in the ground, and all types of mining each year add 2,500 to 3,000 tonnes of new gold to the global stock (World Gold Council 2019c). Estimates vary about how much of this gold comes from ASGM. Many ASGM miners are reluctant to disclose how much gold they extract (Fold et al. 2014). *Gold supply chain participants* include local buyers who procure gold from different sources, including from ASGM miners who lack mining licenses or cannot document the gold's origin. Some gold from ASGM is smuggled across borders, sometimes linked with money laundering, before being refined in Switzerland and other hubs (Berne Declaration 2015; Werthmann 2017; International Peace Information Service 2019). Edward B. Swain and colleagues (2007) estimated that ASGM was responsible for 500 to 800 tonnes of gold in the early 2000s. Kevin H. Telmer and Marcello M. Veiga (2009) put the production in the ASGM sector a bit lower at 350 tonnes per year during that same time. In the mid-2010s, ASGM was believed to produce 600 to 650 tonnes of gold per year (UNEP 2017), which would account for about a quarter of all new gold produced. Participants at the end of the gold supply chain include central banks, jewelry buyers, technology producers, and private investors.

Data on the global scope of current ASGM are sparse and uncertain (IGF 2017). Countries with relatively large numbers of ASGM miners include Indonesia, Ghana, Colombia, Peru, Bolivia, China, Ecuador, and Sudan (UNEP 2017). One estimate puts the total number of ASGM miners at just over 20 million (IGF 2017). If this estimate is correct, it means that roughly

90 percent of all people who work in gold mining do so in the ASGM sector, with only the remaining 10 percent employed in large-scale, mechanized mining operations. Another estimate puts the total number of people who work in ASGM at 10 to 15 million miners, including 4.5 million women and 1 million children (UNEP 2017). Accounting for the total population of miners and *other ASGM community members* who both directly and indirectly depend on mining, at least 100 million people are believed to rely on this sector for their livelihoods in lower income economies in more than one third of the world's countries (UNEP 2017).

Mercury used in ASGM comes from local mercury suppliers who buy *mercury in commerce* from *mercury supply chain participants* on *mercury markets*, which are often international. The most recent estimate is that between 872 and 2,598 tonnes of mercury go into this sector annually (UNEP 2017). This large uncertainty range in global estimates of mercury use in ASGM is a result of substantial gaps in national data; although mercury use in ASGM remains legal in many developing countries, data on its use are incomplete and unreliable at best. Some other countries where licensed ASGM is legal, like Brazil, Colombia, French Guiana, Mongolia, and Peru, have banned the use of mercury in ASGM, but undocumented mercury use continues. Other countries like China have outlawed both ASGM and related mercury use, but ASGM activities using mercury still take place there (Spiegel 2009; Spiegel and Veiga 2010; Fritz et al. 2016). Much evidence exists of growing mercury use in ASGM in countries where it is not permitted, but a lack of data makes it difficult to calculate exact amounts (UNEP 2017).

Even though ASGM miners use rudimentary techniques to extract gold, modern technology plays an important role. *Transportation and communication infrastructure*, such as roads and mobile phone and data networks, provides access to mining areas and connects ASGM communities with mercury sellers and gold buyers (Canavesio 2014). ASGM miners use simple *mining and amalgamation equipment*, but many have access to basic power tools to extract the ore. ASGM largely exploits three different kinds of ore deposits (UNEP 2012). In alluvial deposits, free gold particles can be found in river sediments. Gold can also be found in saprolites (weathered bedrock) and hard-rock deposits. Panning for gold in alluvial deposits and extracting gold from saprolites and hard-rock deposits requires very different mining equipment. The application of *ASGM mercury capture devices* such as fume hoods and retorts, which are often small and round enclosures that allow for closed circuit burning of the amalgam, can capture mercury before it is emitted to air. *Gold processors* who work on mining sites or in nearby communities help miners separate the gold from the mercury and bring it to market.

Knowledge about the *health impacts from mercury exposure* to ASGM miners and other community members, including gold processors, has become more widespread over the past few decades. In turn, knowledge of *health protection techniques* is critical to reducing these impacts. Much of the mercury that is used in ASGM is released into land and water, but the amounts of these releases are uncertain globally (see chapter 3). Because of these releases into local environments, many *ecosystems near ASGM sites* contain high levels of mercury. Knowledge about *mercury concentrations in the environment* and *environmental impacts from mercury discharges* is thus necessary for understanding local environmental consequences of mercury use in ASGM. A large fraction of the mercury that is used in ASGM is emitted into the atmosphere (UNEP 2019). Some of the mercury emitted to air travels long distances before depositing in *ecosystems far from ASGM sites* and affecting *people living far from ASGM sites.*

Institutions at different spatial scales influence the ASGM sector. Many countries have adopted *national and local laws and regulations* on resource extraction and mining rights, which control who can legally mine where, and whether mercury use in ASGM is permitted or banned. The implementation and effectiveness of these laws, however, vary dramatically across countries. ASGM also occurs in the context of supply and demand dynamics on international mercury markets as well as *gold markets.* The commercial prices of mercury and gold influence how much miners have to pay for the mercury that they use, and how much they are paid for the gold that they extract. These prices may fluctuate substantially over time. *International certification schemes* set mining-specific standards, including for mercury use, for participating ASGM miners. Additional *standards set by international organizations* on tracing the origin of gold affect the ability of ASGM miners to bring their gold to international markets. Other international institutions focused on addressing the use of mercury in ASGM include specific guidelines formulated under the *Global Mercury Partnership* and mandates included in the *Minamata Convention.*

Interactions

ASGM and its associated mercury use results in both benefits and harms for human well-being. Figure 7.2 shows interactions in the mercury and ASGM system: we have selected three interactions in that matrix to focus on in this section (the three items in bold type in boxes 1-1 and 2-2); we then trace the pathways that influence them, which we summarize in figure 7.3 (where the bold boxes correspond to the selected interactions). First, people enter the ASGM sector in search of employment (box 1-1), influenced by ecological deterioration and infrastructure, as miners participate in markets with gold and mercury supply chain participants (boxes 1-1, 3-1, and 2-1). Second, mining and amalgamation equipment and techniques affect mercury use in ASGM (box 2-2), as the types of ore at ASGM sites determine

	1. Human	2. Technical	3. Environmental
		Knowledge	
		Institutions	
1. Human	(1-1) **People enter the ASGM sector in search of employment;** ASGM miners participate in markets with gold and mercury supply chain participants; **ASGM miners and community members experience mercury-related health impacts**	(1-2) ASGM miners purchase and use mercury, mining and amalgamation equipment, and mercury capture devices	(1-3) Mercury users directly discharge elemental mercury into ecosystems
2. Technical	(2-1) Infrastructure affects ASGM miners' ability to reach mining areas and buy and sell mercury and gold; Mining and amalgamation equipment and processes lead to elemental mercury exposure	(2-2) **Mining and amalgamation equipment and techniques affect mercury use in ASGM;** Mercury used in ASGM affects the amount of mercury in commerce	(2-3) Amalgamation equipment leads to elemental mercury discharges into ecosystems
3. Environmental	(3-1) Ecological deterioration affects people's decisions to become miners; Methylmercury in nearby ecosystems affects ASGM miners and community members; Methylmercury in ecosystems far from ASGM sites affects people	(3-2) Ore types at ASGM sites affect miners' choice of mining and amalgamation equipment and amount of mercury used in ASGM	(3-3) Ecosystem processes transport mercury and lead to production of methylmercury

Figure 7.2
Interaction matrix for the mercury and ASGM system.

a) Employment in ASGM: Environmental, technological, and socio-economic factors drive employment and local conditions in the ASGM sector

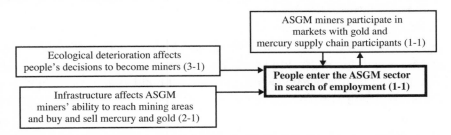

b) Use of mercury in ASGM: ASGM miners, ore types, and mining and amalgamation equipment and techniques affect mercury use in ASGM

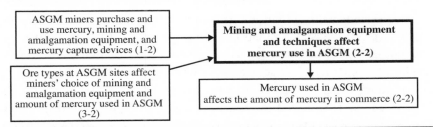

c) Health consequences of ASGM mercury use: Mercury used in ASGM affects ecosystems and human health

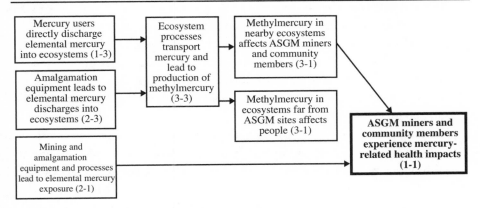

Figure 7.3

Pathways of interactions in the mercury and ASGM system. Bold box indicates focal interaction for each subsection.

the miners' choice of mining and amalgamation equipment and mercury capture device; these choices have an impact on the amount of mercury in commerce (boxes 3-2, 1-2, and 2-2). Third, ASGM miners and community members experience mercury-related health impacts (box 1-1), where the use of mining and amalgamation equipment leads to elemental mercury exposure and discharges into ecosystems; some of this mercury is transformed into methylmercury, which affects ASGM miners and community members as well as people living far from ASGM sites (boxes 2-1, 1-3, 2-3, 3-3, and 3-1).

Employment in ASGM

People enter the ASGM sector in search of employment for multiple reasons (box 1-1). Since the current ASGM boom started in the 1980s, many people have become ASGM miners because of poverty and a lack of other job opportunities. ASGM has relatively low barriers to entry in terms of formal education and financial resources, and thus offers an opportunity for people who struggle to find other kinds of employment to support themselves and their families (Gamu et al. 2015; Langston et al. 2015; Hilson 2016). ASGM miners can be less likely to be in poverty compared with other people who work in different sectors in the same area (Fisher et al. 2009). An expansion of ASGM may have further economic benefits for miners' families and other ASGM community members. Income from ASGM, for example, may allow parents to continue to send their children to school (Hilson et al. 2013; Jenkins 2014). An estimated six jobs per ASGM miner are created downstream, as other community members may work as drivers, cooks, clothing merchants, repair technicians, bookkeepers, and accountants (Hilson and McQuilken 2014).

Changes in international commodity prices and agricultural policy decisions in other countries can influence the income of small-scale farmers, pushing them to enter the ASGM sector (Hilson 2016). For example, when cotton dropped in price on the global market in the mid-1990s, the European Union (EU) and the United States decided to maintain their relatively high levels of domestic subsidies to protect their own cotton farmers from revenue losses. As a result of the price drop and the continued EU and US farm subsidies, 3 million people in Mali who fully or partly depended on cotton production were hit extremely hard, as the domestic cotton market plunged and they struggled to sell their harvests. Many of these farmers, who were no longer able to make a living growing and selling cotton, became ASGM miners. Switching from cotton farming to ASGM provided a means to buy food and lifesaving medicines as well as opportunities for rural families to buy insecticides, fertilizers, and cattle to allow for continued part-time farming (Hilson 2012).

People's decisions to enter and stay in the ASGM sector are affected by the market price of gold, as ASGM miners participate in markets with gold and mercury supply chain participants (box 1-1). The market price of gold between 2000 and late 2012 increased by more than a factor of six, to USD 1,746.50 per ounce, attracting many new miners. The global demand for gold in 2018 reached just over 4,300 tonnes, which at an average price of USD 1,250 per ounce equals a market value of roughly USD 190 billion. The demand for gold from different sectors changes from year to year, but in 2018, the main sectors were: jewelry (50 percent); investment products including bars and coins (27 percent); central banks and other institutions (15 percent); and the technology industry (8 percent) (World Gold Council 2019b). China and India combined for 58 percent of the demand for jewelry and 43 percent of the demand for bars and coins. Many central banks in emerging markets are increasing their gold reserves. The US government, however, remains the world's largest holder of gold at over 8,100 tonnes (US Department of the Treasury 2018). In late 2019, gold sold for just over USD 1,500 per ounce.

High gold prices attract people from other mining sectors to ASGM. This can be further influenced by policy measures on other minerals. A temporary ban on the export of diamonds from Ghana was introduced in 2006 together with the implementation of production and export controls under the Kimberley Process Certification Scheme (Nyame and Grant 2012). These measures, related to the issue of conflict diamonds, led to increased extraction costs and reduced opportunities for small-scale diamond miners to sell to foreign markets. Many diamond miners were unable to find other employment outside the mining sector. With gold prices rising sharply, miners who had the skills to take advantage of a favorable geological terrain switched to mining gold. Gold has been mined in Ghana for more than 1,000 years: before Ghana gained independence in 1957, it was the British colony called the Gold Coast (G. Hilson 2002; Wilson et al. 2015). However, the events of 2006 led to an expansion in ASGM. It is estimated that ASGM produced 34 percent of all gold in Ghana in 2013, involving 1 million people out of a total national population of just over 25 million (IGF 2017).

Many ASGM miners and other community members are vulnerable to different forms of exploitation. Much ASGM mining takes place in the informal sector without worker or social protections. Many mining communities are affected by armed non-state groups, organized crime, high levels of violence and the extortion of protection money and bribes, smuggling of drugs and other goods, and high rates of infectious diseases such as malaria, tuberculosis, and HIV/AIDS (Rees et al. 2010; International Peace Information Service 2019). In some mining areas, there is widespread human trafficking and an extensive sex industry involving both adults and children, leading to much human suffering (Goldstein forthcoming). Already-dire social situations in mining communities can be made worse by widespread government corruption involving local officials. Sometimes tense conflicts over land and mining rights can erupt into violent and deadly conflicts between people associated with large mining firms and ASGM miners (Global Witnesss 2016).

Ecological deterioration can affect people's decisions to become miners (box 3-1). Subsistence farmers may enter the ASGM sector after worsening environmental conditions affect their ability to farm and provide enough food for their families. Many small-scale farmers and herders in Africa became ASGM miners after prolonged periods with little or no precipitation (Gamu et al. 2015). Contemporary ASGM in Burkina Faso started in the 1980s during the major drought in the West African Sahel and Savanna zones. ASGM then became a last resort for many farmers who lost their fields and animals and faced starvation; mining has since been the primary source of income for many of these former farmers (Werthmann 2017). Accelerating climatic changes and other environmentally destructive forces, such as soil erosion and desertification, may have an even greater impact in the future on crop and animal farming in developing countries. This may result in a growing number of people leaving the agricultural sector to enter the ASGM sector in the coming decades.

Infrastructure affects ASGM miners' ability to reach mining areas and buy and sell mercury and gold (box 2-1). This can be seen in the case of the 2,600 kilometer–long Interoceanic Highway that connects the Andes with the Amazon, completed in 2011 as part of a large transportation infrastructure agreement intended to promote economic development between

Brazil and Peru. Madre de Dios quickly became known as the "El Dorado" or "El Wild West" among the locals and the migrant workers because the new highway facilitated the explosion of new mining into areas that were previously too remote to access (Goldstein forthcoming). The highway furthermore spurred the building of secondary and tertiary roads, opening up even greater access to areas with unsettled land tenure rights. Widespread use of the internet and mobile phones made it easier for miners to exchange information about new gold discoveries and to bring the extracted gold to domestic and international markets. This combination of road building and ASGM has led to significant and long-lasting damages to local forests and biodiversity (Elmes et al. 2014).

Transportation and communication infrastructure also facilitates cross-border movements of people. For example, tens of thousands of Chinese nationals moved to rural parts of Ghana in the early 2000s to become ASGM miners. This migration followed in the wake of a large increase in Chinese investments in extractive industry projects in Ghana and other parts of Africa. Chinese-owned service companies in Ghana often actively prompted this movement of miners from China (Hilson et al. 2014). More recently, Ghanaian authorities have repatriated thousands of Chinese miners in the informal sector back to China (Zinsuur 2018). In another example, many ASGM miners in Burkina Faso come from neighboring countries, such as Niger, Mali, Ivory Coast, Ghana, Benin, or Nigeria, and many miners who are from Burkina Faso in turn go to mine in those same countries (Werthmann 2009). Similarly, people in the Amazon frequently move across the national borders of Brazil, Peru, and Bolivia to find work as ASGM miners. A large number of Brazilian miners also operate in Guyana and Suriname.

Use of Mercury in ASGM

The use of mining and amalgamation equipment and techniques affects mercury use in ASGM (box 2-2). Miners predominantly use one of two mercury-based amalgamation techniques to separate gold from ore (UNEP 2012). During whole ore amalgamation, miners mix 100 percent of the ore with liquid mercury (in steel drums, for example) where the mercury forms an amalgam with the gold as the ore is further crushed. This rarely captures more than 30 percent of the gold, and requires quantities of mercury that often range from 4 to 20 parts mercury for every part gold. During concentrate amalgamation, miners instead first concentrate the gold-bearing ore into a smaller mass by, for example, pouring a mixture of ore and water into a sluice box that consists of a tray with a rough surface (such as a carpet) positioned at an inclined angle and open at both ends. The gold-containing particles that get trapped in the carpet, called the concentrate, are washed in a basin. Miners mix liquid mercury into the concentrate to draw the particles into a heavy liquid. Finally, they use a simple panning technique to separate the gold-containing mercury before filtering it through a cloth to get an amalgam of roughly 50 percent gold and mercury. This allows for less mercury use, generally at a 1:1 to 1.3:1 ratio with gold.

Ore types at ASGM sites affect miners' choice of mining and amalgamation equipment and the amount of mercury used in ASGM (box 3-2). The combination of alluvial deposits, saprolites, and hard-rock deposits varies greatly between mining areas, requiring different extraction methods and tools. ASGM is by definition smaller and less mechanized than large-scale commercial gold mining, but the introduction of relatively cheap power tools facilitates more aggressive resource extraction, especially from saprolites and hard-rock deposits. Katja Werthmann (2017, 422), focusing on Burkina Faso, noted: "Some artisanal mines have changed into semi-mechanized operations with cemented shafts, winches and crushers. Ventilators powered by solar panels blow air into the shafts. Instead of donkey carts and bicycles, tricycles (motorized three-wheeled vehicles from China) now transport water, ore and goods as well as passengers." More extraction thus requires more mercury.

ASGM miners' purchase and use of mercury, mining and amalgamation equipment, and mercury capture devices are essential features of ASGM (box 1-2). Suppliers sell elemental mercury to ASGM miners through both legal and illegal channels. In turn, miners must sometimes sell their gold to the same mercury suppliers, often at reduced prices, in order to buy more mercury in a continuing cycle of dependency (Spiegel et al. 2018). Domestic mining provides the mercury that is used in ASGM only in very few countries, including China and Indonesia. In most other countries, the mercury going to the ASGM sector either comes from domestic recycling from other sources, is imported for other legal uses such as dental amalgam and then sold illegally to miners, or is smuggled across national borders. The increase in ASGM-related mercury demand encouraged the resurgence of illegal primary mercury mining from previously closed mines in Mexico and Indonesia (UNEP 2017). Data on the legal trade in mercury are incomplete, and the mercury that comes from illegal mining or enters ASGM through smuggling is not included in any formal statistics (Camacho et al. 2016; Spiegel et al. 2018).

Mercury used in ASGM affects the amount of mercury in commerce (box 2-2). ASGM is currently a leading sector that contributes to the demand for mercury, and thus it has a large impact on international mercury markets. Many of the main mercury-exporting firms were located in the EU (mainly in Spain and the Netherlands) and the United States until the early 2000s. This situation shifted quickly as the EU and US elemental mercury export bans entered into effect in 2011 and 2013. These export bans by two actors that had previously been major mercury suppliers restricted the sources and amounts of mercury available on international markets, especially excess mercury from the chlor-alkali industry and other recycled mercury (since primary mercury mining at that time was already phased out in the EU and the US). ASGM became an important driver in the international demand for mercury following these supply restrictions, as mercury use in ASGM was increasing while other commercial uses of mercury were declining in many regions. The adoption of the export bans by the United States and the EU resulted in mercury price dynamics that are unique to different markets—in contrast to gold, there is no longer a global price for elemental mercury (see chapter 3).

Health Consequences of ASGM Mercury Use

Many ASGM miners and community members experience mercury-related health impacts (box 1-1). The body burden of mercury in some miners is affected by the use of mining and amalgamation equipment and processes leading to elemental mercury exposure (box 2-1). ASGM miners and other community members are exposed to dangerous levels of elemental mercury vapor when burning the amalgam, especially when working indoors in poorly ventilated gold processing centers. People have been found to breathe air with mercury concentrations higher than 50 ug/m^3, which is 50 times higher than the maximum public exposure guidelines set by the World Health Organization (WHO) (Swain et al. 2007). A synthesis of studies from gold-mining communities in 19 different countries demonstrated that measured urinary mercury concentrations (a biomarker of elemental mercury exposure) were well above WHO health guidance values (Gibb and O'Leary 2014).

Mercury users directly discharge elemental mercury into ecosystems (box 1-3). This happens through accidental spills in the field. Use of amalgamation equipment also leads to elemental mercury discharges into ecosystems (box 2-3), but the use of mercury capture devices can reduce the amount of mercury discharged. The process occurs as follows: Miners heat up the amalgam to evaporate mercury. This leaves them with a porous product of sponge gold, which still contains 5 to 10 percent mercury. Open-air burning of the amalgam results in substantial losses of mercury from vaporization, but the use of retorts and other mercury capture devices can capture up to between 75 and 95 percent of the mercury, which can then be reused (UNEP 2012). The sponge gold is further melted to produce a solid bar of unrefined gold (called a gold dore). Miners either do this themselves, or take the sponge gold to nearby gold shops for further processing. This takes out the remaining mercury, which can also be reused if captured. In gold shops, the dore is refined by gold processors up to a 24-karat level of purity, which is considered to be 100 percent gold but in reality is only 99.9 percent pure.

The health of ASGM miners and other community members can be affected by ecosystem processes that transport mercury and lead to production of methylmercury (box 3-3). These processes can occur in local water bodies near ASGM sites. Once methylmercury has formed in ecosystems near ASGM sites, it builds up in living organisms and can affect ASGM miners and community members (box 3-1). This happens through the consumption of local fish (Gibb and O'Leary 2014). Yet, many of the health damages resulting from both elemental mercury and methylmercury may be unknown to those who are affected. Local health care workers may lack not only the knowledge of mercury's dangers but also the ability to distinguish the symptoms of mercury poisoning from those of infectious diseases prevalent in many mining communities. Elemental mercury discharges from ASGM can also travel worldwide, and then be converted to methylmercury in ecosystems far from ASGM sites, where methylmercury affects people (box 3-1).

Interventions

A number of actors with a variety of goals have intervened to change the mercury and ASGM system, including international organizations, national and local governments, international non-state standard-setting bodies, and experts. The matrix in figure 7.4 shows interventions for the mercury and ASGM system. First, some interventions have targeted the ASGM sector by aiming to legalize ASGM and by advocating for better working conditions, thus helping to improve human well-being for miners and community members (box 1-1). Second, international standards shape decisions by ASGM miners on how and where to mine for gold (box 1-1). Third, other interventions specifically target mercury use in ASGM, with the goal of minimizing its use and associated health and environmental damages, largely through targeting mining and amalgamation techniques and equipment (boxes 1-2 and 2-2).

Knowledge Institutions

	1. Human	2. Technical	3. Environmental
1. Human	(1-1) National and local governments take legal actions that influence ASGM activities; International organizations and international non-state standard-setting bodies target gold supply chains	(1-2) International non-state standard-setting bodies formulate rules about mercury use for certification; Experts and international organizations design and disseminate new mining and amalgamation equipment and mercury capture devices; National governments ban mercury use	(1-3)
2. Technical	(2-1)	(2-2) Experts design efficiency improvements in mercury capture and amalgamation techniques; National governments ban the import of mercury for ASGM	(2-3)
3. Environmental	(3-1)	(3-2)	(3-3)
Interveners			
International organizations; National and local governments; International non-state standard-setting bodies; Experts			

Figure 7.4
Intervention matrix for the mercury and ASGM system.

Legal Approaches to ASGM

National and local governments have taken several kinds of legal actions that influence ASGM activities (box 1-1). Legal interventions affecting the ASGM issue in many developing countries are related to the often very close connections between small-scale and large-scale mining. ASGM was largely absent from national and international discussions on poverty alleviation and economic development until the late 1990s (Hilson and Gatsinzi 2014). Many developing-country governments instead focused on large-scale mining as a way to attract more foreign direct investment. To this end, governments rewrote national laws and fiscal policies beginning in the 1970s to open up access to mining operations for foreign firms, with strong encouragement from the International Monetary Fund, the World Bank, and other multilateral donor agencies (World Bank 1992; Hilson and Potter 2005; Hilson et al. 2019). These new policies sometimes led to a pronounced legal separation between formal and informal mining (Perks 2013). Policies favoring mining companies in many cases also ended up pushing more ASGM miners into the informal sector (Hilson and Potter 2005).

Governments have sometimes taken legal actions in the ASGM sector that have had major negative implications for working conditions and levels of mercury use. Governments have occasionally initiated punitive law enforcement actions against ASGM miners operating without permits, as the majority of the world's ASGM miners do (Spiegel et al. 2015; Spiegel 2015). These actions have included steps to outlaw informal ASGM. Unlicensed ASGM miners are then deemed to be criminals, and the police move in with force to shut down mining operations for the purpose of driving miners and other community members from the area. This happened, for instance, in Madre de Dios, as discussed above. The government of Ghana also introduced a ban on ASGM in April 2017, aiming to arrest miners and confiscate equipment (Arthur-Mensah 2018). Approaches that make ASGM mining illegal may also include actions by governments, sometimes with the assistance of the media, to vilify miners as contaminators, or as being contaminated themselves (Goldstein forthcoming).

Bans on unlicensed ASGM can lead to escalating conflicts and deadly confrontations over land tenure and mining rights where ASGM intersects with larger gold mining operations (Patel et al. 2016; IGF 2017; Marshall and Veiga 2017; Werthmann 2017; Bebbington et al. 2018). In some cases, these conflicts are the result of ASGM miners moving into areas where large mining companies have already been given mining concessions, which may have been the result of corruption by government officials. In some cases, these concessions were granted without consulting local communities living on or near mining sites. In other cases, large mining companies move into areas in search of deposits where ASGM miners are already operating without permits. During violent confrontations, governments often side with the larger mining companies over the interests of ASGM miners. A confrontational approach can push miners into even further social marginalization, leading to additional unregulated mercury use and the associated environmental and human health risks that accompany it (Spiegel et al. 2015).

Some governments, in contrast, have taken legal steps to formalize the ASGM sector (Perks 2013). Political support for formalization has grown over the past two decades (G. M. Hilson 2002; Hilson 2016). Formalization makes it easier for governments to collect taxes and other fees from ASGM miners who are licensed by public authorities. It can also create more transparency and afford greater regulatory control with respect to mercury use (Hruschka 2011; Fritz et al. 2016). Bringing more ASGM miners into the formal sector often requires creating simpler licensing procedures. Yet, formalization is often a slow and difficult process in many countries, and is held back by bureaucratic hurdles and weak state structures (Marshall and Veiga 2017; Rochlin 2018). National governments in many countries that have close political and financial ties to large mining firms have also been reluctant to change their attitudes and approaches to ASGM, in part because the mining firms want to protect their land claims and mining rights over those of ASGM miners (Sippl 2015).

The Minamata Convention provides support for measures to bring more ASGM miners into the formal sector. Parties with "more than insignificant" ASGM must include steps to facilitate the formalization and regulation of the ASGM sector in their national action plans. This may involve developing new national policy frameworks and strategies for integrating ASGM and smallholder farming (Hilson 2016). ASGM miners in some countries have formed organizations to facilitate their negotiations with governments and mining companies (Werthmann 2017). Awarding miners official mining rights can help domestic authorities tackle pervasive socio-economic problems—such as child labor and other forms of forced labor that occur in many mining communities—because addressing these kinds of issues is more difficult when large numbers of people are operating without documents in the informal sector (International Labour Office 2006; Hilson and Osei 2014). Greater government involvement can also improve the social and economic status of women and girls, provide better tools to fight the sex trade and human trafficking, and reduce the risks from the outbreak and spread of infectious diseases (Jenkins 2014).

National governments can also take political and economic actions aimed at human development by encouraging people to voluntarily move out of the ASGM sector. This requires counteracting the decades-long trend of more people entering ASGM. Because ASGM provides short-term economic benefits to many people who may otherwise fall into (greater) poverty, longer-range efforts must involve the promotion of alternative livelihoods that address problems of unemployment and poverty while supporting sustainable development (Sippl and Selin 2012). Revised government policies can aim to provide alternative means of support for human development so that miners leave the ASGM sector for employment in other sectors (Hilson 2016). In Madre de Dios, for example, fish farming and the harvesting of Brazil nuts may provide income that is comparable to ASGM with much less damage to the environment and human health (Fisher et al. 2018). Forest conservation can also be supported by implementing payments for ecosystem services programs (Agencia AFP 2019). It can, however, be difficult to successfully implement alternative livelihood programs (Hilson and Banchirigah 2009). In addition, some observers argue that a more sustainable form of ASGM can provide an important source of viable livelihoods (Tschakert 2009).

International Standard Setting

International organizations and international non-state standard-setting bodies target gold supply chains in multiple ways (box 1-1). One initiative by an international organization is the due diligence guidance for responsible supply chains of minerals set by the Organisation for Economic Co-operation and Development (OECD). This guidance, intended to prevent gold buying firms from contributing to local conflicts when they decide which suppliers to use, affects the ability of ASGM miners in conflict-ridden areas to sell their gold on international markets (Organisation for Economic Co-operation and Development 2016). The EU in 2017 mandated that EU importers of gold (and other minerals) follow the OECD guidance, and China has also worked with the OECD to translate the guidance into domestic policy. In the United States, the 2010 Dodd-Frank Act requires that US companies determine whether several minerals including gold come from the Democratic Republic of Congo or an adjacent country. If so, companies must review the supply chain to determine if those minerals help fund local armed groups. Several non-state initiatives also promote transparency in the mining sector, including documenting the origin of the gold that enters international markets (Auld et al. 2018).

International certification schemes by international non-state standard-setting bodies—including efforts by the Alliance for Responsible Mining (ARM) and Fairtrade International (FLO)—are another form of standard setting that targets ASGM. These civil society organizations aim to make ASGM mining more socio-economically and environmentally sound by creating financial incentives for miners to change mining practices through a price premium on the gold they sell, and by convincing consumers to pay more for jewelry made from ethically sourced gold. Building on efforts started by ARM in 2004 with the small mining cooperative Oro Verde in Colombia, ARM and FLO launched a joint pilot project in 2011 that covered the entire gold supply chain from miners, through cooperatives, traders, refiners, manufacturers, and retailers, to consumers of jewelry. ARM and FLO decided to go their separate ways after the pilot project ended in 2013, and now operate separate schemes. The ARM and FLO schemes are slightly different, but for both, in order to have their gold certified, miners must take actions such as organizing democratically, gaining legal mining permits, preventing child labor, and making other behavioral changes to support community development and protect the environment (Sippl 2016).

The design and implementation of the ARM and FLO certification schemes have been criticized, however. John Childs (2014, 129) found a "substantive gap" between how fairness in trade is characterized in the international ARM- and FLO-led discourse compared to the on-the-ground situation for many ASGM miners who participate in these certification schemes. In Tanzania, the fair trade price offered to miners was often significantly lower than local market prices, and many miners mistrusted such external interventions. Gavin Hilson and colleagues (2016) furthermore argue that certification schemes often fail to do enough when it comes to helping the poorest and most marginalized miners. Because certification organizations only work with the relatively small number of miners who have organized collectively and have formal mining permits, they often ignore those miners who are less networked, and who thus need assistance and opportunities the most. Many of the miners excluded from the certification schemes are also those who face the most significant political, economic, and administrative obstacles in obtaining formal mining licenses—and because it is difficult for them to join a certification scheme, they are stuck in the informal sector (Sippl 2020).

In 2018, ARM launched the Code of Risk mitigation for Artisanal and small-scale mining engaging in Formal Trade (known as the CRAFT Code), in large part as a response to criticism of its certification scheme and its low participation by miners (Sippl 2020). The CRAFT Code is designed to help miners document the origin of their gold, and thus relates to efforts by the OECD, the EU, the United States, and others to restrict the trade in conflict minerals. It is intended to apply to artisanal gold but also to be adaptable for other minerals. The code is based on self-reporting without the requirement of external audits that come with the certification schemes, and as such is a less demanding alternative for miners who struggle to meet the certification requirements. This documentation is intended to help ASGM miners continue to sell their gold on international markets. However, the code is less stringent about formal permits than earlier certification schemes, and fulfilling the code does not provide the miners with a price guarantee, as do the certification schemes. ARM has nevertheless expressed a dual hope for the code: that it will provide an incentive for consumers to purchase non-conflict gold at a premium, and that it may attract miners to eventually join a more stringent but also more lucrative certification scheme (Sippl 2020).

Addressing Mercury Use in ASGM

International non-state standard-setting bodies formulate rules about mercury use for certification (box 1-2). The certification schemes that are operated by ARM and FLO specifically focus on gradually reducing and phasing out mercury use in ASGM by offering higher economic incentives. Miners who have been certified and who are able and willing to go mercury free can go beyond the standard price premium and upgrade to the ecological standard, which further increases the price that they are paid for their gold (Sippl 2016). Both ARM and FLO pay the miners 95 percent of the London Bullion Market Price for fulfilling their basic standards (which is higher than the 70 percent that they would normally get on the gold market).

While FLO offers a 15 percent premium on the market price for gold that meets their ecological standard (Sippl 2020), ARM instead offers a fixed premium of USD 6,000 per kilogram for the miners who upgrade to their ecological standard. Both of these ecological standards require mercury-free mining. The 15 percent premium offered by FLO equals a larger dollar amount than the USD 6,000 premium given by ARM when the gold price is above USD 40,000 per kilogram (USD 1,134 per ounce) (Sippl 2016). The market price of gold has almost always been above this level since 2011. The CRAFT Code is less stringent in restricting mercury use than the ARM and FLO certification schemes.

Experts and international organizations design and disseminate technological solutions to mercury use in ASGM in the form of new mining and amalgamation equipment and mercury capture devices (box 1-2). Experts also design efficiency improvements in mercury capture and amalgamation techniques (box 2-2). Challenges associated with the introduction of such equipment vary from country to country, and sometimes even from mining site to mining site (Sousa and Veiga 2009; IGF 2017). The often-significant variations in situations across ASGM communities make it impossible to design a single universal approach to effectively address all aspects of mercury use, discharges, and exposure in ASGM (Sousa and Veiga 2009). It thus becomes important for governments and other interveners to recognize country-specific and site-specific conditions across different mining sites, and to take bottom-up and participatory, multi-stakeholder community-based approaches when engaging individual miners and mining communities (Spiegel 2009; Spiegel and Veiga 2010; Fritz et al. 2016). This has often not been the case in the past.

The Minamata Convention calls for parties to take actions to eliminate whole ore amalgamation. It also mandates parties to develop strategies to encourage mercury-free mining methods. This can involve using cyanide instead, which can allow miners to recover more of the gold from the ore than using mercury-based methods (Veiga et al. 2009). But cyanide-based methods are often more expensive, and they can be difficult and dangerous to use in the field. Mercury and cyanide that are used in the same area may also interact and increase the bioavailability of mercury in the environment, potentially leading to more mercury methylation (Spiegel 2009; Spiegel and Veiga 2010; Spiegel et al. 2018). Several other mercury-free technologies also allow miners to yield more gold, particularly the use of various gravity-based methods, including Gemini (or gold shaking) tables, sluice boxes, and centrifuges (Davies 2014; Veiga et al. 2015). Yet some mercury-free alternatives can be more expensive and complicated to introduce because of cultural and technological factors (Spiegel and Veiga 2010; UNEP 2012; Amankwah and Ofori-Sarpong 2014).

Other mercury-focused interventions engage experts who design efficiency improvements in mercury capture and amalgamation techniques (box 2-2). A central component of such a strategy, which is supported by provisions in the Minamata Convention, involves expanding the use of more effective retorts and other mercury capture devices, and instructing miners how best to use them. The correct use of mercury capture devices yields health benefits as well as economic benefits, as it results in much higher levels of mercury recycling and reuse, and thus reduces the frequency with which miners need to spend money to buy new mercury. There can nevertheless be technical and behavioral barriers to successfully implementing retort use in mining communities. Such challenges require careful attention from experts, particularly those from outside mining communities, when they aim to introduce novel technologies and techniques (Spiegel et al. 2015). A related intervention aims to reduce, and eventually eliminate, open-air burning of amalgam and all burning of amalgam in residential areas, including in gold processing centers (Veiga et al. 2018).

Technology-focused interventions to reduce mercury use are more likely to succeed when they are coupled with initiatives to improve knowledge of mercury risks (Clifford 2014; Spiegel et al. 2015; Veiga et al. 2015). Several international programs have been involved in such twin efforts, including the Global Mercury Project, which was launched in 2002 by the United Nations Industrial Development Organization (UNIDO) with support from the United Nations Development Programme (UNDP) and the Global Environment Facility (GEF) (Chouinard and Veiga 2008). The Global Mercury Partnership has had ASGM as one of its main partnership areas since 2009. Partnership efforts on ASGM shaped the negotiations of the Minamata Convention as well as early treaty implementation (Sun 2017). The Global Mercury Partnership mobilizes knowledge by providing education and raising awareness about amalgamation techniques and mercury-minimizing techniques. For these programs to work, they must often apply a bottom-up approach and be implemented alongside other measures and provisions that ensure the availability of finance and access to equitable gold markets, and in the context of other efforts to reduce human health risks and maximize benefits of mining to local communities (Spiegel 2009; Spiegel et al. 2015; Veiga et al. 2018; Stocklin-Weinberg et al. 2019).

Many miners and other people who live around mining areas and work in gold processing centers are unaware of the health risks of breathing in mercury vapor. Successful science-based risk communication often goes further than just identifying mercury-related environmental and health problems; it offers miners specific health protection measures to address at least some of these problems (Zolnikov 2012). It is important to develop community-targeted information and communication strategies when generating and sharing information on how to reduce mercury use and exposure in ASGM. People in mining communities, especially in former European colonies, can be skeptical of external information about the dangers of mercury; they may fear it comes in the guise of "green colonialism" where outside actors try to impose their environmental protection ideas at the expense of local communities (Zaitchik 2018; Goldstein forthcoming). Without active involvement by local governments, many technical, financial, and social support efforts on awareness raising and training stand a greater chance of failing (Fritz et al. 2016).

Other mercury-focused interventions have addressed the transnational supply of mercury going to ASGM communities. A growing number of national governments have banned the import of mercury for ASGM (box 2-2). Such governmental regulatory actions are also connected to the argument that curtailing the legal international trade in mercury will result in price increases that in turn will lead to reduced use—especially in ASGM, where miners who typically have limited financial resources are sensitive to price changes in mercury (Hylander 2001). However, restrictions on the legal trade in mercury can lead to short-term increases in domestic stockpiling and illegal trade, which can be difficult to monitor and prevent (Greer et al. 2006). This shows the risks of interventions by international organizations and governments that focus on mercury supply without corresponding efforts to reduce demand. In addition, some national governments have banned mercury use in ASGM (box 1-2). Rapid phaseouts and bans on mercury uses in ASGM are also supported by some international organizations and donor-led initiatives.

The introduction of governmental bans on mercury use has rarely (if ever) completely stopped ASGM miners from using mercury. Miners will often look instead to the black market for buying mercury. This continuing reliance on unlawfully smuggled mercury into many ASGM communities is thought to have been a major driving force behind the resurgence of illegal mining in Mexico and Indonesia (Camacho et al. 2016; Spiegel et al. 2018). Most of this increase in the informal supply of newly mined mercury is likely used by ASGM miners, either domestically or once it has been exported illegally to other countries. In a situation where so much mercury use is unregulated, it can be more difficult to set up and implement abatement programs. In the absence of such programs, mercury-related problems in and around ASGM sites can grow, even where there is an official mercury-use ban. A ban on mercury use can also lead to a greater dependency on illegal gold buyers, who provide socially and economically vulnerable miners with smuggled mercury in exchange for exclusive gold buying rights, sometimes at exploitative prices (Spiegel 2009).

Insights

The story of the miners of Madre de Dios illustrates how small-scale gold mining activities are part of a broader ASGM system that occurs in multiple countries simultaneously and is linked with national laws and international markets. In this section, we draw insights from the mercury and ASGM system. First, ASGM is characterized by place-based and global-scale interactions and rapid adaptation to change. Second, past interventions have emphasized an incremental transition that minimizes harm without addressing fundamental questions of mining. Third, effective governance approaches will require coordination of institutions at multiple scales and attention to issues of power and justice.

Systems Analysis for Sustainability

Place-based and global-scale interactions combine to influence not only the use of mercury in ASGM but also the environmental and human health impacts associated with it. The way in which miners operate in ASGM—by looking for gold in alluvial deposits, saprolites, or hard rock deposits; engaging in whole ore amalgamation or concentrate amalgamation; and carrying out open air burning or closed circuit burning of the amalgam—affects the amount of mercury they use and the resulting environmental discharges as well as human exposure levels. Discharges can lead to elevated concentrations of methylmercury in nearby as well as distant bodies of water. International gold markets and commodity chains connect local gold miners and sellers to gold buyers and consumers in faraway places. International codes and certification schemes seek to change local gold production methods, including practices of mercury use. International markets for mercury link individual ASGM miners with foreign supply sources and transnational networks of mercury smugglers. Support from the Global Mercury Partnership and international organizations affects the ability of many governments to reduce socio-economic and environmental problems in ASGM.

The mercury and ASGM system has shown a capacity to adjust rapidly to changes in local and global conditions. This has had mixed implications for human well-being. The steep increase in the international price of gold has encouraged a growing number of people to become miners, allowing them to make more money. Ecological deterioration, policy actions on conflict diamonds, and price changes in the gold market have also pushed more people into the ASGM sector in search of an income. This has resulted in an increase in mercury use, exposure, and discharges. Recently, some ASGM miners have taken measures to meet international standards on conflict-free gold to allow them to continue selling their gold on international markets. Some miners have also joined transnational gold certification schemes, which can sometimes (but not always) contribute to both higher incomes and less mercury use. At the same time, mercury suppliers and miners have responded to mercury export bans and other supply and use restrictions by finding alternative, often illegal ways of transporting mercury into ASGM communities. This has solidified mercury's central role in ASGM, even as mercury continues to be a major human health and environmental problem.

The high degree of locally concentrated use and discharge of mercury in ASGM shows the variability of mercury impacts at different scales. Mercury from ASGM is thought to be the largest global source of emissions to the atmosphere, but it remains uncertain how much of the mercury that is discharged from ASGM stays in the local environment and how much travels to other regions. There are variations in human exposure to mercury both within and across ASGM communities. Based on their choice of mining area and extraction techniques, some miners will use more mercury than others and thereby face a higher degree of exposure. The conditions under which the amalgam is burned off to refine the gold also have major implications for environmental discharges and human exposure. Open air burning without the use of retorts leads to atmospheric mercury emissions, and indoor burning can result in very high concentrations of mercury vapor in poorly ventilated gold shops. All of these factors can result in variable and dangerous local environmental and human health conditions that may not be visible when looking at trends in total emissions or global average mercury levels.

Sustainability Definitions and Transitions
A more sustainable ASGM sector would use less or no mercury, but some analysts and advocates would argue that mining itself is unsustainable by definition. Mining, including ASGM, can be seen as inconsistent with definitions of sustainability that stress the importance of not depleting stocks of non-renewable resources. Eliminating mining would remove all mercury use in ASGM, but few actors who work on addressing socio-economic and environmental problems in ASGM argue for this solution. The long-term goal of the Minamata Convention—to phase out all mercury use in ASGM to protect human health and the environment—is linked to a perspective that sees mercury use as inherently dangerous and undesirable. Many short-term efforts, however, look to reduce (not eliminate) mercury use and implement measures that better protect people in mining communities and limit discharges of mercury to the environment. This is based on the recognition that mercury use in ASGM has some positive value, as it is integral to the ability of many miners to earn a basic livelihood to support themselves and their families. It would also be extremely difficult to implement and enforce an immediate ban on mercury use in all the world's ASGM sites.

A gradual transition away from mercury use in the ASGM sector, with the aim of ultimately phasing out all use, requires considering trade-offs in resource allocation. It is often necessary to decide how much limited human and financial capital should be spent on short-term fixes such as the introduction of retorts, medium-term efforts that focus on the promotion of mercury-free mining techniques, or long-term restructuring that looks toward getting people out of the ASGM sector by providing alternative livelihoods. Expanding the use of retorts reduces the short-term exposure risks and discharges of mercury, but it allows for continued mercury use that does not reduce exposure and discharges to zero. Many mercury-free mining alternatives are more expensive and require practices unknown to miners, making it difficult to immediately expand their use. Phasing out ASGM would address the mercury problem, but the ability for ASGM miners to find other work that provides a livable income depends on the creation of those jobs as well as the miners having the necessary skills to fill those positions. It is also important that those jobs do not create other forms of social and environmental problems.

International institutions and knowledge play important roles in supporting a transition away from mercury use in ASGM. The Global Mercury Partnership, the negotiations of the Minamata Convention, and the creation of gold certification schemes were instrumental not only in highlighting mercury-related and other problems in the ASGM sector that had been largely overlooked by earlier international development efforts, but also in pushing action to reduce mercury use, environmental discharges, and human exposure. The design of new technologies and the formulation of information campaigns on awareness raising, alternative mining techniques, and behavioral change, all of which are essential to this transition, are underpinned by engineering, natural science, and social science knowledge. Greater use of retorts and expanded information sharing about the dangers of mercury exposure in mining communities have had positive effects in some cases. At the same time, the transition toward a mercury-free ASGM sector is severely hindered by a steady supply of mercury, the lack of a simple and cheap alternative to mercury use, and the economic profitability for ASGM miners of a high price of gold.

Sustainability Governance

The Minamata Convention sets out a global approach to address mercury use in ASGM, but much national implementation and on-the-ground change will be up to domestic actors. The requirement for parties with "more than insignificant ASGM" to develop their own individual national action plans is important for institutional fit in that it provides important country-level flexibility. Many large-scale drivers of the recent expansion in ASGM—such as lingering poverty, a continuing lack of alternative employment opportunities, and a high price of gold—are similar across Latin America, Africa, and Asia. Yet, there are significant legal, political, economic, cultural, and environmental variations among countries with ASGM, as well as across individual mining sites, that affect national and local governance efforts.

As a result, countries will identify partly different domestic problems and design different solutions in the national action plans. Moving forward, it is important that global efforts on ASGM under the Minamata Convention and supported by the Global Mercury Partnership are designed to reinforce national-level efforts in ways that recognize variations in domestic conditions.

Governance of mercury use in ASGM involves a combination of approaches to address various leverage points in different ways. The Minamata Convention supports efforts on formalization as an important way to bring miners out of the informal sector. This can allow for greater protection of ASGM miners and better controls on mercury use. Evaluations of field programs show that efforts to address mercury use, exposure, and discharges in ASGM are often more likely to have the desired impact when they consist of a combination of interventions that address behavior and those that introduce new technology. It is often easier to get a miner to change from open air burning to closed circuit burning of the amalgam if a program both provides the miner with an opportunity to acquire a relatively cheap but effective retort and includes an educational component about how to use that retort to recapture mercury for subsequent reuse. International certification schemes seek to harness the power of gold buyers throughout the commodity chain to provide economic benefits in the form of a price premium to ASGM miners who move to mercury-free mining.

Governance of ASGM is deeply connected to socio-economic issues of poverty and marginalization. ASGM miners and other people who live in mining communities are often vulnerable to different forms of exploitation, including forced labor, child labor, human trafficking, organized crime, and actions by mercury smugglers and gold buyers who take advantage of miners' exposed legal and economic situation. In many past interventions by governments and international organizations, the representation of the poorest and least well-organized miners was often ignored. Their voices may also be excluded in conflicts with large mining firms and governments. This may increase the use of mercury because marginalized miners often operate without access to technology and other resources that would help reduce mercury use and thus protect human health. This makes it important for those looking to improve the well-being of the estimated 100 million people who directly and indirectly gain an income from the ASGM sector to pay close attention to dynamics of justice and power.

For societies beholden to Mercury, the god of finance, gold remains a desirable financial instrument. Resource extraction continues to power the world economy, but the process by which people turn gold in geological reservoirs into a resource is associated with multiple social and environmental problems. The use of mercury, which continues to be a valuable extracted resource for this purpose, is a key ingredient in ASGM. This mercury use helps people who are often marginalized and poor make a basic income from gold mining. Mercury used in ASGM also damages human health and the environment: it affects miners, gold processors, and other local community members, and simultaneously contributes a substantial portion of global mercury emissions and releases. Both local conditions and transnational forces drive ASGM, and thus influence efforts to address its negative environmental and human health effects. Different behavioral and technical measures target mercury use in local mining communities, but simultaneous changes in international prices of and trade in mercury and gold have a major impact on the decisions and ultimately the health and well-being of millions of miners, their families, and their communities.

III   Lessons for Sustainability

8   Sustainability Systems: Seeing the Matrix

Analytical frameworks help researchers from various fields and backgrounds examine and compare the structures and functions of different kinds of systems. Identifying the main components of systems, understanding how these components interact, and assessing how actors can intervene to effect change, are topics of broader interest to analysts who focus on a range of sustainability issues. Findings relating to the components of the mercury systems, their interactions, and past interventions drawn from part II can inform systems-oriented studies related to sustainability. Our analysis shows that important operations and properties of the mercury systems can be examined by applying the human-technical-environmental (HTE) framework and a matrix-based approach to identify and map relationships between human, technical, environmental, institutional, and knowledge components. A similar approach could also yield promising insights for other sustainability issues.

We argued in chapter 1 that advancing understanding of sustainability requires in-depth examination of additional empirical cases to further generate knowledge and test theories. With that purpose in mind, the five topical mercury systems detailed in part II provide a rich set of empirical material covering a long period of human history. People, over centuries and millennia, have interacted with mercury and in the process developed new scientific knowledge, technologies, and products with the goal of improving prosperity and well-being. Human uses and discharges of mercury, however, also modified ecosystems and harmed human health. We used the HTE framework to analyze each of the five mercury systems in part II as an individual system within its own boundaries and with its own interactions among human, technical, environmental, institutional, and knowledge components. Yet the mercury systems are connected to one another, not only by the presence of the element mercury, but also by other common system components, and these components also link to other systems relevant to sustainability.

The HTE framework and the matrix-based approach that we introduced in chapter 2 provide a way for an analyst to "see" system components (as referenced in the title of this chapter) and study their dynamics as part of complex adaptive systems. Examining a system involves determining system boundaries and components for analysis, identifying causal pathways of interactions among system components, and assessing the potential for changing system operations. In this chapter, we draw lessons from across the empirically grounded findings from the mercury systems to answer the first three research questions we posed in chapter 1. First, what are the main components of systems relevant to sustainability? Second, in what ways do components of these systems interact? Third, how can actors intervene in these systems to effect change? We conclude the chapter by discussing the utility of the HTE framework and the matrix approach for analyzing other systems relevant to sustainability.

The Matrix Revisited: System Components, Interactions, and Interventions

The five chapters in part II are intentionally structured in an identical way to follow four consistent steps. The first three steps make up the matrix-based approach: first, the identification and classification of system components; second, the examination of pathways of selected system interactions; and third, the assessment of interventions toward greater sustainability. In this section, we discuss and compare the findings from these three steps and associated research questions. We discuss findings relating to the fourth step, drawing insights from the system analysis, in chapter 9.

System Components

Three findings based on our analysis of the five mercury systems connect to our first research question, about the main components of systems relevant to sustainability. First, the mercury systems can be usefully described as comprising a combination of human, technical, environmental, institutional, and knowledge components. Second, the mercury systems can be analyzed by identifying a relatively small number of each type of component, and different system descriptions require differing levels of specificity for system components. Finally, while some components are unique to a single mercury system, others are common across several mercury systems, and these may also be important to other sustainability-relevant systems.

All of the individual components from the mercury systems are summarized in figure 8.1 (with numbers in parentheses to indicate the chapters they correspond to, 3 through 7). The components are listed in the order in which they appear in those chapters, but they are grouped based on the level of empirical detail with which we have chosen to represent each component. For example, as human components, we treat workers and employers in mercury-using sectors as a single component in the health system and the products and processes system. In other systems, however, we disaggregate human components at a finer level of empirical detail. The artisanal and small-scale gold mining (ASGM) system includes ASGM miners as well as gold processors, whereas the health system involves medical

Human components	Technical components	Environmental components

Workers and employers in mercury-related sectors (4,6)
• Miners (3)
  -ASGM miners (7)
• Medical professionals (4)
• Gold processors (7)
Producers and consumers of energy and (industrial) goods (3,5)
• Producers and consumers of mercury-added products and other goods made using mercury (6)
  -Medical patients (4)
• Consumers of food (4)
• Producers of commercial market food (4)
• Gold supply chain participants (7)
• Mercury supply chain participants (7)
People living near mercury discharges or contaminated sites (6)
• Consumers of food (4)
• Non-commercial harvesters (4)
• People living near industrial point sources (5)
• Other ASGM community members (7)
People living far from mercury discharges (6)
• Consumers of food (4)
• Non-commercial harvesters (4)
• People living far from industrial point sources (5)
• People living far from ASGM sites (7)
	Extracted mercury (3)
	• Mercury in commerce (3,6,7)
	• Mercury in stockpiles and landfills (3,6)
	Extracted geological materials containing mercury (3)
	• Extracted fossil fuels (6)
	  -Extracted coal (5)
	Mercury-added products (3,6)
	• Mercury-containing medicines and medical treatments (4)
	Mercury in production processes (3,4,6)
	• Mercury used in ASGM (7)
	Energy and industrial goods (5)
	• Other goods made using mercury (6)
	Industrial point sources of mercury discharges (6)
	• Industrial point sources of mercury emissions (5)
	Air pollution control devices (5)
	Transportation and communciation infrastructure (7)
	Mining and amalgamation equipment (7)
	ASGM mercury capture devices (7)
		Geological reservoirs (3,6)
		• Coal in geological storage (5)
		• Ore at mining sites (7)
		Atmosphere (3,5)
		Land (3)
		Oceans (3)
		Terrestrial and aquatic ecosystems (3)
		• Ecosystems near mercury (emission) sources/uses (4,5,6)
		  -Living organisms (3)
		    -Fish, shellfish, and marine mammals (4)
		  -Rice (4)
		  -Ecosystems near ASGM sites (7)
		• Ecosystems far from mercury (emission) sources/uses (4,5,6)
		  -Living organisms (3)
		    -Fish, shellfish, and marine mammals (4)
		  -Ecosystems far from ASGM sites (7)
		Contaminated sites (6)

Institutional components	**Knowledge components**	
Markets for energy and goods (5) • Mercury markets (3,6,7) • Markets for mercury-added products and goods made using mercury (6) • Gold markets (7) Regional treaties (3) • International air pollution agreements (5) Global Mercury Partnership (3,5,6,7) Minamata Convention (3,4,5,6,7) Trade controls (3) National and local laws and regulations (4,5,6,7) Dietary recommendations (4) Cultural norms (4) International certification schemes (7) Standards set by international organizations (7)	Forms of mercury (3,4) Properties of mercury (3,4,6) Long-range transport (3,5) Mercury concentrations in the environment (3,7) • Mercury concentrations in the atmosphere (5) Quantities of mercury in stockpiles and trade (3) Exposure routes (4) Mercury concentrations in people (4) Health impacts from mercury exposure (4,6,7) Health protection techniques (4,6,7) Mercury deposition (5) Techniques for air pollution control (5) Mercury-based product development and production techniques (6) Individual and societal beliefs (6) Environmental impacts from mercury discharges (6,7) Mercury-free product development and production techniques (6) Gold extraction techniques (7)	

Figure 8.1

Unique and common components of the mercury systems in part II (chapter numbers in parentheses).

professionals, all of whom are also included in the larger group of workers in mercury-related sectors. Figure 8.1 thus illustrates the relative level of detail for the individual components in each mercury system, and the relationship of these components to each other.

Categories of System Components   The system components we chose as the ones most useful for describing the individual mercury systems cover all five categories of components: the material human, technological, and environmental components, and the non-material institutional and knowledge components. Humans are present as workers, employers, producers, consumers, and people living near and far from mercury discharges. Technical components include extracted mercury, mercury-added products, industrial point sources of mercury discharges, and air pollution control devices. Environmental components include the atmosphere, land, oceans, and contaminated sites. Institutional components include markets for energy and goods, the Minamata Convention, and national and local laws and regulations. Knowledge components include information about different forms of mercury, health impacts from mercury exposure, and mercury-free product development and production techniques.

Our decision to classify components among the five predetermined categories undoubtedly influenced our selection and categorization process. Nevertheless, we believe that two findings based on the individual components that we identified are valid and useful. First, we were not able to describe and analyze any of the five mercury systems without including human, technical, environmental, institutional, and knowledge components. This strengthens the assertion at the core of our systems perspective: analyses that only address some of these five sets of components might omit important parts of systems relevant to sustainability. Second, we did not identify any important components that did not fit (or any that fit very imperfectly) into one of the five categories of system components. This suggests that we are not missing an important component category by focusing on these five categories in our analytical framework.

Our categories of system components are similar, but not identical, to those found in other system descriptions such as social-ecological systems, coupled human-natural systems, and socio-technical systems (Liu et al. 2007; Ostrom 2009; de Weck et al. 2011). Several of these systems would capture some of the same individual components that we identify, but sometimes in different categories. For example, what we call technological components would mostly be included in the human category in coupled human-natural systems research (Liu et al. 2007). In our analysis, we find it useful to draw a clearer distinction between technological and human components for the subsequent analysis of interactions and interventions. Some of our components may also be treated as external to systems in other literatures. For example, in engineering systems literature focusing on socio-technical systems, environmental components are often treated as boundary conditions.

In our framework, we differentiate material (human, technical, and environmental) from non-material (institutional and knowledge) components, which helps us connect more explicitly to efforts to model and simulate systems, as discussed below. Our categories of components also overlap with efforts to assess the importance and value of different kinds of capital assets that are viewed as fundamental building blocks of human well-being in sustainability analysis. These capital assets can include human capital, natural capital (most environmental components), manufactured capital (e.g., infrastructure and buildings), social capital (including institutions), and knowledge capital (Polasky et al. 2015). The overlap between our categories and those used in other system descriptions and studies of capital assets suggests that human, technical, environmental, institutional, and knowledge components are indeed important constituents of sustainability.

Number and Specificity of System Components   The total number of human, technical, environmental, institutional, and knowledge components that we identify for each component category in the five mercury systems ranges from two to seven. The total number of system components for each of the mercury systems ranges from 20 to 27. The identification of this relatively limited number of individual components nevertheless made it possible to capture the properties and dynamics of each mercury system that we needed for our analysis. It would of course have been possible to identify a larger number of components for each mercury system. This would have given us the ability to describe each mercury system in greater empirical detail. However, we believe that using a greater number of system components would not have had a meaningful positive impact on our analysis of system interactions and interventions; we found that the aggregated descriptions of the selected components captured the necessary degree of detail in each system.

Each system description contains components that vary in specificity and spatial scale. Ecosystems near mercury sources or uses, for example, are included as an environmental component in the health system, atmospheric system, and products and processes system, encompassing a broad range of ecosystems in many regions across the world. Ecosystems near ASGM sites in the ASGM system are a specific sub-category of ecosystems near mercury sources. Similarly, extracted fossil fuels in general are addressed as a technical component in the products and processes system, but extracted coal in particular is identified as an individual component in the atmospheric system. This shows that components need not be described at the same level of detail in system descriptions. With respect to spatial scale, environmental components could be a small contaminated site or as big as the oceans. Technical components can be small and limited to a specific location, such as a mercury capture device that is used in ASGM, or diffused across the world, such as mercury in commerce. Institutions can take the form of locally specific dietary recommendations, or be global in scale like the Minamata Convention.

Clearly identifying the varying levels of specificity and spatial scale needed to describe system components can assist those who model and simulate systems. The modeling community refers to the development of model simulations at differing levels of scale and complexity as a "hierarchy" or "spectrum" of models (Claussen et al. 2002). The term "hierarchy" was initially used to capture the existence of models ranging from simple to more detailed, but could be construed to imply that high-resolution models are better. The term "spectrum" avoids this implication, acknowledging that the scale of modeling should fit its intended purpose. If system descriptions are useful at varying levels of specificity, different models should be developed to answer different questions. The use of mercury models ranging from global biogeochemical cycle models to detailed ecosystem-specific models helps the scientific community to understand interactions occurring on varying scales across time and space (Obrist et al. 2018). The climate modeling community similarly has used a spectrum of models to address interactions between human activities and the climate system at different levels of resolution (Claussen et al. 2002).

Uniqueness of System Components Some of the identified components are unique to one of the five mercury systems. Unique human components include consumers of mercury-added products and other goods made using mercury in the products and processes system, and ASGM miners in the ASGM system. Mercury-containing medicines and air pollution control devices are unique technical components in the health and atmospheric systems, respectively. Contaminated sites are a unique environmental component in the products and processes system, and rice is an environmental component in only the health system. Gold markets and international certification schemes are institutional components only in the ASGM system. Knowledge of technology-based air pollution controls is part of just the atmospheric system, whereas knowledge of mercury-based and mercury-free production techniques is unique to the products and processes system.

Several material components are common across two or more of the mercury systems, and are thus important for examining connections among those systems. Ecosystems far from sources of mercury discharges are affected by mercury emissions and releases from a combination of sources, including industrial point sources, ASGM, mercury-added products, and mercury in production processes. People living far from mercury sources are similarly affected by mercury from all these sources, as some discharged mercury travels long distances in the environment before it is converted into methylmercury. Mercury in commerce links mercury-added products and mercury in production processes to ASGM through supply and demand dynamics.

Common institutional and knowledge components also link several of the mercury systems. The Minamata Convention covers issues that are central to all five mercury systems through its lifecycle focus on mercury management. Other institutions such as the Global Mercury Partnership link societal efforts to control mercury emissions in the atmospheric system, to move to mercury-free alternatives in the products and processes system, and to phase out mercury-based amalgamation methods in the ASGM system. National and local laws and regulations on mercury discharges connect initiatives to protect the environment and human health from several different sources of mercury in the atmospheric, products and processes, and ASGM systems. Scientific knowledge about the properties of mercury influences its use in the products and processes and ASGM systems. Medical knowledge about the human health effects of mercury exposure shapes mercury use in the health, products and processes, and ASGM systems, and also informs the formulation of dietary guidelines targeting particularly vulnerable groups in the health system.

Some of the mercury system components are also relevant to other issues, suggesting that they may be important to sustainability more broadly. Components relevant to other issues fall within all our categories of system components, both material and non-material. For example, workers and consumers are central in the human health, atmospheric, products and processes, and ASGM systems (as miners and supply chain participants). They are also essential components of different types of production and consumption systems as well as energy issues. Extracted fossil fuels are a key technical component for the products and processes system and the atmospheric system (in the form of extracted coal), and are an important component for other air pollution, climate change, and energy issues. Ecosystems far from mercury sources or uses, present in several of the mercury systems, may be areas of particular concern for other environmental issues such as biodiversity. Knowledge of health protection techniques that is important in the health system is also relevant to pesticides and industrial chemicals. In addition, national and local laws and regulations on pollution are key elements in many sustainability-relevant systems. The relevance of components across several domains relates to the concept of nodes in network analysis, which has been applied in studies of social-ecological systems (Janssen et al. 2006). Common system components could be considered as nodes, and interactions considered as links, if the matrix were represented as a network map.

System Interactions

To address our second research question on interactions, we focus on the 14 pathways of interactions that we identified across the five mercury systems. These pathways are summarized in figure 8.2, grouped into four different types based on which material interactions the pathways include. We identify three major findings related to the ways in which the material components interact, in the context of the two non-material components. First, most (but not all) pathways involve interactions among all three categories of material components. Second, pathways that coexist with each other vary with respect to the number of interactions that they include and how interactions are connected to one another. Third, pathways cross

Human and Technical (HT) Pathways

HT1: Occupational exposure and health impacts: Workers in different sectors come into contact with mercury, and different forms of mercury affect their health (Chapter 4, boxes 2-1, 1-1, 1-2: T-H, H-H, H-T)

HT2: Medical use and health impacts: Medical professionals prescribe and patients use mercury-containing medical treatments, which affect patients (Chapter 4, boxes 2-1, 1-2: T-H, H-T)

Human and Environmental (HE) Pathways

HE1: Mercury discharges and global biogeochemical cycling: People discharge mercury, which cycles through ecosystems and converts to methylmercury (Chapter 3, boxes 1-3, 3-3: H-E, E-E)

HE2: Climate and ecosystem changes, mercury cycling, and methylmercury production: People alter ecosystems in ways that lead to changes in mercury cycling and methylmercury production (Chapter 3, boxes 1-3, 3-3: H-E, E-E)

Technical and Environmental (TE) Pathways

TE1: Mercury cycling in society: Mercury in commerce is traded across borders and leads to emissions and releases (Chapter 3, boxes 2-2, 3-2, 2-3: T-T, E-T, T-E)

Human, Technical, and Environmental (HTE) Pathways

HTE1. Dietary exposure to methylmercury: Mercury-contaminated ecosystems provide food for consumers, affecting their health (Chapter 4, boxes 3-1, 2-3, 3-3, 1-3, 1-1: E-H, T-E, E-E, H-E, H-H)

HTE2. Industrial production and air pollution: Production of energy and industrial goods benefits society and leads to emissions of air pollutants, including mercury (Chapter 5, boxes 1-1, 1-2, 2-1, 3-2, 2-3: H-H, H-T, T-H, E-T, T-E)

HTE3. Atmospheric transport of air pollutants: Point sources affect air pollution transport and mercury distribution in ecosystems (Chapter 5, boxes 2-3, 3-3, 3-1: T-E, E-E, E-H)

HTE4. Pollution control and mercury emissions: Air pollution controls reduce mercury emissions, but incur economic costs to producers and consumers (Chapter 5, boxes 2-2, 2-3, 2-1: T-T, T-E, T-H)

HTE5. Commercial mercury benefits and harms: Economic and social conditions and geological reservoirs containing mercury prompt the development of consumer goods that provide human benefits but also cause harms (Chapter 6, boxes 2-1, 3-2, 2-2, 1-1,1-2: T-H, E-T, T-T, H-H, H-T)

HTE6. Commercial mercury and the environment: Commercial mercury enters ecosystems, where it affects human health (Chapter 6, boxes 2-3, 3-3, 3-1, 1-3: T-E, E-E, E-H, H-E)

HTE7. Employment in ASGM: Environmental, technological, and socio-economic factors drive employment and local conditions in the ASGM sector (Chapter 7, boxes 1-1, 3-1, 2-1: H-H, E-H, T-H)

HTE8. Use of mercury in ASGM: ASGM miners, ore types, and mining and amalgamation equipment and techniques affect mercury use in ASGM (Chapter 7, boxes 2-2, 3-2,1-2: T-T, E-T, H-T)

HTE9. Health consequences of ASGM mercury use: Mercury used in ASGM affects ecosystems and human health (Chapter 7, boxes 1-1, 2-1, 1-3, 2-3, 3-3, 3-1: H-H, T-H, H-E, T-E, E-E, E-H)

Figure 8.2

Summary of interaction pathways from chapters in part II, grouped by the types of material component interactions that the pathways include. The underlined box number in each pathway identifies focal interaction.

spatial scales from local to global, and temporal scales from minutes to millennia.

Types of Interaction Pathways   None of the 14 pathways that we identified in the mercury systems is composed exclusively of interactions that involve only one type of material component. Some parts of pathways, however, involve interactions among material components of the same type. These parts of pathways involve interactions between human components in the form of producers and consumers in the atmospheric system and the products and processes system, and the cycling of mercury among different environmental components and its conversion to methylmercury in the global cycling system. Separate literatures characterize interactions among one type of component. Economic models document interactions among people in markets. Lifecycle assessment models in industrial ecology simulate the flows of materials in industrial production processes. Ecological models of mercury cycling capture ecosystem interactions such as those between different species in a food web. Because the 14 pathways involve at least two different types of material interactions, they fall into four characteristic types: human and technical (HT); human and environmental (HE); technical and environmental (TE); and human, technical, and environmental (HTE).

Pathways involving human and technical interactions: These types of pathways include interactions between human and technical components that are noted in boxes 1-1, 1-2, 2-1, and 2-2 in the interaction matrices in part II (figures 3.3, 4.2, 5.2, 6.2, and 7.2). We identify two of these human and technical pathways, both in the health system. In the first pathway, workers in different sectors come into contact with mercury, and different forms of mercury affect their health (HT1). In the second pathway, medical professionals prescribed mercury-containing medical treatments to patients for a wide range of illnesses that in turn affect the health of these patients (HT2). The appearance of these two human-technical pathways in the health system is not altogether surprising, as health care systems are increasingly examined as socio-technical systems. Socio-technical perspectives are also common in studies of innovation (Geels 2004). In the case of mercury, however, several other pathways that involve innovation, for example in the products and processes system, include interactions involving environmental components as well (e.g., HTE5 and HTE6).

Pathways involving human and environmental interactions: These types of pathways include interactions that are in boxes 1-1, 1-3, 3-1, and 3-3 in the interaction matrices in part II. Two of these types of pathways are present in our analysis, both in the global cycling system. These pathways are where people discharge mercury that cycles through ecosystems and converts to methylmercury (HE1) and where people alter ecosystems in ways that lead to changes in mercury cycling and methylmercury production (HE2). Methods for analyzing pathways involving human and environmental components can be drawn from the literature on coupled human-natural systems (Liu et al. 2007; Chen 2015). Interactions where environmental components influence human components relate to the concept of ecosystem services in which nature provides functions for society (Daily 1997). The Intergovernmental Platform on Biodiversity and Ecosystem Services use similar concepts to link biodiversity to human well-being (Díaz et al. 2015). We only consider pathways to be human-environmental in character if they do not include interactions with an identified technological component. The highly technological nature of the mercury systems likely explains why we found it useful in most cases to explicitly identify technological components that mediate interactions between human and environmental components. For example, discharges of mercury in ASGM are mediated by amalgamation equipment (HTE8).

Pathways involving technical and environmental interactions: These types of pathways include interactions that are located in boxes 2-2, 2-3, 3-2, and 3-3 in the interaction matrices in part II. One pathway in the global cycling system is of this type. In the identified pathway, mercury in commerce is traded across borders and leads to emissions and releases (TE1). A variety of methods have been developed to track flows of different metals, including mercury, in anthropogenic systems, and most of these methods focus on quantifying flows through technical components (Müller et al. 2014). Emissions inventories track the amount of pollutants entering the environment from technological sources (e.g., Muntean et al. 2018). In the majority of cases that we identify in the mercury systems, however, technical-environmental dynamics were closely coupled with human components. For example, in the ASGM system, pathways that involved interactions between amalgamation equipment and the environment also involved its effects on human health (HTE9).

Pathways involving human, technical, and environmental interactions: Pathways that include interactions that involve all three human, technical, and environmental components are the most common type that we identified across the mercury systems (that is, they include combinations of interactions involving all three material components). These interactions are found in all boxes in the interaction matrices in part II, and 9 out of the 14 pathways fall into this category. These pathways are present in the health, atmospheric, products and processes, and ASGM systems. The predominance of pathways that include human, environmental, and technical interactions suggests that the sustainability-relevant dynamics of most interest—those we chose as focal interactions in each mercury system—often involve influences across all the three material components.

Pathways involving human, technical, and environmental interactions include many different components. In the health system, mercury-contaminated ecosystems provide food for consumers, affecting their health (HTE1). In the atmospheric system, production of energy and industrial goods benefits society and leads to emissions of air pollutants, including mercury (HTE2). Point sources affect air pollution transport and mercury distribution in ecosystems (HTE3). Air pollution controls reduce mercury emissions, but incur economic costs to producers and consumers (HTE4). In the products and processes system, consumer goods that are produced using mercury provide benefits but also cause harms to human health (HTE5). Commercial mercury also enters ecosystems, where it affects human health (HTE6). In the ASGM system, environmental, technological, and socio-economic factors drive employment and local conditions in the ASGM sector (HTE7). ASGM miners, ore types, and mining and amalgamation techniques affect mercury use in ASGM (HTE8). The mercury used in ASGM affects ecosystems and human health (HTE9).

Few previous system analyses have assessed human, technical, and environmental components and their interactions in ways that give each component type equivalent emphasis. Literatures on human-environment systems and socio-technical systems, as noted above, cover subsets of human-technical-environmental pathways. Scientists have developed methods to explore dynamics of environmental and human components, including the environmental cycling of mercury and its effects on wildlife and people. Some engineering approaches explore socio-technical dynamics including human and technical components. For example, the development of emission scenarios for mercury involves projecting economic activity as well as the application of control technologies. The prevalence of pathways in the mercury systems that involve interactions among all three types of material components, however, shows how important it is to integrate perspectives from different disciplines when analyzing interactions among material components of systems related to sustainability.

Some existing studies offer promising ways forward to address system interactions across human, technical, and environmental components in a more integrated and comprehensive manner. For example, one modeling study in China links underlying economic drivers (interactions among producers and consumers in markets) with flows of mercury through technologies and the atmosphere, resulting in the estimate quoted in chapter 5 that 33 percent of China's mercury emissions and releases for 2010 were related to the production of goods for export (Hui et al. 2016). There is, however, a need for more interdisciplinary studies and modeling that focuses on societal and environmental flows of mercury and other hazardous substances. Further combining social science, natural science, and engineering studies toward this end holds much promise for advancing sustainability analyses.

Scope and Complexity of Interaction Pathways   Pathways vary with respect to their scope (the number of interactions that they contain), and their complexity (whether they are linear or involve multifactor causality and/ or feedbacks). Some pathways involve few linked interactions, such as HT2, which consists of only two linked interactions. In contrast, other pathways consist of a greater number of linked interactions, as illustrated by HTE9, involving seven linked interactions. Some pathways are linear, including in the global cycling system where people discharge mercury into ecosystems, which then cycles through the atmosphere, converts to methylmercury, and then affects living organisms (HE1). Other pathways involve multifactor causality. This is the case in the ASGM system, in which both ecological deterioration and the availability of transportation and communication infrastructure affect the choice of potential miners to enter the ASGM sector (HTE7). Some pathways include feedbacks in the form of reciprocal interactions, such as in the health system where workers use mercury that in turn affects their health (HT1). A larger feedback loop is present in the products and processes system in the pathway involving commercial mercury benefits and harms (HTE5). In this pathway, market interactions between producers and consumers lead to the sale of mercury-added products that provide benefits or harms to consumers, thus influencing their purchasing decisions about mercury-added products.

Pathways of varying numbers of interactions and degrees of complexity coexist and interact with each other. Pathways that include a larger number of interactions caused by multiple factors became more common as a result of the growing use and dispersal of mercury during the industrial era. The global distribution of mercury through the atmosphere—combining mercury from multiple sources—became important once emission levels increased enough to pose risks to faraway ecosystems (HTE3). Some pathways that emerged earlier in history, however, continued to operate alongside the more complex pathways that became more important later on. In the health system, workers in different sectors for centuries came into contact with mercury that affected their health (HT1). The importance of this pathway has declined recently with the reduced commercial use of mercury, but it existed for a long time even as longer pathways emerged from the industrial use of mercury. New knowledge can also reveal the existence of previously unknown pathways. Only recently have scientific studies been able to trace the transport of mercury across environmental components, and identify the importance to environmental levels of previously unrecognized sources such as mercury in commercial products (HTE6).

The high prevalence of non-linear pathways with larger numbers of interactions poses substantial challenges for scientific efforts to document causality and for researchers developing and refining methods to better understand the mercury systems. For example, isotopic analysis makes it possible to identify mercury in an environmental sample that has a specific atomic mass, which can help determine its source (Blum et al. 2014). This holds much promise to account for environmental interactions, but these methods cannot be easily integrated to track causality across human, technical, and environmental components. Environmental monitoring and analysis techniques may at some point be able to distinguish between different sources of historically and recently released mercury in air or water. However, they would not be able to identify whether that mercury was less or more prevalent because of a regulatory policy, the application of a technology, or broader economic forces such as changes in production and trade patterns. This suggests that new interdisciplinary methods are needed to capture the full complexity of interactions in human-technical-environmental systems.

Spatial and Temporal Scale of Interaction Pathways   Pathways often comprise interactions that cross multiple spatial scales. The spatial extent of the mercury issue seems global when focusing on global-scale cycling, but environmental and human-driven transport of mercury creates regional and local dynamics as well. In part II, we identified several examples of pathways across local to global levels of spatial scale in the mercury systems. Whale consumers in the Faroe Islands are harmed by mercury released far from their borders (HTE1). But their local culture and dietary habits—the decision to continue the *grindadráp* and to consume whale meat and blubber—mediates their exposure and determines its ultimate impacts on their health. ASGM has many of the hallmarks of a traditional place-based case study of sustainability, but it is fundamentally linked to global-scale forces (HTE7). The international gold market sets the economic conditions for ASGM, and much of the mercury that is used in ASGM is imported (both legally and illegally) from other countries. Multinational mining firms also interact with national authorities and local miners to create complex political situations and socio-economic conditions in the ASGM sector.

All five types of components are involved in interactions crossing spatial scales, from the local to the global. Some people (human components) have moved across countries and even continents to enter the ASGM sector based on conditions in global markets (HTE7). Some mercury in commerce (a technical component) is traded internationally before it is emitted and released (TE1). Mercury that is discharged into ecosystems (environmental components) cycles globally (HE1). The degree to which mercury emissions from industrial point sources travel to affect both local and faraway ecosystems depends on the site-specific application of air pollution controls (HTE4). The application of control technologies involves local technical knowledge as well as institutions such as domestic laws and the global Minamata Convention that mandate technology-based control approaches. Not all interactions, however, occur across spatial scales. Many people who live near contaminated sites that are caused by mercury discharges from local sources are affected by their close proximity to these sites for long periods of time (HTE6). Local exposure to mercury also remains common in ASGM (HTE9).

Interactions involving the five types of components operate at different temporal scales that combine through pathways to determine the timescales of each system. The environmental cycling of mercury continues over centuries and longer (HE1). In contrast, many interactions involving technology and humans involve shorter timescales. Exposure to mercury and related health impacts for workers in mercury mining and other sectors can happen on a timescale of hours to months (HT1). Similarly, interactions involving air pollution controls may occur on timescales of minutes to months, where control devices can be installed and turned on and off and thus affect emissions rapidly (HTE4). Institutions and knowledge play a substantial role in determining the timescales on which interactions occur and change. Although the aspects of exposure in pathway HT1 are short-term, these exposures can persist for much longer because of the influence of institutions and knowledge. For example, while the average life of a miner in Huancavelica was only a few months, colonialism and conscription allowed for mining to continue over centuries. For the pathway HTE4, development of new knowledge such as techniques for pollution controls may take years or longer. The process of changing laws and regulations for pollutant control can further stretch from years to decades, and is influenced by developing scientific understanding of mercury impacts on the environment and human health.

The fact that pathways cross both spatial and temporal scales can separate impacts in space and time from their causes, which challenges efforts to link specific sources to particular observed effects. This can be seen in pathway HTE1 in the health system. Tracing this pathway backward, consumers of commercial market seafood are affected by the concentrations of methylmercury in their entire diet of different species of fish and shellfish, and much commercial seafood is traded across borders. These fish and shellfish spent their various lifetimes (short to long, depending on the age of the harvested seafood) in different freshwater and saltwater basins where they had varying diets, depending on the structure of the ecosystem and their position within the food web. The methylmercury in the fish and shellfish came from a combination of mercury discharges from local and distant sources that was originally discharged during both historic and present times. This methylmercury accumulated in biota over timescales that may have ranged from months to decades depending on ecosystem characteristics.

System Interventions
For the third research question on interventions, we focus on four aspects of how actors can intervene in complex adaptive systems to effect change

toward greater sustainability. First, some interveners took actions targeting mercury specifically, while other interventions that affected mercury use and discharges were taken with other goals in mind. Second, interventions addressed both material and non-material components at different leverage points across pathways, and many of the more effective strategies combined different types of interventions. Third, interventions occurred at different spatial and temporal scales, but they often propagated across scales. Fourth, a broad range of actors with varying levels of power and influence was able to prompt system-level changes. Figure 8.3 summarizes the interveners for the five mercury systems, and identifies the material interactions that different interveners targeted.

Goals of Interventions   In the mercury systems, the goals of interventions fall into two main categories: those that primarily targeted mercury and those that mainly involved other issues but also affected mercury. Some interventions were taken to protect human health and/or the environment from mercury. Bans of mercury in teething powder and medical applications were introduced to protect small children from acrodynia. Mercury emission controls such as the US Mercury and Air Toxics Standards were set to mitigate human health damages. Bans and restrictions on the use of mercury-containing pesticides such as Panogen in Sweden, the United States, and elsewhere were designed to protect the environment and human health simultaneously. The EU's recent calls to establish a mercury-free economy are motivated by environmental and human health concerns. The elimination of the very small remaining uses in products, such as mercury in button-cell batteries, is related to this goal. In ASGM, phasing out mercury use is a vitally important piece of efforts to reduce environmental discharges and to protect the health of miners, gold processors, and other community members.

Other interventions that reduced or eliminated the use and/or environmental discharges of mercury were largely driven by non-mercury concerns, but had the positive side effect that they also addressed different aspects of the mercury problem. Governments instituted air pollution controls beginning in the 1970s and 1980s that mainly targeted sulfur emissions from industrial point sources, but these controls also helped reduce mercury emissions as an additional benefit. Economic forces and technological innovation, rather than environmental and human health concerns, drove the development of many mercury-free manufacturing processes because

Knowledge
Institutions

	1. Human	2. Technical	3. Environmental
1. Human	MC Article 7 (3); National and local governments (7); International organizations (7); International non-state standard-setting bodies (7)	MC Articles 4–6, 7 (3); Industries (4); National and local governments (4,6,7); Professional organizations (4); International non-state standard-setting bodies (7); Experts (7); International organizations (7)	MC Article 11 (3); National and local governments (4)
2. Technical	MC Article 16 (3); Industries (4); National and local governments (4)	MC Articles 3, 8–9, 10 (3); National and local governments (5,6,7); Industries (5,6); International bodies (5,6); Experts (6,7)	MC Articles 8–9 (3); National and local governments (5,6); International bodies (6)
3. Environmental	MC Article 16 (3); National and local governments (4,5)	MC Article 3 (3); International bodies (5); National and local governments (5); Industries (5)	MC Article 12 (3); National and local governments (6)
Interveners			

Minamata Convention parties (3); Minamata Convention bodies (3); Experts (6,7); Global Mercury Partnership participants (3); Industries (4,5,6); National and local governments (4,5,6,7); Professional organizations (4); International bodies (5,6); International organizations (7); International non-state standard-setting bodies (7)

Figure 8.3
Interveners and interventions in the five mercury systems discussed in part II. Minamata Convention (MC) provisions are specified by article, consistent with chapter 3.

they were cheaper and more efficient for producers. The discovery of penicillin, and its effectiveness in treating syphilis, led to the phaseout of this longstanding medical use of mercury before there was a more sustained effort to phase out mercury use in medicine based on health concerns. The higher energy efficiency of LED lighting technology is a central driver of supplanting the minimal use of mercury in compact fluorescent bulbs, which are still allowed under the Minamata Convention.

Some interventions had unintended damaging effects to present and future human well-being as a result of interveners acting with incomplete knowledge of system-level processes and interactions. In some cases, early efforts to communicate mercury-related risks from seafood consumption resulted in pregnant women eating less seafood overall, which was potentially dangerous to fetal development given the nutritional benefits of fish consumption. Switching to cyanide technology is one option for moving away from mercury amalgamation in ASGM, but if done improperly, this can result in even more mercury methylation in local environments (and also creates cyanide management–related problems). Phaseouts of mercury in chlor-alkali production, especially before the US and EU elemental mercury export bans, resulted in the diversion of mercury stocks into ASGM, where much of the mercury was discharged into the environment. Efforts to ban ASGM and related mercury use often result in greater harms to miners and other people who live in mining communities when the activities continue in the informal sector.

Mercury-specific interventions had both positive and negative spillover effects on other sustainability issues. The introduction of mercury emission controls on coal-fired power plants in the United States and Canada, which likely triggered some older plants to close, thus had larger positive effects on abating other types of air pollution and climate change. These effects may not have been completely unintended by those who supported the mercury emission standards, however, given that governments and other public authorities are engaged in broader efforts to improve air quality and reduce carbon dioxide emissions. In contrast, advanced power plants with emission controls can sometimes have a longer lifetime, continuing to emit carbon dioxide. A growing literature aims to better assess links between different interventions and their sustainability-relevant impacts. Many studies focus on connecting actions on climate change to local and regional air pollution, achieving near-term benefits for human health simultaneously with longer-term climate benefits (Nemet et al. 2010; Anenberg et al. 2012). Researchers have also identified the existence of both synergies and trade-offs among actions to meet targets under the Sustainable Development Goals (Pradhan et al. 2017).

Targets of Interventions   Interventions in the mercury systems targeted all nine types of interactions among the material components (as illustrated by figure 8.3, where interventions are present in all nine boxes of the intervention matrix). Different interveners addressed different types of interactions:

this reflects that interveners may have at least partially different goals and capacities when they act to modify specific interactions. National and local governments targeted all nine types of interactions—that is, they appear as interveners in all nine boxes—illustrating the many critical roles that public authorities play in protecting and promoting greater human well-being. The Minamata Convention also addresses all nine types of interactions, consistent with its goal to target the entire lifecycle of mercury. Other interveners targeted more specific types of interactions: for example, industries targeted three of the types of interactions that involve technical components.

Some interventions sought to directly alter interactions among material components, while others instead focused on non-material components in order to change material interactions. Interventions involving material components aimed to reduce discharges of mercury by altering technological components (such as deploying air pollution control devices) and also to physically modify attributes of environmental components (such as by cleaning up contaminated sites). Interventions that modified institutional and knowledge components aimed to alter the rules and parameters for interactions among the material components. National and local governments passed laws to regulate the sale of mercury and mercury-added products and to control mercury emissions and releases. International non-state standard-setting bodies formulated rules about mercury use for certification. These certification schemes and codes shape decisions by ASGM miners on mercury use as well as choices by gold buyers throughout the commodity chain. International bodies collected and diffused knowledge of techniques for air pollution control, affecting implementation of standards through technical components.

Many of the more successful interventions to protect the environment and human health from mercury addressed material and non-material components simultaneously. The deployment of air pollution control devices, a technical intervention, occurred concurrently with efforts to modify institutions to set emission standards at national and international levels. The design and dissemination of mercury capture devices such as retorts in ASGM have been more effective when they are combined with information sharing and education of miners. The role of interventions in influencing material and non-material components relates to ongoing debates on the relative importance that behavioral and technical change has in contributing more broadly to sustainability (Fischer et al. 2012). For example, much debate surrounds the role of individual choices relative to more structural approaches in preventing dangerous climate change (Fragnière 2016; Wynes and Nicholas 2017). A longer-term perspective drawn from the mercury systems supports the claim that interventions that aim to modify behavior together with introducing new technology can reinforce each other in ways that are beneficial for human well-being.

Interventions differed based on where the targeted interactions (as leverage points) were located along the pathways of causal influence. Some "upstream" interventions addressing mercury uses targeted interactions early in a causal pathway, while other more "downstream" interventions on mercury discharges and exposure focused on interactions located later in a pathway. Upstream interventions have different capacities to influence system operations and outcomes than do those that are taken further downstream. For example, controls on discharges occur further upstream than efforts to provide guidelines on dietary consumption within the pathway involving exposure to methylmercury from food sources (HTE1). Limits on discharges can have long-term, comprehensive positive impacts, but can be slow to propagate through the system, especially through environmental components, and it can be difficult to evaluate their impact. In contrast, interventions further downstream can have more immediate desirable impacts on human well-being, but leave the underlying processes causing mercury discharges unmodified.

Analysts (and decision-makers) increasingly acknowledge that upstream and downstream interventions often need to occur simultaneously to maximize their impact. To protect the environment and human health from mercury it is necessary to address mercury use, discharges, and exposure in a comprehensive manner. This idea is related to the concepts of mitigation and adaptation in climate change, with mitigation looking to reduce greenhouse gas emissions and adaptation referring to efforts to adjust to a changing climate. Climate change experts increasingly recognize the need for societies to adapt to climate change while concurrently acting to curb emissions (Schipper 2006). How to "best" balance mitigation and adaptation efforts can be controversial; adaptation efforts may be viewed as taking away from the urgency to reduce pollution. Evidence from the mercury systems shows that conducting simultaneous upstream and downstream interventions can have greater influence on human well-being in the near and long term than either approach in isolation.

Spatial and Temporal Scales of Interventions   Interventions in the five mercury systems ranged in spatial scales from local to global. The introduction of measures to protect workers from mercury exposure, such as wearing gloves and other protective clothing in workplaces or increasing ventilation in laboratories, addressed mercury exposure in a specific place. The cleanup of contaminated sites is also locally focused, as is publishing local guidelines on dietary consumption for seafood consumers. At the other extreme, much recent action on mercury has occurred in global forums, with government representatives from all over the world negotiating the provisions of the Minamata Convention and collaborating with other stakeholders under the Global Mercury Partnership. Interventions can occur at different scales than the interactions they ultimately aim to change. Interventions to protect local consumers of seafood, for example, can involve setting global emission standards under the Minamata Convention. Many interventions that affect local mercury use in ASGM emerged in the context of international development efforts and global-scale mercury policy, including bans on mercury mining and restrictions on reuse and trade in elemental mercury under the Minamata Convention.

Interventions can propagate across scales through both top-down and bottom-up dynamics. Recent global top-down interventions under the Minamata Convention prompted efforts to ban all remaining primary mercury mining and to phase out mercury use in chlor-alkali production worldwide, to be implemented and enforced by each party within its own jurisdiction in specific geographical places. In contrast, some efforts to ban specific mercury-containing products prior to the adoption of the Minamata Convention were driven by bottom-up efforts. Individual US states led many local efforts to ban mercury in several products, such as batteries, that were then adopted nationally. These initiatives prompted transnational private sector technological development as well as learning across different jurisdictions. In another example of interventions spreading from a smaller to a larger scale, the thresholds incorporated in the Minamata Convention on maximum allowable mercury content in specific consumer products such as batteries and light bulbs are the same as the thresholds already set by many national authorities and the EU, who advocated for their adoption at the global level.

Institutions and knowledge had substantial impacts on the degree to which interveners were able to effect change across different spatial scales.

The emergence of international institutions on mercury facilitated efforts to address domestic and regional mercury problems and also spurred global action. The Global Mercury Partnership provided a mechanism for supporting and coordinating mitigation efforts in different regions of the globe toward a common goal of reducing mercury use, emissions, and releases (but allowing for variations in domestic implementation). Networks of experts in one partnership area initiated the exchange of information on best available techniques (BAT) for controlling mercury emissions from coal-fired power plants in ways that later facilitated collective action under the Minamata Convention. Experts in another partnership area similarly supported the dissemination of information on ways to reduce mercury use and exposure in the ASGM sector. Governmental regulations in different regions also encouraged the development and international diffusion of mercury-free alternatives for products and manufacturing processes.

Interventions occurred over very different temporal scales. Initiatives to protect workers in mining and manufacturing evolved over centuries, going back to at least the 1500s. Government actions starting in the 1900s often developed over years to decades. For example, Swedish authorities prohibited the use of Panogen in the 1960s, but did not ban all forms of mercury pesticides until the 1980s. In many countries, regulations mandating the application of air pollution control technology were gradually strengthened over time. The political process that led to the negotiations of the Minamata Convention started with the global mercury assessment that was finalized in 2002; the treaty negotiations were completed in 2013, and the Minamata Convention entered into force in 2017. Several of its provisions on mercury mining and use as well as emission controls will only come into effect 5 to 10 years after its entry into force for a party. In contrast, some interventions are more immediate. Dietary advice to pregnant women or women who might become pregnant can change exposure rapidly. This can have relatively rapid benefits because the lifetime of mercury in the human body is a few months.

Role of Power and Influence in Interventions   Interventions in the mercury systems varied depending on which actors were involved, and how they exerted power and influence. An intervener's influence can be assessed by analyzing that actor's ability to effect change. Governments initiated some interventions in the mercury systems, while the private sector took the lead in other instances. Most emission controls were government-led. In contrast, some of the earliest phaseouts of mercury in medicine and products and processes were initiated by the private sector. Many of the actions of governments and industry during the past 70 years intersected and reinforced one another, suggesting that each actor alone may not have sufficient power to comprehensively alter a system. Anticipation of future government controls, such as the phaseout of mercury in batteries, drove some industry action. Global voluntary partnerships beginning in the early 2000s both implemented interventions (for example, in reducing mercury use and emissions using retorts in ASGM) and facilitated the introduction of legal requirements (for example, the diffusion of BAT standards for point sources).

Non-state actors other than industry, and individuals acting in groups, also facilitated change in the mercury systems. Protests and advocacy on behalf of people affected by Minamata disease, and working through the Japanese judicial system, forced an eventual response by public officials and the Chisso Corporation as well as Showa Denko. Similarly, public protests in Tamil Nadu resulted in the government shutting down the mercury-using thermometer factory in Kodaikanal. Conservationists and ornithologists in Sweden first raised public concerns about the use of mercury-based pesticides, prompting the authorities to investigate and eventually take regulatory action. Public awareness and concerns about the dangers of mercury spurred governmental action in other cases as well. Discussions of mercury hot spots in the United States related to efforts by the US Environmental Protection Agency (EPA) to control mercury emissions from power plants in the early 2000s resonated with a broader public, prompting states and environmental groups to challenge the proposed federal regulation in court. Community concerns about methylmercury exposure shaped the formulation of dietary recommendations in many places, including for locally caught seafood.

Some interventions reflected learning among actors across different contexts and forums. In some instances, one political entity set a standard or adopted a law that was later used as a basis for passing a similar or identical standard or law in another jurisdiction. For example, China's current standard for mercury emissions from coal-fired power plants is modeled after (and set at the exact equivalent of) Germany's emission standard. China, South Korea, and California used the EU regulations on mercury and other hazardous substances in electronic and electrical equipment to adopt similar measures within their own jurisdictions. Yet there are also examples where a lack of knowledge possibly delayed interventions. Researchers in Kumamoto who were seeking an answer to what caused the strange disease in Minamata in the 1950s did not initially have access to information from Europe about methylmercury's dangers to human health. If this information had been available in Japan earlier, it may have helped the researchers to more quickly identify the link to methylmercury. This may (or may not) have resulted in a quicker political response.

Matrix Reloaded: Applying the Matrix Approach to Other Sustainability Challenges

The HTE framework and matrix-based approach that we present and apply in this book is not specific to mercury. Further research could therefore use the same framework and matrix approach to examine other human-technical-environmental systems beyond mercury. Selections of additional empirical cases might draw from the idea of focusing on "action situations" in the Institutional Analysis Development framework (Ostrom 2009; Ostrom 2011). Researchers can use the HTE framework to gain a deeper understanding of other important sustainability issues, and to identify similarities and differences among different cases through comparative studies, thereby further developing and modifying theories about the structure and functions of complex adaptive systems relevant to sustainability. The application of the HTE framework together with the matrix approach would allow for an analytically consistent comparison across different empirical systems.

The fact that the HTE framework centers on material interactions makes it a particularly useful tool for analyzing material flows, including those of pollutants, goods, nutrients, water, or energy. Applying the framework to other types of pollutants would push analysts to consider social and technical factors that interact with the environmental cycling of pollutants. Lead, for example, has a similarly long history of human use and exposure when compared to mercury, and is another element that continues to pose significant environmental and human health challenges in many local communities—even as its major modern emission source, leaded automobile fuel, has been phased out in the vast majority of countries worldwide.

Problems related to many persistent organic pollutants (POPs)—both pesticides and industrial chemicals—stem from their extensive use and transport in both the environment and international trade. Many different forms of plastics circulate in society, where they provide benefits to people but also cause much environmental harm, especially in aquatic ecosystems.

The applicability of the HTE framework is not limited to analyzing flows of materials across spatial scales; it can also be used to examine place-based systems. In chapter 2, we used the example of Minamata disease around Minamata Bay to introduce the framework and illustrate how we apply the matrix approach. This type of locally centered analysis could be extended further. In another mercury-focused example, the matrix approach can be used to analyze a specific community in which ASGM is practiced, such as in Madre de Dios. Applications of the HTE framework and the matrix approach to place-based case studies, particularly where interactions have been previously studied with other frameworks, might be useful to compare insights across different analytical frameworks and disciplines. If the HTE framework were to be applied to common-pool resource issues such as forest or fisheries management, for example, its focus on material interactions and technology might result in insights distinct from those gleaned from studies using a social-ecological systems framework (e.g., Ostrom 2009).

The HTE framework and matrix approach can also inform other analytical efforts to examine sustainability-relevant questions. This includes network analysis (e.g., Bodin et al. 2019), since all different pathways identified using the matrix approach can be displayed in a network diagram to illustrate the same information, as we discussed earlier in this chapter. Representing interactions as networks may help analysts disentangle multiple, complex causal pathways. Here, we chose illustrative pathways, but a full network analysis applied to a detailed matrix would more formally identify key nodes and multiple, interacting causal pathways. Network analysis methods may help those who interpret matrix analyses at higher levels of complexity, multiple levels of scale, and linkages between different systems. The HTE framework and matrix approach may also give some guidance to transitions researchers who seek to better understand system interactions and their dynamics. Transitions researchers have emphasized the need for new analytical approaches to assess transitions involving technological, economic, social, and ecological change (Turnheim et al. 2015).

The HTE framework and matrix approach could be further modified to address different aspects of issues important to sustainability in more detail. One such development may be to explicitly classify different types of interactions, which could be divided between flows of pollutants, energy, money, or information based on categories used in the engineering systems literature (e.g., de Weck et al. 2011). The matrix approach can also be used to categorize cross-scale system connections by classifying component attributes relative to common metrics of spatial or temporal scales (such as kilometers or hours, for example). As we noted in chapter 2, analysts may wish to use the matrix as an organizing principle for helping to design a quantitative systems model where the necessary data are available. The matrix could thus serve as a mechanism for bridging qualitative and quantitative analysis, which has been urged in the transitions literature (Turnheim et al. 2015; Köhler et al. 2017). Analysts could then use a quantitative model based on a matrix analysis to measure systems relative to different sustainability metrics, and their progress over time.

Further empirically grounded theory development is important to sustainability analysis. There is, however, an ongoing debate about what types of theories are most useful. Elinor Ostrom (2007) argues that scholars of social-ecological systems should move "beyond panaceas," looking not for universal solutions to their problems but instead to diagnose the variables and outcomes that affect specific systems in their full complexity. Oran R. Young (2018) similarly suggests that analysts should focus on deriving mid-range theories about the effectiveness of international environmental regimes that can explain a subset of cases rather than looking for those that are universally applicable. Categorizing components, interactions, and interventions across systems in a common structure is one way in which analysts can begin to develop and test such mid-range theories. However, the complexity of the mercury systems illustrates the difficulty in deriving theories that can apply even to the entirety of the mercury problem. This suggests that even mid-range theories might be too ambitious, and that deep empirical engagement and the development of smaller-range theories could advance thinking and scholarship. In this vein, we draw some initial insights relevant to selected areas of scholarship in chapter 9.

The HTE framework coupled with a matrix-based approach offers analysts a valuable tool to see and study complex adaptive systems relevant to sustainability. The

matrix approach can be used to identify system components, examine their inter-
actions, and analyze interventions. Our analysis of the mercury systems shows
that identifying a relatively small number of aggregated system components at
varying spatial scales can describe systems relevant to sustainability in ways that
are analytically useful. Focusing on interactions among material system com-
ponents in the context of institutions and knowledge can help trace important
pathways of linked interactions that shape system dynamics across spatial and
temporal scales. Interveners who have opportunities to influence either material or
non-material components can alter system operations at different leverage points
and geographical scales to effect change. Further studies could extend and apply
the HTE framework and the matrix-based approach to examine additional topical
or place-based systems relevant to sustainability.

9   Sustainability Insights: Earth "Under Pressure"

Human activities have dramatically altered the Earth. Analysts in a variety of academic fields have attempted to better understand the drivers and impacts of human pressures on the Earth's life support systems, how these pressures have changed over time, and the ways in which societies have acted to both address and adapt to them. Reducing the human footprint on environmental processes while protecting and enhancing human well-being is at the heart of sustainability. To this end, some researchers are interested in better understanding the structure, function, and outcomes of complex adaptive systems that are relevant to sustainability. Others focus on defining critical aspects of sustainability or studying the dynamics of transition processes. Still others examine the design, implementation, and effects of governance arrangements for sustainability at and across various geographical scales. Insights of relevance to these different groups of researchers can be drawn from the analysis of the five mercury systems.

Humans have mobilized large amounts of mercury from the Earth's crust, especially over the past five centuries. This has had large environmental and human health consequences, as we detailed in part II. Uses and discharges of mercury occur in parallel with other accelerating human pressures. These include increases in energy use, population, consumption, carbon dioxide emissions, nitrogen pollution, and biodiversity loss (Steffen et al. 2015a). Addressing these human influences on the Earth is an urgent matter. Researchers in different fields share common interests in how changes to the Earth's life support systems occur, and how they threaten human well-being. Many scientists investigate the dynamics and consequences of altered physical and biological processes. Other scholars engage in efforts to better define sustainability, and to explain how transitions toward greater sustainability can be understood and encouraged. Those interested in governance issues examine the processes by which people come together to collectively address human activities that harm the environment and human well-being. All of these fields share a goal of understanding and theorizing about Earth systems that are under pressure from human activities in novel and changing ways.

The idea of human activities pressuring Earth systems relates to the message sent to negotiators of the Minamata Convention when Freddie Mercury and Queen's "Under Pressure" echoed over loudspeakers to urge delegates to address specific aspects of the mercury issue. Insights from the analysis of the mercury systems provide fodder for further developing and testing theories (either mid- or smaller range) about the ways in which systems respond to human pressures, for charting the progress of systems toward sustainability, and for identifying the functions and dynamics of institutions that set out to govern them. In this chapter, we answer our fourth research question—What insights can be drawn from analyzing these systems?—by synthesizing across the three areas that we identified in part I and addressed in each chapter of part II: systems analysis for sustainability, sustainability definitions and transitions, and sustainability governance. We draw these insights with the aim of distilling actionable knowledge to support the formulation of collective, but often contentious, decisions on how to steer societies toward greater sustainability.

Systems Analysis for Sustainability

Multiple sets of literatures apply systems-oriented approaches to examine connections between identified components. Many systems relevant to sustainability are complex and adaptive, and their behavior can be difficult, if not impossible, to predict (Sterman 2011). Different scholarly fields propose varying methods and approaches for analyzing complex adaptive systems (e.g., de Weck et al. 2011; Binder et al. 2013). In this section, we draw three insights that are relevant to those who are interested in better understanding systems and their behavior. First, new systems-oriented analytical approaches could better account for dynamics of human, technical, environmental, institutional, and knowledge components in an integrated way. Second, adaptive capacity stands out as an important dynamic that explains system behavior over time, and it can have both positive and negative effects on human well-being. Finally, concepts that guide systems analysis in the Anthropocene could better capture system-wide variability and changes over temporal and spatial scales.

Understanding Complex Adaptive Systems

Much systems analysis is conducted within individual disciplinary fields. Nevertheless, as we discussed in chapter 8, gaining a deeper understanding of the dynamics of complex adaptive systems relevant for sustainability requires further integration of perspectives and expertise from multiple disciplines. The importance of multidisciplinary analysis is well known, yet few efforts have been able to achieve an integrated perspective that gives different types of system components and their interactions equal attention. Our analysis of the mercury systems illustrates why such an integrated perspective is methodologically and analytically necessary. It also suggests several ways forward for advancing such multidisciplinary analysis of interactions among people, technology, the environment, institutions, and knowledge in order to better understand interactions between different system components, and to explore and assess opportunities through which actors can intervene to promote greater human well-being.

In applying a systems perspective, researchers who study the environmental cycling of materials like mercury could benefit from focusing on the combined impacts of human and natural drivers. For example, biogeochemical cycling analyses aim to quantify the degree to which environmental concentrations of a variety of chemical elements have been increased by human activities; they sometimes do this by calculating global enrichment factors (Sen and Peucker-Ehrenbrink 2012; Schlesinger et al. 2017). Many mercury researchers focus on understanding to what extent global environmental concentrations of mercury have been enhanced due to human activities (Selin et al. 2008; Amos et al. 2013; Outridge et al. 2018). But a growing amount of historically emitted mercury, from both anthropogenic and natural sources, is remobilized from environmental storage by human-induced land-use changes and climate change, and as a result the distinction between natural and anthropogenic flows is becoming increasingly artificial. Thus, a focus on trying to specifically attribute the footprint of human influence may be less relevant to sustainability analysis than more comprehensive systems-oriented approaches.

Systems analysis that better accounts for two-way interactions between environmental processes and human activities is especially relevant to sustainability-focused studies. Uses and discharges of mercury both affect, and are affected by, environmental processes. The prevalence of ASGM and associated mercury use may increase as a result of ecological deterioration, pushing farmers who can no longer make a living growing crops and raising animals into the mining sector. The availability and relative price of coal and oil as feedstocks influenced the amount of mercury that was used in the chemicals industry, leading to discharges of mercury into the environment. Improved scientific understanding of the environmental and human health risks from mercury prompted societal concerns and the adoption of controls on mercury emissions and releases. Treating anthropogenic discharges of mercury as external to a system may be useful for addressing some questions, such as identifying the total amount of present-day emissions and releases. However, studies that do not account for feedbacks between the environment and human activities provide only a partial picture of the material flows that are important to sustainability analysis.

Efforts to quantify how toxic substances cross the globe should account for both societal and environmental transport. The long-range transport of mercury and other toxic substances is substantial regardless of whether it occurs via the atmosphere or international trade. Many quantitative models for mercury and other pollutants nevertheless calculate the impact of one region on another purely based on environmental factors such as long-range atmospheric transport (Fiore et al. 2009; Corbitt et al. 2011). Societal use and flows of substances are often overlooked in models of environmental transport, but they are also relevant in assessing the impact of one region on another. Mercury and many other hazardous substances that move through trade affect human well-being far from the site of their extraction or production. But reliable data on cross-border trade flows are lacking for many commercial substances, including mercury (UNEP 2017). This creates a need for the scientific community to expand efforts to gather further data on environmental cycling together with societal cycling of substances for the purpose of improving sustainability analysis and informing decision-making.

One way forward may be to better integrate biogeochemical cycling studies, environmental modeling, and methods from industrial ecology. A large body of literature in industrial ecology and related fields on material flows through society and the economy employs concepts of industrial, urban, and socio-economic metabolism to better understand the flow of pollutants (Ayres 1989; Fischer-Kowalski and Haberl 2007; Ferrão and Fernández 2013). Studies on heavy metals focus on whether such materials are recovered and recycled, or if they are released or "dissipated" in the environment (Müller et al. 2014). For example, mercury use in a pesticide is considered dissipative because it cannot be recovered, while mercury in a closed product such as a thermometer can, at least in principle, be recycled. Environmental modeling focuses on the results of this dissipation in the form of environmental discharges, transport, and transformation. Some work has been done to link material flow analysis with mercury transport and environmental behavior, for example in China (Hui et al. 2016; Wu et al. 2016). Further combining these perspectives would provide a more complete systems-level accounting of sustainability issues.

The development and use of quantitative models that capture dynamics of human, technological, and environmental interactions simultaneously helps researchers to better understand system behavior and develop related theories. Integrated modeling that accounts for both human and environmental dynamics has a long history (Dowlatabadi 1995). This is especially the case in climate change modeling, where integrated models simulate the economic drivers of fossil fuel emissions and the environmental and human health consequences of climate change. Some researchers have called for further improvements in representing human and environmental dynamics in models for studying climate change, as well as other issues such as forestry, agriculture, and marine ecosystems (Bonan and Doney 2018; Calvin and Bond-Lamberty 2018). Insights from the mercury systems reinforce that any system model intended to explore sustainability-relevant questions would likely neglect important information if it omitted whole categories of human, technical, or environmental components or related interactions. Effectively incorporating all three categories is a challenge, however, and requires model development. Capturing technology development, for example, has proved particularly difficult in integrated models that account for both human and environmental dynamics (Ackerman et al. 2009).

It is important to recognize that adding more complexity to models may not always serve a useful analytical purpose for sustainability-related research. Including specific dynamics may only be relevant for sustainability analysis where they lead to different conclusions and suggest alternate intervention strategies. Analyses of different sustainability issues also require different model capabilities. Including and quantifying specific biological mechanisms for methylmercury uptake in animals and humans may be necessary for modeling a particular ecosystem with the goal of designing better interventions in the form of updated and more detailed dietary recommendations. But that kind of detailed analysis may be less relevant to efforts that seek to better understand and phase out mercury use in production processes with the aim of reducing worker exposure and environmental discharges. In such a case, a model of the use and flows of mercury in a specific type of manufacturing plant may be more analytically relevant. This variation in modeling needs across different aspects of a single issue such as mercury suggests the value of having access to a spectrum of different models, as we discussed in chapter 8.

Adaptation Dynamics and Human Well-Being

Much systems analysis focuses on better understanding how systems reconfigure as a result of disturbances or changes. Previous literature identifies adaptive capacity or adaptability as an important property of systems, including those relevant to sustainability (Walker et al. 2004; de Weck et al. 2012). The concept of adaptive capacity, as the ability of a system to adjust to maintain key functions when under stress, is also related to the idea of resilience, which can be defined as the capacity to maintain a desired set of functions in the face of disturbances (Biggs et al. 2012). Carl Folke (2016) argues that resilience "reflects the ability of people, communities, societies and cultures to live and develop with change, with ever-changing environments." The resilience literature also addresses the important difference between maintaining system function as a positive characteristic, and the kind of negative rigidity in systems that may impede desirable transformations (Olsson et al. 2014). A high degree of adaptation is often seen as positive for maintaining system operations over time, but it is sometimes necessary to actively break down a system because it generates outcomes that undermine human well-being. In such instances, a resistance to change poses an obstacle to interventions that support greater sustainability.

Our analysis of the mercury systems further highlights the importance of understanding the adaptive behavior of complex systems when examining their dynamics over time. Technological innovation emerged as a type of adaptation that was important in the products and processes system, as mercury use over time both increased and decreased with the development of new techniques for producing goods. Innovation was also prominent in the atmospheric system, which reconfigured in response to regulatory pressure to reduce mercury emissions from industrial point sources. For example, the application of emission control technologies resulted in fewer mercury emissions, but the control technology altered the fraction of different forms of emitted mercury that travels short and long distances. Market forces and human social and economic behavior contributed to adaptation as well. Changes in food availability and dietary preferences as a result of mercury contamination affected people's exposure to and impacts from methylmercury. Ecological degradation led to surges in the number of ASGM miners, and supply restrictions on mercury led to the illegal reopening of old mercury mines and trade.

Dynamics of adaptation can have varying and sometimes conflicting impacts on human well-being, leading to benefits as well as harms. Innovation, involving technical and human components, maintained the ability of several mercury systems to provide socially valuable goods over time. This was important for developing new mercury-free products and production processes and end-of-pipe control technologies for industrial point sources. The ability of producers and societies to innovate served to maintain benefits from the consumption of goods and energy while reducing mercury use and discharges. In contrast, adaptive dynamics sometimes proved negative for human well-being. Innovation resulted in new uses of mercury simultaneously with new mercury-free techniques. The supply of mercury to the ASGM sector continues, along with its harms to human health and the environment, despite domestic and international interventions to limit its flow into mining communities. Conversely, an absence of adaptation was also sometimes beneficial for human well-being: a resistance from international organizations, health authorities, and doctors to eliminating mercury use in vaccines has ensured the continuing provision of life-saving health services to many people.

The low ability of interactions involving environmental components to adapt to human-induced pressures related to toxic substances can have particularly negative consequences for human well-being. The sometimes very long timescales of interactions among environmental components can, at times, make it difficult for humans to modify these interactions, complicating both shorter-term and longer-term efforts to promote sustainability. Concentrations of mercury in the atmosphere, land, and oceans are relatively resistant to decreases due to the extended timescales in mercury's global biogeochemical cycle. This makes it more difficult for human efforts to significantly reduce environmental concentrations of mercury by simply addressing current and future emissions and releases. The inability of some aquatic ecosystems to absorb mercury deposition without converting it into methylmercury is a further sign of the low adaptive capacity of environmental components that has negative consequences for the health of both animals and humans.

Interveners often try to enhance the resilience of a system. The fact that a high degree of adaptive capacity can lead to mixed outcomes in terms of human well-being, as we have observed in the mercury systems, makes efforts to enhance resilience a challenge. Encouraging positive adaptation while mitigating its negative elements requires careful and targeted intervention, and the recognition that some components are not easily changed. Where human activities have long-term consequences for environmental components, efforts to mitigate harms to human well-being through other means become important. These can include efforts to reduce human exposure through dietary advice or preventing the harvesting of seafood from certain bodies of water. Such measures are sometimes still partially ineffective—fish-eating birds do not follow dietary recommendations, no matter how sound they are. This suggests that although focusing on overall resilience can provide useful guidance, different aspects of systems may require very different types of interventions to promote sustainability.

Challenges for Planetary-Scale Systems Analysis

Some believe that we live in the epoch of the Anthropocene (Crutzen 2006), although the usefulness of this concept and its potential start date as a new epoch are debated (Malm and Hornborg 2014; Lewis and Maslin 2015; Zalasiewicz 2015). Geologists define epochs by identifying a globally visible "golden spike" or environmental marker in the geological record to determine their beginning. Some analysts argue that the Anthropocene began in the 1900s, around the time that fossil fuel use and consumption, as well as population levels, accelerated (Malhi 2017). Others believe that a much earlier start date is more appropriate, positing that human activities began to have global-scale impact on environmental processes far back in history. Some archaeological evidence suggests that global human-induced land-use change began 3,000 years ago (Stephens et al. 2019). In 2019, an expert working group under the International Commission on Stratigraphy, a scientific body of the International Union of Geological Sciences, voted in favor of establishing the Anthropocene as a formal geological epoch with a mid-twentieth-century start date. If this proposal is accepted by the International Commission on Stratigraphy, it will be forwarded to the Executive Committee of the International Union of Geological Sciences, which will make the final decision (Subramanian 2019).

The human influence across regional and global scales involving both environmental and societal cycling of mercury goes back at least to the Spanish colonial silver and gold mining in Latin America starting in the 1500s. This is consistent with an argument that a New–Old World collision that began in 1492—which resulted in a large human population shift, the expansion of trading networks linking Europe, China, Africa, and the Americas, and a mixing of previously separate biotas—is a critical point in the history of human impact on the planet (Lewis and Maslin 2015). Mercury was a central element in this often environmentally and socially destructive process, due to its substantial trade and importance in gold and silver mining. The global transport of mercury escalated at that time, although the ultimate environmental fate of this mercury remains contested. Measurements of mercury in environmental archives, as discussed in chapter 3, do not all show a globally identifiable "golden spike" during that time. Even if the environmental distribution of mercury remained largely local, however, societal flows of mercury increased with the advent of transatlantic trade, as much mercury from Almadén and Idrija was shipped to the Americas.

Defining the Anthropocene based only on the geological record does not account for the full scope and influence of societal processes: environmental archives and geological records alone cannot identify all global-scale anthropogenic influences that are driven by interactions between humans and technology. Mercury has been traded across societies for millennia, going back at least to Roman times when cinnabar and mercury extracted from mines in Almadén was an important commodity. Evidence of this is found in written records of how much mercury different mines produced and in logs of mercury trades. Defining when dynamic global-scale interactions characteristic of the Anthropocene began, and identifying lessons for present-day systems-oriented sustainability analysis, thus requires consideration of societal flows in addition to environmental data. The concept of the Anthropocene is used not only by geologists but also by a broader group of researchers who may define and use the term differently from those who consider it only as a marker of global-scale influence measurable everywhere in environmental compartments. Scholars applying the Anthropocene idea to sustainability, especially those who define sustainability with a focus on human well-being, should therefore carefully define what they mean by the term and how it applies to their analysis.

Researchers have developed the concept of planetary boundaries to assess the ultimate limits of the Earth system based on the identification of global-scale environmental markers. They argue that going beyond these planetary-scale boundaries can have detrimental, or even catastrophic, consequences for humanity's ability to continue to live in a relatively stable environment (Rockström 2009; Rockström et al. 2009). Scientists have also attempted to quantify the outer limits of specific planetary boundaries for critical Earth-system processes; for example, identifying a range of "safe" atmospheric carbon dioxide concentration levels that should not be exceeded to avoid catastrophic climate change. The idea of a planetary boundary is envisioned to ensure that society stays a safe distance from dangerous thresholds and tipping points in Earth systems. Some experts argue that some critical boundaries have already been reached or breached, in particular in the areas of biodiversity loss, climate change, and nitrogen cycling (Steffen et al. 2015b). Nevertheless, the concept of planetary boundaries is contested (Nordhaus et al. 2012).

Applying the concept of planetary-scale boundaries is challenging for toxic substances like mercury. This is grounded in the fact that pollution involves a large number of substances from multiple sources, in different places, with different toxicities, and in varying amounts (Diamond et al. 2015). Johan Rockström and colleagues (2009), however, argue that there are two possible approaches to identifying planetary boundaries for chemical substances. One is to focus on substances that have global environmental distribution, and the other is to focus on those that pose unacceptable long-term and large-scale effects on living organisms. The latter approach may require a combination of efforts: setting a range of sub-boundaries based on the effects of many individual substances, and identifying specific effects on sensitive organisms. When smaller-scale processes occur simultaneously in different places around the globe, their impacts may aggregate to a point that a common planetary threshold or tipping point is crossed. Matthew MacLeod and colleagues (2014) further argue that three conditions must be met for chemical pollution to pose a planetary boundary threat: (1) it has a disruptive effect on a vital Earth system process; (2) the disruptive effect is not discovered until it is a problem, or inevitably will become one, at a planetary scale; and (3) the disruptive effect is poorly reversible.

Defining a global boundary for mercury and related chemical pollutants is difficult, however. It may also turn out to be both impossible and undesirable in the context of sustainability. First, there is no clear way of empirically identifying a single planetary boundary for a substance such as mercury that is dispersed across the entire planet by both environmental and societal processes at highly varying concentrations and forms in air, oceans, and land. This variability also makes it difficult to determine from environmental data alone when a particular threshold has been crossed at global scale. Second, much harm to human health and the environment from bioaccumulative toxic substances like methylmercury can be highly locally concentrated, yet nevertheless is affected by global forces. Mercury discharges affect nearby as well as distant places, with major differences in human impacts that depend on local levels of mercury use and discharges, ecosystem processes, and seafood consumption patterns. In addition, individual variations in genetic characteristics are increasingly understood to affect the sensitivity of individuals to mercury and other toxic exposures. These factors combine to make the link between severe local harms from mercury and any global boundary tenuous at best.

Sustainability Definitions and Transitions

Sustainability is both a contested concept and an urgent planetary goal. How sustainability is defined has important implications for analysis, as it influences which specific factors are seen as essential to sustainability, in both theory and practice. The choice of definition in turn affects analyses of important drivers and factors that hinder or promote transitions toward greater sustainability. We identify three overall insights concerning sustainability definitions and transitions. First, different values attributed to benefits and risks complicate analysis of human well-being across populations and over time, challenging efforts to define sustainability. Second, many transitions toward sustainability were characterized by incremental change, but some nevertheless had substantial benefits for human well-being. Finally, different patterns and modes of transitions that had interacting dynamics occurred simultaneously.

Sustainability and Human Well-Being

We elected in our sustainability-oriented analysis of the mercury systems to focus on human well-being, rather than another factor, such as the maintenance of particular natural resource stocks at certain levels over time—a perspective that would have been more aligned with a "stronger" sustainability definition. Our choice of an anthropocentric focus on sustainability shaped our analysis by centering it on factors and processes that improve or detract from human well-being. Applying a stronger sustainability criterion of natural resource maintenance to the mercury systems would have been conceptually challenging given mercury's extensive use over history. Analysts have disagreed about the extent to which sustainability requires non-renewable resources to be maintained (and not depleted), and the degree to which different resources are able to substitute for one another (Ayres 2007). We allowed for the possibility that one resource can substitute for another in providing for human well-being, and we empirically examined where and when this was possible. The mercury systems provide several examples of how different people, in different places and at different times, have perceived these substitutions and associated trade-offs.

Insights from the mercury systems show that preventing extraction of a non-renewable resource like mercury does not always benefit contemporaneous and future human well-being. While mercury led to many harms, uses of mercury provided many tangible benefits to people all over the world for millennia. In more recent times, mercury's unique properties facilitated innovation, including improving measurement precision and enabling the development of fundamental scientific knowledge that paved the way for new technology. The use of the mercury thermometer, mercury amalgam in dentistry, and very small doses of mercury as a preservative in life-saving vaccines made it possible to provide better health care for billions of people over many generations. Mercury use in chlor-alkali production enabled the manufacturing of chlorine to rapidly expand, and resulting societal benefits included disease prevention by chlorinating drinking water to make it safer for human consumption. Modern uses of mercury in compact fluorescent bulbs helped reduce energy demand from fossil fuels, and even reduced net mercury emissions in some places by decreasing the demand for electricity produced by coal.

The concept of a circular economy, which focuses on the recovery and recycling of materials, is gaining increased attention from analysts and policy-makers (Ghisellini et al. 2016). The establishment of a more circular economy can help better manage flows of substances, but implications of these flows on sustainability can be mixed and complex. Some analysts consider a circular economy a necessary condition for sustainability, others assess it as beneficial but not required, and still others argue that it has both pros and cons from a sustainability perspective (Geissdoerfer et al. 2017). Among those analysts who have critiqued the circular economy perspective, Julian M. Allwood (2014) points out that implementing a circular economy for materials can increase energy demand for recovery and recycling, potentially resulting in decreased overall sustainability depending on the impacts of the sources of this energy. This underscores the importance of not just studying societal and environmental flows of a particular substance such as mercury in isolation, but also considering how those flows affect, and are affected by, the flows of other substances.

Examples from the mercury systems highlight the complexity of assessing the impact of material flows. Ensuring sustainable use of materials can defy simple solutions (Olivetti and Cullen 2018). A growing number of countries are adopting supply restrictions on mercury to curb its use and trade, including the adoption of the US and EU export bans on elemental mercury in the 2000s. However, this has resulted in an increase in illegal mercury mining together with the smuggling of mercury into local ASGM communities in many countries. This mercury is sold on the black market and subsequently discharged into the environment from amalgamation processes and accidental spills. Yet not all uses and environmental discharges of mercury may be net negative for human well-being. Mercury in a vaccine enters the human body and is then excreted into the environment. Most analysts would argue that the benefits of safe and effective vaccines for human well-being far exceed the harms of this miniscule "loss" of mercury dissipated from human bodies into the environment.

A better understanding of the long-term challenges stemming from early and more recent uses of mercury could nevertheless have prevented some past, present, and future harms to human health. Assessing the impacts of mercury use on human well-being requires understanding the complexity of how it affects humans over time and space. The perceived benefits and harms of mercury use changed dramatically over time as a result of new scientific and medical knowledge and societal norms. Mercury as a commodity, once a valued asset, is now increasingly seen as a liability. Even today, though, mercury remains a valuable commercial product, not least because of its extensive use in ASGM. At the same time, public authorities and firms have spent billions of US dollars on cleanup where mercury has contaminated local environments, and more money will go toward addressing contaminated sites in the future. Much is also spent on health care costs related to mercury exposure as well as public information campaigns about the dangers of mercury in food and the formulation of dietary recommendations. Estimates of the monetized harms of mercury pollution to health can far exceed expected costs of mitigating emissions (Giang and Selin 2016; Sunderland et al. 2016).

Accounting for the value of assets that might become liabilities in the future is a major challenge for efforts to define sustainability. Following the literature on inclusive wealth, an analyst from the not-so-distant past might have valued mercury positively as a depletable asset and regarded underlying stocks of mercury as contributing to human well-being. Mercury, despite its contribution to environmental damage, still maintains a positive commodity price and is traded both legally and illegally. But through most of the history of its human mobilization, the commercial stock of mercury represented a yet unknown future claim on human well-being. This is not just a matter for history: a recent version of the World Bank's Changing Wealth of Nations report, detailing benchmarks on comprehensive wealth, includes the value of mineral reserves for lead, another toxic heavy metal (Lange et al. 2018). A similar argument can be made today for including reserves of coal and oil in such accounting. Of currently identified reserves, a third of oil, half of gas, and more than 80 percent of coal need to remain unused if global average temperature increases are to stay within a 2-degree Celsius target (McGlade and Ekins 2015).

The existence of trade-offs in benefits and costs of mercury uses also reinforces the importance of considering equity and power when defining and analyzing sustainability. The way benefits and harms are valued is contingent on what, and who, matters, and to whom. Many Arctic indigenous communities where people consume large quantities of seafood are shouldering the human costs of being affected by methylmercury that originates from distant economic activities—and are receiving few of the related benefits. Responses to contamination incidents such as those in Minamata, Grassy Narrows, and Kodaikanal show that mercury exists in the context of other, competing economic values, such as the worth of an industrial manufacturing plant to a local community and factory owners. The lack of value that Spanish colonial rulers in South America gave to the human lives of indigenous peoples allowed for mining, and its health and environmental hazards, to persist for centuries. In contemporary ASGM, the use of mercury adds to other challenges present in poor communities where many miners and other community members operate in the informal sector, vulnerable to exploitation.

Incremental versus Fundamental Transitions

Scholars disagree about the importance of incremental relative to fundamental transitions toward sustainability, with some analysts putting more faith in incremental steps than others. The development and stepwise application of more environmentally friendly technology is central to moving toward greater sustainability in the literature on ecological modernization (Mol 2003). However, critics argue that the concept of ecological modernization relies too heavily on the future promise of technological salvation while not fully recognizing the necessary scope and depth of behavioral change (Gibbs 2006). Some research on sustainability transformations instead emphasizes that radical changes are necessary: Derk Loorbach and colleagues (2017) note that major transitions are not merely technological, but are accompanied by socio-cultural change that has a deep effect on institutions, routines, and beliefs. Technological and behavioral changes are, of course, not mutually exclusive, but can take place in tandem. Yet, perspectives that emphasize incremental solutions often focus more on technological improvements, whereas those who argue that fundamental change is needed often stress the need for behavioral changes.

Transitions in the mercury systems show that incremental changes can have substantial benefits for the well-being of both present and future generations. The gradual implementation of end-of-pipe control technology on large industrial point sources, such as on waste incinerators and coal-fired power plants in Europe and North America, mitigated local harms by cutting mercury emissions, and relatively quickly reduced exposure levels to nearby populations. These controls also prevented the mobilization of mercury that would otherwise have contributed to future mercury cycling in the environment for decades to centuries. Similarly, the application of retorts in ASGM that reduce mercury emissions and exposure has large benefits for the health of miners, people working in gold processing shops, and other community members as well as for nearby and faraway consumers of seafood. The introduction of increasingly stringent worker protection measures in other mercury-using sectors also prevented much health damage. In addition, piecemeal efforts that reduced levels of mercury use in industrial processes lessened mercury releases to the environment and their short-term and long-term damage.

The introduction of new technology had multifaceted implications for sustainability transitions in the mercury systems, with both positive and negative elements. Compact fluorescent bulbs are an improvement over the incandescent bulb with respect to energy efficiency and, in many cases, associated mercury emissions from power generation. However, they still contain mercury. Increased technological efficiency of production processes, such as in the European chlor-alkali plants that still relied on the mercury-based production method, reduced mercury use and discharges but delayed progress toward fully eliminating mercury use by extending the lifetime of the old production method. The expanded use of retorts on local mining sites has clear benefits to human health and the environment, but lessens the urgency to phase out mercury use in ASGM. The development of more efficient pollution control technology, capturing mercury that would otherwise be emitted to air from industrial point sources, significantly reduces the amount of mercury that enters the environment. This technology, however, allows for the building of additional coal-fired power plants as long as any mandated pollution controls are implemented, contributing to climate change.

The mercury systems provide further evidence that a substitution for a known hazardous substance can lessen one kind of damage, but may create new problems. For example, the environmental and human health implications of discharges from the production and use of membranes made of PFAS, toxic and long-lasting substances that substituted for mercury-based technology in chlor-alkali plants, are unclear. This was also the case in the past when early substitutes for DDT were less persistent but more acutely toxic (Walker et al. 2003). Furthermore, some substitutes for ozone depleting substances are potent greenhouse gases (Wallington et al. 1994). The toxicity of different heavy metals that are needed to make alternative sources to fossil fuel energy (such as photovoltaics and batteries) may (or may not) be less than that of mercury, and their environmental and human health effects locally and globally are at least partly unknown. The use of such heavy metals in energy technologies also changes where and for how long they are present in the atmosphere, land, and water. This suggests that analyses of benefits and costs to human well-being should be approached with humility given the limitations of scientific knowledge.

Our analysis of the mercury systems also shows that incremental changes can add up to support more fundamental change. The gradual phaseout of mercury in medicine over the last 100 years profoundly altered how mercury is used in medical treatments; the main remaining uses are in dentistry and vaccines. The stepwise controls of mercury in products facilitated a global ban of the production and sale of most currently existing mercury-added products under the Minamata Convention. The strengthening over time of mandates for air pollution control technology on large stationary sources in some instances both reduced mercury emissions and impacted fossil fuel production. The US Mercury and Air Toxics Standards and the Canadian mercury regulations likely contributed to the early closure of old coal-fired power plants, where it did not make economic sense for the owners to install new mercury-control technology. This interplay between stepwise and fundamental change suggests the need for more critical analysis of positive and negative consequences of incremental versus comprehensive approaches to transitions. In this context, some analysts have called for "incremental change with a transformative agenda," or "radical incrementalism" (Najam and Selin 2011, 453).

The fact that fundamental change can sometimes come from cumulative incremental steps makes it difficult to empirically distinguish between different degrees of transitions. Several kinds of transitions that had aspects of both incremental change and fundamental disruptions occurred in the mercury systems. Because technological change involving many mercury-added products and mercury-based production processes were not obviously associated with large-scale societal changes, some analysts may classify these as incremental changes. But a closer look reveals that there were substantive institutional developments in markets and domestic and international regulations as part of these changes. Consumers and governments put pressure on companies to reduce mercury use in products, and regional and global agreements codified maximum allowable thresholds for mercury concentrations in these products. There were also major changes in scientific knowledge and societal perceptions about the environmental and human health risks from mercury. Analysts must thus look at a combination of technological, economic, political, legal, social, and knowledge factors when examining the scope and depth of a change process.

Drivers of Transition Dynamics

Analysts are discussing the uniqueness of the drivers and dynamics of sustainability transitions compared to other types of societal transitions. Frank W. Geels (2011) posits that sustainability transitions differ from many other types of previous socio-technical transitions. However, Björn-Ola Linnér and Victoria Wibeck (2019) argue from a historical and comparative perspective that contemporary efforts to move toward sustainability can learn from other major changes that societies have undergone in the past, including the Industrial Revolution and the abolition of slavery. Transitions that are relevant to sustainability are deeply social and political processes that often are characterized not only by cooperation, but also by contestation and confrontation (Avelino et al. 2016). Systems can change both slowly and abruptly, but many system transitions related to sustainability are characterized by strong path-dependencies and lock-ins, as established practices are intertwined with individual choices and lifestyles, cultural traditions, business models and economic systems, technology, regulations and other policies, and organizational and political structures (Markard et al. 2012).

Transitions in the mercury systems reflected both economic and technical drivers (e.g., the development of cheaper and better ways to make large mirrors) and scientific advances (e.g., penicillin replacing previous remedies). The transition literature on niches, regimes, and landscapes helps analyze the multi-scale nature of these processes (Geels 2002). Many technical innovations that allowed for the phaseout of mercury use in products and processes occurred in niches that supported experimentation. For example, mercury-free technologies for chlor-alkali production emerged in different places in different times, sometimes from an interplay between environmental factors (such as the availability of brine wells) and national regulatory regimes. Later mercury phaseouts were driven by interactions between innovation processes and government-led regulatory measures focused on protecting the environment and human health. The landscape of the Global Mercury Partnership and the Minamata Convention accelerated some of these phaseouts. The prevalence of economic and technical drivers suggests that increased attention to socio-technical transition dynamics is useful in sustainability analysis.

Distributions of agency and power, sometimes embedded in institutions, can play a substantive role in driving transitions. Many employers had a much greater influence on the use of mercury than did their employees, in both mining and manufacturing. Similarly, doctors who prescribed mercury-based treatments were in a position of power over their patients. This gave employers and doctors, and later also governments, a dominating role in driving transitions away from these mercury uses. There are also instances where private sector actors successfully fought to delay or stop government actions, including for setting maximum allowable mercury concentrations in commercial fish, banning phenylmercury compounds in paints, and the introduction of pollution controls on stationary sources. All ASGM is locally concentrated, but many choices by potential and active miners (including those related to the use of mercury) are influenced by domestic laws on mining rights and by international markets for gold and mercury, over which individual miners have very little influence. These examples highlight the importance of assessing power relations among different actors in transition analyses.

Sustainability Governance

Much research on sustainability is motivated by the goal of identifying opportunities for more effective interventions that benefit human well-being. The challenge of governance, involving institutions at a variety of spatial scales, is in itself a complex system problem. We identify three main ways in which the many efforts to govern mercury-related issues over centuries illuminate how governance for sustainability could become more effective. First, ensuring that institutions fit the physical problems that they are designed to address involves paying close attention to material system components and their interactions. Second, governance strategies can look to address multiple sustainability issues simultaneously through institutional design. Third, evaluating institutional effectiveness requires simultaneously considering environmental and societal factors that shape outcomes.

Problem Structure and Institutional Fit

Many scholars argue that polycentric governance structures are often a better institutional fit with complex environmental and sustainability issues than more monocentric and traditional top-down governance structures

(Victor and Raustiala 2004; Ostrom 2010; Keohane and Victor 2011; Abbott 2012). Polycentric governance structures involve a large number of actors and policy instruments that are not organized in a strictly hierarchical way, and governance happens in a large number of forums at multiple geographical scales simultaneously. This polycentricism can create dynamic opportunities to address different aspects of large and complex issues in separate places at the same time, which can add up to better overall governance than attempting to accomplish all goals in the same venue. A polycentric governance approach also allows for greater policy experimentation, learning, and diffusion across venues and jurisdictions (Hoffman 2011). It is, however, important that governance efforts in different forums and across different geographical scales do not inhibit or contradict each other, but rather create synergistic benefits (Selin 2010).

Our analysis of the mercury systems shows the benefits of designing a polycentric governance approach across global, regional, national, and local scales. Efforts to limit the risks of methylmercury exposure resulting from the consumption of seafood benefit from targeted, locally specific advice, but much seafood is traded on regional and global markets, requiring some coordination across governance scales. The effectiveness of pollution-control technologies for capturing mercury emissions from industrial point sources depends on technical characteristics that may be specific to an individual plant. The Global Mercury Partnership and cooperation under the Minamata Convention facilitates the development and diffusion of pollution standards globally. The Minamata Convention sets out ambitious provisions for the phaseout of mercury in products and processes; the implementation and enforcement of these depend on actions at the national and local level. Addressing mercury use in ASGM involves a combination of community-level capacity building, changes to national mining laws, and efforts to change the behavior of transnational gold market participants.

The governance literature tends to focus on political, economic, and technical factors when characterizing the structure of environmental problems. From this perspective, environmental problems have been classified on a spectrum from "benign" (relatively easy to address) to "malign" (much harder to solve) (Miles et al. 2002). Other problem categories include "wicked" and "super-wicked" problems (Levin et al. 2012; Head and Alford 2015; Grundmann 2016). Super-wicked problems have been defined as comprising four key governance features: (1) time is running out; (2) those who cause the problem also seek to provide a solution; (3) the central authority needed to address it is weak or nonexistent; and (4) policy responses discount the future irrationally (Levin et al. 2012). These types of characterizations of problem structure, however, are much too general and simplified to fully capture the dynamics of many sustainability issues, including the dangers of mercury to human well-being, and to inform any deeper understanding around interactions and interventions.

It is important that a governance approach to an individual sustainability issue take into consideration its unique biophysical as well as societal characteristics. A key aspect of many chemicals issues is how substances cycle through the environment and society. But there are important differences even between chemical substances, which means that individual substances need their own detailed description of biogeochemical cycling and of societal uses and flows. Factors that influence governance can also vary among different aspects of a single substance, which we see across the mercury systems. Governments can, if they have the political support to do so, set pollution controls on a smaller number of major stationary sources relatively readily (as effective control technology already exists). It is much harder to address the activities of tens of millions of ASGM miners and community members operating in the informal sector. Designing institutions thus requires a detailed engagement with different aspects of what might seem to be a single issue.

The Minamata Convention attempts to engage different aspects of the mercury issue by setting out a global-scale legal framework for action on the full lifecycle of mercury (Selin et al. 2018). The negotiations and early implementation of the Minamata Convention drew political attention to mercury, and triggered some countries to take administrative and regulatory actions that they otherwise would not have taken (or at least not as early as they did). International law, including multilateral treaties such as the Minamata Convention, however, is only a "thirty-percent solution" to environmental problems (Bodansky 2010, 15). The general language in several Minamata Convention articles creates a need for further specificity, such as the formulation of national emission standards and national action plans on ASGM. Initiatives outside the scope of the Minamata Convention such as international certification schemes and codes are also important governance instruments. Stakeholders range from individual consumers of skin-lightening creams to ASGM miners and gold market participants to operators of fossil fuel–based energy infrastructure. This means that future governance will require efforts by a large number of public, private, and civil society actors, in a multitude of global, regional, national, and local forums, to realize the objective of protecting human health and the environment from mercury.

Governance Strategies

The literature on governance strategies for complex systems suggests that the identification of components and interactions that are located at different places within a system can help guide efforts toward effecting system change (Meadows 1999; Abson et al. 2017). These efforts include attempts by interveners to intentionally move systems toward greater sustainability. Many early efforts to address air and water pollution of mercury and other hazardous substances focused on ways to prevent emissions and releases through the application of end-of-pipe technologies. In contrast, a growing number of policy initiatives over the past few decades emphasizes the importance of interventions further upstream through modifications to underlying production processes that produce pollution (Browner 1993; Mayer et al. 2002). Upstream approaches are particularly influenced by discussions about the precautionary principle that focuses on preventing harm before it occurs, even when scientific knowledge about environmental and human health impacts of pollution may be incomplete or uncertain (Harremoës et al. 2013).

Insights from the mercury systems show that a combination of interventions is often necessary to enhance human well-being, for both present and future generations. Approaches such as recycling and environmentally sound disposal helped reduce mercury discharges. Actions further upstream that prevent mercury discharges completely by eliminating all mercury uses also remain critical to reduce mercury-related problems. However, the further upstream the intervention, the slower it may be to propagate through the system. People have altered the environmental cycling of mercury to such an extent that environmental concentrations will not return to natural levels on human-relevant timescales. Thus, while technology-based approaches that avoid generating pollution can help prevent further contamination, they will need to be paired with other actions to protect human well-being. This includes other kinds of downstream initiatives such as the issuing of global recommendations for maximum allowable mercury concentrations in food by the World Health Organization (WHO), together with more nationally and locally tailored dietary recommendations for specific populations and geographical locations.

Different actors with varying abilities to effect change engage in multiple ways in governance for sustainability. The mercury systems feature large and well-organized interests in energy generation and industrial production, and more diffuse groups such as consumers, medical patients, and ASGM miners. As in other issue areas, less powerful groups of actors often find it difficult to make their voices heard, as seen in Minamata, Grassy Narrows, Kodaikanal, and elsewhere. Many narratives surrounding pollution focus on populations who are on the receiving end and have no control over the processes that create it. The same is true in many cases for mercury, including for subsistence fishers and others who depend on local fish for important nutrients. At the same time, whale consumers in the Faroe Islands have much individual freedom to make their own dietary choices, and Arctic indigenous peoples collaborate in transnational networks to shape regional and global institutions, including the Minamata Convention. This is consistent with studies that show that victims of pollution may have opportunities to both individually and collectively exercise power and influence (Fernández-Llamazares et al. 2020).

It is sometimes possible to design interventions on one sustainability issue so that they positively affect another one as well. The application of pollution-prevention technology on industrial point sources can address sulfur, particulate, and mercury emissions simultaneously, having multiple benefits for environmental quality and human health. Reducing dependency on fossil fuels addresses both mercury and carbon dioxide emissions; slowing down temperature increases limits the negative impacts of climate change and may lead to less remobilization of legacy mercury from environmental storage. The same strategy for addressing land use change might benefit biodiversity, prevent runoff of nitrogen and hazardous substances into waterways, and create conditions that reduce the formation of methylmercury in aquatic ecosystems. Increasing resources for climate change adaptation can help prevent conditions that drive people to become ASGM miners, whose actions lead to mercury discharges. All of these examples illustrate the importance for analysts (and decision-makers) to consider opportunities for enhancing cross-sectoral synergies when designing governance structures and strategies to support a move toward greater sustainability.

At the same time, it is important to recognize that efforts to enhance governance synergies across sustainability issues can have drawbacks as well. Global cooperation on mercury may not have resulted in a legally binding treaty if solutions to mercury emissions were linked to restrictions on coal burning (necessary for addressing climate change). Enhancing linkages between the Minamata Convention and the Basel, Rotterdam, and Stockholm Conventions has been a topic of much discussion. Supporters argue that it would create political and administrative synergies, but skeptics fear that it may result in reduced attention to the Minamata Convention. Analyses of the efforts to link the Basel, Rotterdam, and Stockholm Conventions show that such efforts can facilitate policy-making and implementation as well as transfer political disagreements from one forum to another (Selin 2010; Allan et al. 2018). This suggests that efforts to craft "win-win" solutions to sustainability governance should consider political as well as material connections among issues. Less-coordinated actions can have a greater influence on human well-being than carefully optimized strategies where decision-making and implementation are contested or incomplete. A realistic governance approach may include both a higher-level and a longer-term focus on sustainability transitions in combination with an "honest recognition of the realities of near-term incrementalism at the same time" (Patterson et al. 2017, 4).

Evaluating Effectiveness

Evaluating whether governance strategies enhance present and future human well-being is analytically challenging. Analyses of the effectiveness of international environmental treaties like the Minamata Convention that focus on institutional dynamics include qualitative case studies, quantitative case studies, and large-number quantitative analyses (Young 2017). A nuanced understanding of institutional fit and problem structure is needed not just when designing institutions (as we discussed earlier), but also when examining the effectiveness of environmental treaties. Paying more careful attention to the role of technical and biophysical systems, however, poses significant research design challenges (Young 2011). Accounting for problem structure raises the issue of endogeneity, in which an environmental treaty is influenced by problem structure and in turn aims to influence it (Mitchell 2006). This makes it analytically difficult to study environmental treaties—and their impacts and effectiveness—as a dependent variable. Many effectiveness evaluations also rely on indicators that do not capture the data necessary to trace key elements linking policy to environmental outcomes (Wöhrnschimmel et al. 2016).

Institutionally focused evaluations of treaty effectiveness can use combinations of outcome and process indicators (Selin et al. 2018). The use of outcome indicators, focusing on environmental conditions, requires having the scientific ability and data to document complex pathways of interactions. First, it is necessary to determine whether a particular intervention caused a change in behavior or operation. For example, was it mercury-specific regulations that caused the mercury emission reductions that were observed in the United States in the 2000s? Or were these reductions the result of other air pollution controls, or market forces that made the burning of coal less economically profitable? Finding the answer requires expertise in policy analysis and economics, coupled with a deep understanding of the technologies involved. Next, it must be determined whether the decrease in mercury emissions have resulted in changes in the amount of mercury entering nearby and remote ecosystems. Measuring deposition trends in the context of both short-range and long-range transport of emissions as well as ecosystem variability and environmental change is difficult, and attributing deposition trends requires integrating environmental measurements with models and counterfactual analysis.

The challenge of attributing environmental change to its causes is much discussed in the scientific literature (Deser et al. 2012; Selin et al. 2018). Time trends for mercury in biota may not track concurrent trends in atmospheric concentrations or deposition because of the strong influence of other drivers on conversion to methylmercury and uptake in biota (Wang et al. 2019). It is well appreciated that many interactions, especially in the environment, are complex and often poorly understood. Much knowledge is still developing about the environmental behavior of different forms of mercury. For example, uncertainty remains about the chemical reactions that mercury undergoes in the atmosphere, the degree to which it is taken up by land ecosystems, and the timescales of its cycling in the environment (Selin 2009). Even more uncertainty surrounds links between environmental mercury cycling and climate and other large-scale environmental changes (Obrist et al. 2018). In addition, there is a growing appreciation that mercury, its behavior, and its impacts are variable in time and space (Eagles-Smith et al. 2018; Giang et al. 2018; Hsu-Kim et al. 2018).

The use of process indicators offers an additional way to evaluate treaty effectiveness. Process indicators provide a way to gather complementary information to the typically incomplete environmental and human health data that are generated through application of outcome indicators. For process indicators, the focus is not on whether a specific intervention has had a direct impact on mercury concentrations in fish in a particular body of water, for instance, or in a specific group of people, but rather if practical steps that are likely to have had a positive impact have been taken. In the earlier example of emissions in the United States in the 2000s, it can be assumed that measures such as the closure of old point sources that burn coal and/or the greater application of emission control technology will result in the avoidance of at least some mercury emissions that otherwise would have entered the atmosphere. This does not come nearly as close to tracking the effects from specific policy measures as using outcome indicators, but the latter may not be scientifically feasible at present. Process indicators can thus provide a rough approximation with much more straightforward analysis.

David C. Evers and colleagues (2016) suggest a combination of outcome and process indicators to evaluate the effectiveness of the Minamata Convention. Their evaluation framework consists of specific metrics for measuring changes in mercury levels that are derived from environmental monitoring, including of hot spots (outcome indicators), as well as a suite of short-, medium-, and long-term metrics that are related to the control articles in the Minamata Convention (process indicators). The use of process indicators also makes it possible to highlight multiple aspects of international cooperation ranging from the number of countries that have taken regulatory measures to the levels of financial flows in support of treaty implementation. Work on the first Minamata Convention effectiveness evaluation, scheduled for 2023, is examining a variety of outcome and process indicators. Hybrid approaches that associate process indicators with proxies for impact can also be developed. Based on experiences from the Montreal Protocol on stratospheric ozone depletion, it has been suggested that indicators estimating the impact of present-day mercury emissions on ecosystems in the future could help policy-makers better account for the timescales of the mercury problem when considering different policy options (Selin 2018).

The fact that planet Earth is under much (and growing) pressure from human activities makes sustainability an urgent challenge. For those researchers who engage in systems analysis, insights from the mercury systems underscore the need to apply multidisciplinary perspectives when considering human, technological, environmental, institutional, and knowledge components together, to account for adaptation in understanding sustainability-relevant systems, and to critically consider the applicability of concepts like the Anthropocene and planetary boundaries. For those who aim to advance definitions of sustainability and who explore sustainability transitions, human interactions with mercury over time and space offer fodder for thinking about definitions of sustainability and the dynamics of incremental change versus fundamental disruptions. For those who focus on governance challenges related to sustainability, analysis of the mercury systems shows how environmental and societal factors together affect institutional design, and how efforts to evaluate institutional effectiveness require paying attention to both types of factors.

10   Sustainability Champions: "We'll Keep on Fighting …"

A major goal of sustainability research is to better understand how people can intervene to actively steer complex adaptive systems toward greater sustainability. This is a significant challenge in many human-technical-environmental systems. Efforts to intervene to effect change in support of human well-being, such as those being conducted for mercury under the Minamata Convention and in other forums, often build from prior experience, where it may be possible to draw lessons from the past. Lessons can provide both motivation and ideas for change, and can help people understand causal relationships among components that will affect critical interactions within complex systems. Lessons from this book can help mercury researchers, decision-makers, and thoughtful citizens in efforts to mitigate the harms caused by mercury pollution, and thus help societies make progress on sustainability challenges.

Future mercury stories remain largely untold. Interventions on mercury go back centuries, and most of these were carried out long before the Minamata Convention entered into force in 2017. Many of the early interventions in the mercury systems that we discussed in part II helped safeguard the environment and human well-being from mercury use and pollution; however, work to address the full range of the mercury problem is far from over. Many of the necessary on-the-ground efforts to meet the objective of the Minamata Convention have only just started. The Minamata Convention and other initiatives to address mercury-related problems are also implemented in a broader context of promoting sustainability globally. The United Nations Development Programme (UNDP), as we mentioned in chapter 3, highlights linkages between its efforts to support the Minamata Convention with the seven Sustainable Development Goals on poverty, hunger, health and well-being, energy, work, consumption and production, and life below water (UNDP 2016).

The sustainability challenge centers on a human struggle to live and thrive across generations on a finite planet. Our analysis of the mercury systems illustrates the multifaceted nature of this endeavor. Important dynamics in the mercury systems occur in different places and at different times, but share common patterns across history. Mercury miners in Almadén, Idrija, Huancavelica, and elsewhere suffered greatly from mercury poisoning. Risks from using mercury in silver and gold mining connect colonial era miners in South America with present-day artisanal and small-scale gold mining (ASGM) miners. Arctic indigenous populations and other vulnerable communities look to the Minamata Convention to address mercury emissions from across the world. Several governments see opportunities through this global treaty to expand their regulatory authority on mercury, or look to global action as a vehicle for receiving financial and technical support.

When the Minamata Convention was adopted, as we mentioned in chapter 1, the music that played during the negotiations, Freddie Mercury and Queen's "Under Pressure," switched to another of the band's songs, "We Are the Champions." That song's lyrics highlight an important goal of research on systems relevant to sustainability: informing the actions of interveners—or champions—with the goal of promoting human well-being now and in the future. Efforts toward sustainability, however, occur within a large landscape of actors who have different interests and varying levels of power and access to resources. The mercury issue involves individuals who face different challenges, and who have different goals and priorities about how to balance competing demands and tradeoffs toward protecting and enhancing human well-being. Our analysis of the mercury systems underscores the slow and deliberate process of coming to consensus as a society about matters that influence people in different places, now and in the future.

The line from "We Are the Champions" that we have used in the title of this final chapter—"We'll keep on fighting …"—is an appropriate charge for current and future champions. In this chapter, we discuss lessons for readers who support the aim of the Minamata Convention of protecting human health and the environment from mercury. The first set of lessons is addressed toward researchers in the natural and social sciences and engineering who seek to understand the properties and behavior of mercury and related societal issues. The second set of lessons speaks to decision-makers and others, including public officials, representatives of industry or non-governmental organizations, or expert advisers, who work to craft and implement mercury-related policies and actions. The third set of lessons is relevant to thoughtful citizens who are concerned about the widespread use and dispersal of mercury and its implications.

Lessons for Researchers

Researchers from many disciplines focus on better understanding the environmental behavior of mercury, its ecological and human health impacts, and efforts and opportunities to mitigate those impacts. Many work, for example, as scientists who measure and model mercury in the environment, as health professionals who aim to prevent or treat mercury exposure, as designers of pollution control equipment or soil remediation technology, or as scholars of international treaty-making and other efforts to address mercury-related problems. Roughly a thousand participants from a wide range of fields attend the biennial conferences on mercury as a global pollutant, continuing the series of conferences that started in Gävle, Sweden, in 1990. We identify three main lessons for the diverse community of mercury researchers: (1) *consider mercury in a larger context*; (2) *work across disciplines*; and (3) *develop and communicate relevant knowledge.*

Consider Mercury in a Larger Context
Much early research on mercury had a singular focus on specific aspects of the behavior of this one element, but it is now well established that mercury use, discharges, transport, and exposure take place in a larger environmental and societal context. The scientific community as a result increasingly addresses connections between mercury and other sustainability issues. The mercury systems that we analyzed in this book furthermore illustrate just how closely the element mercury is linked to a broad range of other components of human, technological, and environmental systems. A recent synthesis of the state of mercury science drew attention to the fact that human-influenced environmental processes, such as climate and land-use changes, are increasingly affecting the cycling of previously discharged mercury from environmental storage in land and oceans (Chen et al. 2018). As a result, researchers will not fully understand many important aspects

of the mercury problem unless they consider it together with other factors including drivers of local and global change.

The range of examples of how mercury is connected to other sustainability challenges across part II illustrates some of the challenges in studying linked mercury-related issues. Technologies to control mercury emissions from point sources often overlap with, and interact with, those that address other air pollutants such as sulfur dioxide and particulates. This highlights for the engineering community the importance of designing effective multi-pollutant control technologies. The formation and accumulation of methylmercury in food webs depend on ecosystem characteristics and structure. As a result, ecological factors sometimes determine where and when this highly toxic form of mercury will reach its highest concentrations. Health impacts of methylmercury are shaped not only by exposure to the substance through dietary intake, but also by genetic factors that vary among individuals, as well as cultural traditions around food. Further analyzing these interconnected issues requires a comprehensive approach.

Scientists in different fields have greatly advanced research on mercury, but there remain important knowledge gaps about mercury's environmental cycling. This creates a need for more sampling of mercury in air, water, soil, and biota to further study its environmental behavior. Mercury could be increasingly measured together with other pollutants, rather than through separate monitoring networks, and it is important to design new techniques that make it possible to measure mercury without using expensive equipment. Modeling efforts could focus on further development of a spectrum of models to address the full complexity of how mercury moves and transforms in the environment and society. These different kinds of mercury-related studies could be better integrated as a part of larger global change research efforts. In addition, mercury researchers should come from a broad range of fields and geographical areas, and not just a few well-known mercury-focused research groups in a small number of mainly industrialized countries.

Work Across Disciplines

The complex and interacting environmental and societal dimensions of the mercury issue underscore the need to reach across disciplinary boundaries in the natural sciences, social sciences, and engineering to better examine system interactions and interventions from a sustainability perspective. Analysis of the biogeochemical and societal cycling of mercury requires understanding how mercury flows through environmental and technical components as well as humans. Analyses of mercury pollution and its consequences should consider the importance of economic and technological factors that influence coal-fired power plants and industrial production, as well as the local and long-range atmospheric transport of mercury. It is also important to compile other kinds of information, such as on the costs and effectiveness of pollution controls. Studies of the impacts of mercury use in ASGM should look at a wide range of factors, including the role of poverty, the influence of mining laws, technology use during the extraction process, and mercury and gold market forces.

One important argument in favor of more interdisciplinary research is that researchers, including those who study different aspects of the mercury issue, may reach incomplete or even incorrect conclusions relevant to sustainability if analyses are conducted within one discipline in isolation. Those scientists who are interested in the atmospheric transport of mercury may be able to better capture how political and economic forces affect the operation and distribution of point sources if working with colleagues in the social sciences. Mercury-capture technologies for use in ASGM designed by engineers working alone are less likely to be effective if behavioral aspects of ASGM production are not accounted for in the design process. Governance scholars drawing lessons from institutional efforts to control air pollutants may make inappropriate conclusions about the match—or fit—between the scope of these institutions and the environmental processes they are designed to govern if they do not consider technological factors and the long-term environmental dynamics of mercury and other substances.

Many reports and studies call for more interdisciplinary research (Brewer and Lövgren 1999; National Academy of Sciences et al. 2005; Repko 2008; Shaman et al. 2013). Such calls often specifically note the need to reach across traditional disciplines in addressing sustainability challenges. The continuing difficulties of achieving interdisciplinary collaborations are already well known (Rhoten and Parker 2004). Much research is still organized based on traditional disciplines, and conventional reward structures for promotion in academia do not typically value interdisciplinary work (Spangenberg 2011). We do not purport to have new solutions from our examination of the mercury systems to the more general challenge of how to better support and reward interdisciplinary collaborations, but researchers may draw on examples from this book to argue that interdisciplinary collaborations hold much promise in generating more nuanced and societally relevant results.

Develop and Communicate Usable Knowledge

It is often critical for researchers to develop and share new and usable knowledge in support of interventions. This involves applying transdisciplinary approaches that engage those outside of academia (Brandt et al. 2013). Researchers in the past few decades have been increasingly called on to pay more attention to crafting usable knowledge, including by working with stakeholders (Clark et al. 2016). Examples from the mercury systems show how such usable knowledge has been influential. The diffusion of knowledge about how to make mercury-free goods and how to design mercury-free manufacturing processes helped phase out much intentional mercury use. Dietary guidelines related to methylmercury levels in food must be tailored based on how people in different locations interpret them, to ensure people keep gaining the benefits of other nutrients in seafood. ASGM miners are more likely to adopt new extraction and amalgamation techniques when these techniques fit local needs and take into account behavioral and cultural factors.

Contemporary mercury debates illustrate the continued importance for scientists to generate and diffuse authoritative information in partnership with non-experts. Two of these debates concern the use of mercury amalgam in dentistry and the use of thimerosal in vaccines. The longer-term goal is to phase out both of these areas of mercury use, but doing so prematurely is likely to cause more harm than good to human health. Mercury amalgam sometimes remains the best alternative for repairing teeth, as recommended by the World Health Organization (WHO). The surge of an anti-vaccine movement in North America, Europe, and elsewhere poses a serious public health problem. Efforts by the WHO, the GAVI Alliance, and others during the Minamata Convention negotiations helped communicate science-based knowledge on the health benefits of using thimerosal-containing vaccines. Researchers can learn from this experience in thinking critically about benefits and harms, including unintended effects and perceived risks, in partnership with stakeholders.

Producing usable knowledge, even with the participation of stakeholders, may not by itself be sufficient to lead to change. Knowledge about mercury is not always evenly shared and distributed. There are many historical examples of how knowledge about the dangers of specific forms of mercury were documented in some places and largely unknown or contested in others. Some European medical associations already warned against the use of mercury in medicine in the sixteenth century, but doctors prescribed many different mercury-based treatments well into the twentieth century. Those who treated early patients of the "strange disease" in Minamata did not know of the experience of Hunter-Russell syndrome and associated knowledge of organic mercury poisoning. Knowledge can also be suppressed: the Chisso Corporation knew it was releasing dangerous quantities of methylmercury into Minamata Bay, but it did not publicly share this information. Understanding how information is generated and disseminated, including aspects of power and influence, is thus critical.

Lessons for Decision-Makers

The negotiations and early implementation of the Minamata Convention helped to elevate the political awareness and perceived importance of mercury abatement globally. Decision-makers and public authorities in different countries, however, have taken action on mercury for much longer, with more modern controls on mercury use and discharges dating back to the 1960s. As we mentioned in the introduction to this chapter, the implementation of the Minamata Convention has only just begun, and much more action by decision-makers will be needed to meet the treaty's objective of protecting human health and the environment from mercury. We believe that decision-makers taking actions in support of this goal are well advised to consider three important points: (1) *intervene in different ways and at multiple scales*; (2) *focus on high-impact interventions*; and (3) *consider long-term impacts*.

Intervene in Different Ways and at Multiple Scales

Many interventions that are needed to comprehensively address the mercury issue must be taken by decision-makers across global, regional, national, and local forums. Mercury is a multi-scale problem in which several factors interact simultaneously, and cross-scale actions are needed to mitigate the various ways that mercury harms human well-being. The protection of people in all regions of the world from mercury's health effects depends on concomitant local-to-global policy action mandating stricter pollution controls on large point sources. Reducing mercury use in consumer products and production processes requires sharing knowledge about mercury-free alternatives across national borders. Greater formalization of artisanal and small-scale mining is an important step toward addressing ASGM-related problems, but it needs to be combined with local interventions addressing mercury use, exposure, and discharges as well as efforts that target mercury trade and gold supply chains.

Global-to-local scale governance efforts to address the human health impacts of mercury are greatly affected by the underlying dynamics of power and influence that shape the differential impacts of mercury use, exposure, and environmental distribution. This creates a need for decision-makers to pay close attention to the situations of populations vulnerable to mercury exposure, including future generations. For example, people in many indigenous communities in the Arctic and elsewhere feel that they are unfairly harmed because of risks from the methylmercury in their traditional diets of fish and marine mammals. Lamenting that some traditional foods consumed for centuries or more are no longer safe to eat, a representative of indigenous peoples during the final negotiation session of the Minamata Convention quoted a line from the song "Under Pressure": "it's the terror of knowing what this world is about" (Earth Negotiations Bulletin 2013b).

Because mercury is linked to other sustainability issues, there is potential for interventions that have multiple and simultaneous benefits for the environment and human well-being. Efforts to reduce the negative environmental impacts of industrial production and wastes can focus on mercury discharges together with other pollutants. Linkages between mercury and climate change mean that addressing mercury can also require actions by decision-makers that at first glance may not seem related to the mercury issue. For example, avoiding deforestation may be an effective way to reduce methylmercury production (Hsu-Kim et al. 2018). Research is still developing on the impact of climate change on the environmental cycling of mercury, but preventing the remobilization of mercury from long-term storage by mitigating climate change could at the same time lessen future human health damages from the consumption of seafood contaminated with methylmercury.

Focus on High-Impact Interventions
Decision-makers operate under resource constraints. Money spent on problems involving mercury is finite, and funding for mercury-related issues competes with the resources needed to address other pressing sustainability

challenges, in both developing and industrialized countries. It is therefore important for decision-makers to consider the relative impact, feasibility, and costs of potential interventions on mercury. Some high-impact interventions may be long-term in scope. For example, initiatives that contribute to increased economic opportunities for miners in regions where ASGM takes place would help address the harms from mining and associated mercury use and exposure. Similarly, phasing out fossil fuel use including coal burning in favor of low-carbon energy sources would have a major impact on mercury emissions. Reducing fossil fuel use would also dovetail with efforts to reduce air pollution and mitigate climate change, and could link to efforts to address economic inequality.

The highest-impact interventions, however, are not always the ones that are the most ambitious. Idealized solutions—such as addressing pollution sources far upstream, including by phasing out coal burning—may not always be politically and practically feasible. Addressing the root causes of the mercury problem would have both shorter-term and longer-term benefits for the environment and human health, but it is important that a focus on such transformative change does not detract from the introduction of more incremental or downstream interventions that would benefit current and future generations. For example, local actions to mitigate human exposure—such as the formulation of dietary advice to vulnerable populations—can occur at the same time as efforts to prevent future discharges to the environment. Simultaneous adaptation and mitigation efforts on mercury, like those on climate change, are necessary to support human well-being where changes are irreversible on human-relevant timescales (N. E. Selin 2014).

Decision-makers should be conscious of tradeoffs in trying to identify high-impact interventions. Moving away from the use of coal would have substantial benefits for the mercury problem, but the Minamata Convention would likely not be in force today had national delegates and policy advocates focused on pushing this most ambitious solution. This would have left mercury emissions from coal burning as well as other aspects of the mercury issue unregulated globally. The ability to respond to changing knowledge is also important. Lessons from the mercury systems suggest that efforts to intervene ought to be conducted with recognition of the extent to which our understanding of both environmental and societal dynamics of the mercury issue have changed over time, and may change again in the future. A first step is to consider the current state of knowledge. Decision-makers can also help researchers design scientific assessments to provide information that both is relevant to their needs and accounts for the perspectives of stakeholders (Cash et al. 2003).

Consider Long-Term Impacts

Mercury's persistence in both the environment and society shows the need for decision-makers to take a long-term perspective. Mercury is an element, and as such it will not go away. Many of the problems that are caused by mercury pollution today come from the activities of the past. Similarly, today's mercury discharges become tomorrow's legacy pollution. A sustainability perspective means that there is a need to consider both present and future generations in decision-making. Stakeholders from future generations (by definition) are not physically present in negotiations about present-day policies, but decision-makers have a moral obligation to pay attention to the long-term impacts of their decisions (to intervene or not to intervene). Governments and others can better account for the perspectives of future generations in decision-making. Making assumptions more explicit, for example in discount-rate analysis of the costs and benefits of pollution controls, could help to better highlight impacts on future generations.

Considerations of the long-term impacts of interventions can be hampered by a lack of scientific knowledge. Many early efforts to address mercury use, exposure, and discharge were based on the dominating perception that mercury was a local problem that could be addressed by local action. This view persisted for centuries. Until the late twentieth century, scientists did not understand that mercury transports long distances through the atmosphere, and cycles between the atmosphere, land, and oceans for generations. There is still scientific uncertainty about the timescales of mercury cycling through the environment and the speed at which concentrations of mercury in the atmosphere and ocean (and methylmercury in different organisms) will respond to changing discharges. This suggests that decision-makers should recognize that decisions today may need to be revisited with further scientific and technical information. Monitoring and evaluation will continue to be necessary to understand both the short-term and long-term impacts of interventions.

Evaluations of the effectiveness of interventions can help to better assess their impact over time. Additional measurements of mercury in the environment would assist in these efforts. Decision-makers should support the establishment of more monitoring programs and stations in all regions of the world, but especially outside North America and Europe. It is critical, however, that environmental data are supplemented with other policy-relevant information. The Minamata Convention effectiveness evaluation process presents a particularly useful opportunity for periodic assessments to explore whether policy actions taken under the treaty are contributing to its objective of protecting human health and the environment from mercury. This involves using outcome and process indicators that couple environmental monitoring data with other types of information about the domestic measures that parties have taken in support of treaty implementation (Selin et al. 2018; Selin 2018).

Lessons for Thoughtful Citizens

Many people across the world are harmed by mercury alongside other stresses that also affect their lives and well-being. Some people are part of societal groups that are highly exposed or particularly vulnerable to mercury exposure and other hazards, such as indigenous peoples, ASGM miners and other community members, and pregnant women and small children. Others are not disproportionately affected by mercury pollution, but may nevertheless be concerned and motivated to take action on addressing a substance that has worldwide impacts. The mercury systems draw attention to the potential influence that thoughtful citizens can have on addressing sustainability challenges, including those involving mercury pollution. For individuals concerned about mercury and its impacts, we suggest three lessons: (1) *consider consumption choices*; (2) *organize to push for change*; and (3) *share sustainability stories*.

Consider Consumption Choices

Many individual behavioral choices involving mercury are related to consumption habits. Everyone, but especially those who are highly exposed or particularly vulnerable to the impacts of methylmercury, should pay attention to dietary guidelines and eat low-mercury seafood. People who consume self-caught fish should be attentive to warnings about mercury and other pollutants specific to local bodies of water. The introduction of mercury-free alternatives makes it easier for consumers to avoid buying mercury-added products, but some products and other consumer goods still contain mercury. People can help minimize future mercury pollution by purchasing mercury-free products where available, and by ensuring that mercury-containing products, such as old thermometers or compact fluorescent bulbs, are disposed of according to local regulations. Using less energy reduces mercury emissions and other forms of pollution and climate change when energy is produced using fossil fuels.

There is an ongoing debate about individual ability to address sustainability issues versus the need for more systemic change. The mercury issue shows that individual consumer choices matter. The effects of certification schemes are debated, but consumers who buy certified mercury-free gold not only send an important market-based signal but also contribute to efforts that look to phase out mercury use in ASGM and improve living conditions in mining communities. Consumers who install solar panels on their homes or buy their electricity from renewable sources help to address mercury emissions as well as other air pollutants and climate change by reducing coal burning. At the same time, systemic change is critical. Much of the demand for gold that drives ASGM comes from large jewelry companies, central banks, and large investment firms. Decisions in many electricity and energy markets are dominated by the fossil fuel industry, which receives much government support. Yet, individual choices can be a catalyst for systemic change, pushing innovation, shifting public opinion, and driving policy development.

Organize to Push for Change

There are many examples in the mercury systems demonstrating that champions, individually and collectively, can have an impact on making people's lives better by organizing and pushing for change. Early advocates for better working conditions in factories, including Alice Hamilton, helped push worker protection efforts on mercury and other hazardous substances. Ornithologists in Sweden alerted the public and the authorities about the dangers caused by the extensive use of mercury pesticides in farming. Advocates in Minamata pressed authorities to remediate the damage to Minamata Bay and helped victims of Minamata disease to get certified and gain compensation. Serious poisoning incidents are not just a thing of the past. Mercury continues to pose human health risks in Grassy Narrows and Kodaikanal, impacting the well-being of community members in complex ways. High levels of mercury exposure and other pressing socio-economic problems in ASGM create further needs for advocacy and local-level community engagement.

Individuals also have power to make a difference on other aspects of the mercury problem. Communities that are concerned about air pollution and "hot spots" can be important sources of advocacy for public authorities to mandate the use of pollution-abatement technology that reduces mercury emissions from industrial point sources. Advocates can continue to raise awareness about the health dangers of using mercury-added products in their communities, for example in places where mercury is still used in cosmetics and skin-lightening creams or religious ceremonies. Individuals can thus help bridge the gap between global-scale interventions such as the Minamata Convention and local action. In addition, concerned citizens can support advocacy and community-based organizations by offering their financial support, participating in collective actions, or providing information (such as joining citizen science projects that aim to collect information on mercury).

Share Sustainability Stories

The Minamata storytellers, by publicly sharing their own and their family members' experiences with Minamata disease, helped raise awareness of the dangers of mercury. Their message resonated with many other people in Japan as well as all over the world. Their stories have also touched others who are affected by mercury pollution, such as the indigenous communities in Grassy Narrows. Their words have echoed at the highest levels of global governance, from the 1972 Conference on the Human Environment in Stockholm to the 2017 meeting of the Minamata Convention Conference of the Parties (COP) in Geneva. Although researchers, experts, and decision-makers have a role to play, the stories of individual people who seek better lives for themselves and for future generations are central to advancing sustainability.

In October 2006, on the fiftieth anniversary of the first official recognition of Minamata disease, a committee of local citizens wrote "Minamata's Pledge" to draw attention to the lessons of the past half century when looking forward to the next 50 years (Minamata-Juku Committee 2006). The pledge acknowledges that Minamata has learned through the failures of history how difficult it is to restore "both a polluted natural environment and a confused social environment." It commits to reconsidering citizens' relationships with nature, ways of living, industrial activities, and community, and creating "spiritually abundant and satisfied lives." This pledge was handed out to delegates at the ceremonial signing of the Minamata Convention in 2013, with a request to translate it into as many different languages as possible; in this way, the story of Minamata and its lessons can continue to reach people worldwide and inspire them to act.

We have done our best in this book to highlight some of the stories of people affected by mercury. People concerned about mercury can share these stories further. They can communicate information about healthy fish consumption during pregnancy, and help others better understand the life-saving benefits of vaccination. By giving voice to those who may not have been listened to in the past, individuals can help make efforts toward sustainability more inclusive for all. Telling stories and hearing the perspectives of others is a critical prerequisite toward building coalitions and advocating for change. Every individual has a story to tell about their experiences, what they value, and what they hope for, for themselves and for future generations. Sharing these stories is a vital part of working toward sustainability.

The urgent need for moving toward greater sustainability underscores the importance of continued actions by champions. Our analysis of the mercury systems highlights the basic challenge for sustainability: to steer often highly complex systems to maintain and enhance the well-being of both present and future generations. With respect to mercury, this requires continued attention to the stories of the most vulnerable, and interventions that weigh the implications of short-term actions alongside their long-term effects. Human activities continue to dominate the Earth in surprising and often damaging ways, and meeting the challenges of sustainability is more than ever dependent on the efforts of people acting alone and in groups to inform and manage a transition toward a more just and sustainable future. As Freddie Mercury sang decades ago—in words that resonate with those who are seeking to address the sustainability challenges of today and the future—we, indeed, are the champions of the world.

References

Abbott, K. W. 2012. The Transnational Regime Complex for Climate Change. *Environment and Planning C: Government and Policy*, 30, 571–590.

Abson, D. J., Fischer, J., Leventon, J., Newig, J., Schomerus, T., Vilsmaier, U., … Lang, D. J. 2017. Leverage Points for Sustainability Transformation. *Ambio*, 46, 30–39.

Ackerman, F., DeCanio, S. J., Howarth, R. B., & Sheeran, K. 2009. Limitations of Integrated Assessment Models of Climate Change. *Climatic Change*, 95, 297–315.

Adger, W. N. 2000. Social and Ecological Resilience: Are They Related? *Progress in Human Geography*, 24, 347–364.

Agencia AFP. 2019. Perú Quiere Que Sus Indígenas Sean Los Primeros Vigilantes De Los Bosques. *Gestión*. https://web.archive.org/web/20190725092126/https://gestion.pe/peru/peru-quiere-indigenas-sean-primeros-vigilantes-bosques-262148

Agnihotri, S. 2016. Kodaikanal Mercury Contamination: Why Unilever Is Paying Settlement to Its 591 Workers. *India Today*, March 9. https://web.archive.org/web/20191224164328/https://www.indiatoday.in/fyi/story/kodaikanal-mercury-contamination-unilever-pays-settlement-workers-312550-2016-03-09

Agocs, M. M., Etzel, R. A., Parrish, R. G., Paschal, D. C., Campagna, P. R., Cohen, D. S., … Hesse, J. L. 1990. Mercury Exposure from Interior Latex Paint. *New England Journal of Medicine*, 323, 1096–1101.

Albright, D., Burkhard, S., Gorwitz, M., & Lach, A. 2017. North Korea's Lithium 6 Production for Nuclear Weapons. Institute for Science and International Security. https://web.archive.org/web/20190203170926/http://isis-online.org/uploads/isis-reports/documents/North_Korea_Lithium_6_17Mar2017_Final.pdf

Allan, J. I., Downie, D., & Templeton, J. 2018. Experimenting with Triplecops: Productive Innovation or Counterproductive Complexity? *International Environmental Agreements: Politics, Law and Economics*, 18, 557–572.

Allwood, J. M. 2014. Squaring the Circular Economy: The Role of Recycling within a Hierarchy of Material Management Strategies. *In:* Worell, E., & Reuter, M. A. (eds.),

Handbook of Recycling: State-of-the-Art for Practitioners, Analysts, and Scientists, 445–477. Waltham, MA: Elsevier.

Amankwah, R., & Ofori-Sarpong, G. 2014. A Lantern Report for Small-Scale Gold Extraction. *International Journal of Environmental Protection and Policy,* 2, 162.

AMAP (Arctic Monitoring and Assessment Programme). 2011. *AMAP Assessment 2011: Mercury in the Arctic.* AMAP, Oslo, Norway. https://web.archive.org/web/2019122915 2151/https://www.amap.no/documents/download/989/inline

AMAP (Arctic Monitoring and Assessment Programme). 2018. *AMAP Assessment 2018: Biological Effects of Contaminants on Arctic Wildlife and Fish.* AMAP, Tromsø, Norway. https://web.archive.org/web/20191229152057/https://www.amap.no/documents /download/3080/inline

Amos, H. M., Jacob, D. J., Streets, D. G., & Sunderland, E. M. 2013. Legacy Impacts of All-Time Anthropogenic Emissions on the Global Mercury Cycle. *Global Biogeochemical Cycles,* 27, 410–421.

Amos, H. M., Sonke, J. E., Obrist, D., Robins, N., Hagan, N., Horowitz, H. M., … Corbitt, E. S. 2015. Observational and Modeling Constraints on Global Anthropogenic Enrichment of Mercury. *Environmental Science & Technology,* 49, 4036–4047.

Anenberg, S. C., Schwartz, J., Shindell, D., Amann, M., Faluvegi, G., Klimont, Z., … Vignati, E. 2012. Global Air Quality and Health Co-Benefits of Mitigating Near-Term Climate Change through Methane and Black Carbon Emission Controls. *Environmental Health Perspectives,* 120, 831–839.

Anonymous. 1924. Germans Again Claim to Make Gold from Mercury. *The Science News-Letter,* 5, 7–8.

Anonymous. 1926. Novasurol. *Canadian Medical Association Journal,* 16, 179–181.

Anonymous. 1945. Mercury Lamp Improves Hangar Lighting. *Electrical Engineering,* 44.

Anonymous. 1949. Fingerprinters Poisoned. *The Science News-Letter,* 56, 231.

Anonymous. 1964. Panogen Vapor Action Seed Treatment. *The National Future Farmer,* 12, 33.

Anonymous. 1970. Mercury in Lake St. Clair. *Science News,* 97, 388.

Anonymous. 2012. One Dead, 32 Hurt as Wildcat Miners Clash with Peru Police. *Reuters,* March 14. https://web.archive.org/web/20190203191108/https://www.reuters .com/article/us-peru-protest-idUSBRE82D15W20120314

Anonymous. 2020. German Sandwich Poisoning Victim Dies after Four Years in Coma. *BBC News,* January 9. https://web.archive.org/web/20200113134804/https:// www.bbc.com/news/world-europe-51050407

Arthur-Mensah, G. 2018. One Chinese Arrested, Six Ghanaians Escape for Illegal Mining. *Ghana News Agency.* https://web.archive.org/web/20190203191151/http://www.ghananewsagency.org/social/one-chinese-arrested-six-ghanaians-escape-for-illegal-mining-139779

Auld, G., Betsill, M., & VanDeveer, S. D. 2018. Transnational Governance for Mining and the Mineral Lifecycle. *Annual Review of Environment and Resources,* 43, 425–453.

Australia National Centre for Immunisation Research and Surveillance. 2009. Fact Sheet: Thiomersal. December 2009. https://web.archive.org/web/20190203155226/http://www.ncirs.org.au/sites/default/files/2018-12/thiomersal-fact-sheet.pdf

Avelino, F., Grin, J., Pel, B., & Jhagroe, S. 2016. The Politics of Sustainability Transitions. *Journal of Environmental Policy & Planning,* 18, 557–567.

Axelrad, D. A., Bellinger, D. C., Ryan, L. M, & Woodruff, T. J. 2007. Dose-Response Relationship of Prenatal Mercury Exposure and IQ: An Integrative Analysis of Epidemiologic Data. *Environmental Health Perspectives,* 115, 609–615.

Ayres, R. U. 1989. Industrial Metabolism. *Technology and Environment,* 1989, 23–49.

Ayres, R. U. 2007. On the Practical Limits to Substitution. *Ecological Economics,* 61, 115–128.

Baccarelli, A., & Bollati, V. 2009. Epigenetics and Environmental Chemicals. *Current Opinion in Pediatrics,* 21, 243.

Baker, J. P. 2008. Mercury, Vaccines, and Autism: One Controversy, Three Histories. *American Journal of Public Health,* 98, 244–253.

Baker, J. R., Ranson, R. M., & Tynen, J. 1939. The Chemical Composition of the Volpar Contraceptive Products. Part 1. Phenyl Mercuric Acetate as a Spermicide. *The Eugenics Review,* 31, 261–268.

Bakir, F., Damluji, S. F., Amin-Zaki, L., Murtadha, M., Khalidi, A., Al-Rawi, N., ... Smith, J. 1973. Methylmercury Poisoning in Iraq. *Science,* 181, 230–241.

Bar-Yam, Y. 1997. *Dynamics of Complex Systems.* Reading, MA: Addison-Wesley.

Barnes, H. J. 1967. Polluted Fish Sale Banned. *Science News,* 92, 564.

Barringer, F. 2006. As a Test Lab on Dirty Air, an Ohio Town Has Changed. *New York Times,* September 27, A18.

Basu, N., Goodrich, J. M., & Head, J. 2014. Ecogenetics of Mercury: From Genetic Polymorphisms and Epigenetics to Risk Assessment and Decision-Making. *Environmental Toxicology and Chemistry,* 33, 1248–1258.

Bebbington, A., Abdulai, A.-G., Bebbington, D. H., Hinfelaar, M., & Sanborn, C. 2018. *Governing Extractive Industries: Politics, Histories, Ideas.* New York: Oxford University Press.

Bell, F. W. 2019. *Food from the Sea: The Economics and Politics of Ocean Fisheries*. New York: Routledge.

Bell, L., DiGangi, J., & Weinberg, J. 2014. *An NGO Introduction to Mercury Pollution and the Minamata Convention on Mercury*. IPEN. May 2014. https://web.archive.org /web/20190723065819/https://ipen.org/sites/default/files/documents/ipen-booklet -hg-update-v1_6-en-2-web.pdf

Bellanger, M., Pichery, C., Aerts, D., Berglund, M., Castaño, A., Čejchanová, M., ... Fischer, M. E. 2013. Economic Benefits of Methylmercury Exposure Control in Europe: Monetary Value of Neurotoxicity Prevention. *Environmental Health*, 12, 3.

Berkes, F. 2017. Environmental Governance for the Anthropocene? Social-Ecological Systems, Resilience, and Collaborative Learning. *Sustainability*, 9, 1232.

Bernard, S., Enayati, A., Redwood, L., Roger, H., & Binstock, T. 2001. Autism: A Novel Form of Mercury Poisoning. *Medical Hypotheses*, 56, 462–471.

Berne Declaration. 2015. *A Golden Racket: The True Source of Switzerland's "Togolese" Gold*. Lausanne. https://web.archive.org/web/20190203191243/https://www.publiceye .ch/fileadmin/doc/Rohstoffe/2015_PublicEye_A_golden_racket_Report.pdf

Bettencourt, L. M. A., & Kaur, J. 2011. Evolution and Structure of Sustainability Science. *Proceedings of the National Academy of Sciences*, 108, 19540–19545.

Beusse, R., Blair, C., Canes, H., Charen, S., Fabrirkiewicz, S., Hatfield, J., ... Van Orden, J. 2006. *Monitoring Needed to Assess Impact of EPA's Clean Air Mercury Rule on Potential Hotspots*. US Environmental Protection Agency. 2006-P-00025. https:// nepis.epa.gov/Exe/ZyPURL.cgi?Dockey=P100URP3.txt

Bharti, R., Wadhwani, K. K., Tikku, A. P., & Chandra, A. 2010. Dental Amalgam: An Update. *Journal of Conservative Dentistry: JCD*, 13, 204.

Biermann, F. 2014. *Earth System Governance: World Politics in the Anthropocene*. Cambridge, MA: MIT Press.

Biermann, F., Abbott, K., Andresen, S., Bäckstrand, K., Bernstein, S., Betsill, M. M., ... Folke, C. 2012. Transforming Governance and Institutions for Global Sustainability: Key Insights from the Earth System Governance Project. *Current Opinion in Environmental Sustainability*, 4, 51–60.

Biester, H., Bindler, R., Martinez-Cortizas, A., & Engstrom, D. R. 2007. Modeling the Past Atmospheric Deposition of Mercury Using Natural Archives. *Environmental Science & Technology*, 41, 4851–4860.

Biggs, R., Schlüter, M., Biggs, D., Bohensky, E. L., BurnSilver, S., Cundill, G., ... Kotschy, K. 2012. Toward Principles for Enhancing the Resilience of Ecosystem Services. *Annual Review of Environment and Resources*, 37, 421–448.

Billings, C. E., & Wilder, J. 1970. *Handbook of Fabric Filter Technology. Volume I. Fabric Filter Systems Study.* National Technical Information Service, US Department of Commerce, prepared by GCA Corporation, Bedford, MA.

Binder, C. R., Bots, P. W. G., Hinkel, J., & Pahl-Wostl, C. 2013. Comparison of Frameworks for Analyzing Social-Ecological Systems. *Ecology and Society,* 18, 24.

Birch, R. J., Bigler, J., Rogers, J. W., Zhuang, Y., & Clickner, R. P. 2014. Trends in Blood Mercury Concentrations and Fish Consumption among US Women of Reproductive Age, NHANES, 1999–2010. *Environmental Research,* 133, 431–438.

Black, J. 1999. The Puzzle of Pink Disease. *Journal of the Royal Society of Medicine,* 92, 478–481.

Bloom, N. S. 1992. On the Chemical Form of Mercury in Edible Fish and Marine Invertebrate Tissue. *Canadian Journal of Fisheries and Aquatic Sciences,* 49, 1010–1017.

Blum, D. 2011. *The Poisoner's Handbook: Murder and the Birth of Forensic Medicine in Jazz Age New York.* New York: Penguin.

Blum, J. D., Sherman, L. S., & Johnson, M. W. 2014. Mercury Isotopes in Earth and Environmental Sciences. *Annual Review of Earth and Planetary Sciences,* 42, 249–269.

Bodansky, D. 2010. *The Art and Craft of International Environmental Law.* Cambridge, MA: Harvard University Press.

Bodin, Ö., Alexander, S., Baggio, J., Barnes, M., Berardo, R., Cumming, G., ... Garcia, M. M. 2019. Improving Network Approaches to the Study of Complex Social–Ecological Interdependencies. *Nature Sustainability,* 2, 551–559.

Bolton, H. C. 1900. *Evolution of the Thermometer 1592–1743.* Easton, PA: The Chemical Publishing Company.

Bonan, G. B., & Doney, S. C. 2018. Climate, Ecosystems, and Planetary Futures: The Challenge to Predict Life in Earth System Models. *Science,* 359, eaam8328.

Boutron, C. F., Vandal, G. M., Fitzgerald, W. F., & Ferrari, C. P. 1998. A Forty Year Record of Mercury in Central Greenland Snow. *Geophysical Research Letters,* 25, 3315–3318.

Braile, R. 2000. FDA Fails to Warn Public of Tuna, Swordfish That Is Laced with Mercury. *Boston Globe,* April 26.

Bramwell, G., Wilson, F., & Faunce, T. 2018. Mercury Pollution from Coal-Fired Power Plants: A Critical Analysis of the Australian Regulatory Response to Public Health Risks. *Journal of Law and Medicine,* 26, 480–487.

Brandt, P., Ernst, A., Gralla, F., Luederitz, C., Lang, D. J., Newig, J., ... Von Wehrden, H. 2013. A Review of Transdisciplinary Research in Sustainability Science. *Ecological Economics,* 92, 1–15.

Brewer, G. D., & Lövgren, K. 1999. The Theory and Practice of Interdisciplinary Work. *Policy Sciences*, 32, 315–317.

Bright, A. A., & Maclaurin, W. R. 1943. Economic Factors Influencing the Development and Introduction of the Fluorescent Lamp. *Journal of Political Economy*, 51, 429–450.

Broad, W. J. 1981. Sir Isaac Newton: Mad as a Hatter. *Science*, 213, 1341–1342.

Brooks, S. C., & Southworth, G. R. 2011. History of Mercury Use and Environmental Contamination at the Oak Ridge Y-12 Plant. *Environmental Pollution*, 159, 219–228.

Brooks, W. E. 2011. *2009 Minerals Yearbook: Mercury*. US Department of the Interior, United States Geological Survey.

Brooks, W. E. 2012. Industrial Use of Mercury in the Ancient World. *In:* Bank, M. S. (ed.), *Mercury in the Environment: Pattern and Process*, 19–24. Berkeley: University of California Press.

Brown, K. W. 2001. Workers' Health and Colonial Mercury Mining at Huancavelica, Peru. *The Americas*, 57, 467–496.

Browner, C. M. 1993. Pollution Prevention Takes Center Stage. *EPA Journal*, 19, 6.

Bulkeley, H., Jordan, A., Perkins, R., & Selin, H. 2013. Governing Sustainability: Rio+20 and the Road Beyond. *Environment and Planning C: Government and Policy*, 31, 958–970.

Butler, T. J., Cohen, M. D., Vermeylen, F. M., Likens, G. E., Schmeltz, D., & Artz, R. S. 2008. Regional Precipitation Mercury Trends in the Eastern USA, 1998–2005: Declines in the Northeast and Midwest, No Trend in the Southeast. *Atmospheric Environment*, 42, 1582–1592.

Cain, A., Disch, S., Twaroski, C., Reindl, J., & Case, C. R. 2007. Substance Flow Analysis of Mercury Intentionally Used in Products in the United States. *Journal of Industrial Ecology*, 11, 61–75.

Cain, A., Morgan, J. T., & Brooks, N. 2011. Mercury Policy in the Great Lakes States: Past Successes and Future Opportunities. *Ecotoxicology*, 20, 1500–1511.

Calvin, K., & Bond-Lamberty, B. 2018. Integrated Human-Earth System Modeling— State of the Science and Future Directions. *Environmental Research Letters*, 13, 063006.

Camacho, A., Van Brussel, E., Carrizales, L., Flores-Ramírez, R., Verduzco, B., Huerta, S. R.-A., Leon, M., & Díaz-Barriga, F. 2016. Mercury Mining in Mexico: I. Community Engagement to Improve Health Outcomes from Artisanal Mining. *Annals of Global Health*, 82, 149–155.

Camuffo, D., & Bertolin, C. 2012. The Earliest Spirit-in-Glass Thermometer and a Comparison between the Earliest CET and Italian Observations. *Weather*, 67, 206–209.

Canadian Council of Ministers of the Environment. 2000. Canada-Wide Standards for Mercury Emissions. https://web.archive.org/web/20190725121011/https://www .ccme.ca/files/Resources/air/mercury/mercury_emis_std_e1.pdf

Canadian Council of Ministers of the Environment. 2006. Canada-Wide Standards for Mercury Emissions from Coal-Fired Electric Power Generation Plants. October 11. https://web.archive.org/web/20190203164533/https://www.ccme.ca/files/Resources /air/mercury/hg_epg_cws_w_annex.pdf

Canadian Council of Ministers of the Environment. 2019. Mercury. https://web.archive .org/web/20190725121254/https://www.ccme.ca/en/resources/air/mercury.html

Canavesio, R. 2014. Formal Mining Investments and Artisanal Mining in Southern Madagascar: Effects of Spontaneous Reactions and Adjustment Policies on Poverty Alleviation. *Land Use Policy,* 36, 145–154.

Carbon Brief. 2018. Mapped: The World's Coal Power Plants. Retrieved February 3, 2019, from https://web.archive.org/web/20190203163817/https://www.carbonbrief .org/mapped-worlds-coal-power-plants

Carey, I. 2017. Swiss to Ban Mercury Exports by 2027, Environmentalists Want It to Stop Now. European Environmental Bureau, October 27. https://web.archive .org/web/20191224164907/https://meta.eeb.org/2017/10/27/swiss-to-ban-mercury -exports-by-2027-environmentalists-want-it-to-stop-now/

Carpenter, S. R., Mooney, H. A., Agard, J., Capistrano, D., DeFries, R. S., Díaz, S., ... Pereira, H. M. 2009. Science for Managing Ecosystem Services: Beyond the Millennium Ecosystem Assessment. *Proceedings of the National Academy of Sciences,* 106, 1305–1312.

Carson, R. 1962. *Silent Spring,* New York: Houghton Mifflin.

Cash, D. W., Clark, W. C., Alcock, F., Dickson, N. M., Eckley, N., Guston, D. H., ... Mitchell, R. B. 2003. Knowledge Systems for Sustainable Development. *Proceedings of the National Academy of Sciences,* 100, 8086–8091.

Chabay, I. 2015. Narratives for a Sustainable Future: Vision and Motivation for Collective Action. *In:* Werlen, B. (ed.), *Global Sustainability: Cultural Perspectives and Challenges for Transdisciplinary Integrated Research,* 51–61. Cham: Springer.

Chauvin, L. 2018. Pope Brings Environmental Crusade to Peru's Amazon, Citing "Defense of the Earth." *Washington Post,* January 19.

Chemical Watch. 2017. China Announces Mercury Timetable, Applying Minamata Convention. https://chemicalwatch.com/asiahub/58274/china-announces-mercury -timetable-applying-minamata-convention

Chen, B., Li, J., Chen, G., Wei, W., Yang, Q., Yao, M., ... Dong, K. 2017. China's Energy-Related Mercury Emissions: Characteristics, Impact of Trade and Mitigation Policies. *Journal of Cleaner Production,* 141, 1259–1266.

Chen, C. Y., Driscoll, C. T., Eagles-Smith, C. A., Eckley, C. S., Gay, D. A., Hsu-Kim, H., ... Thompson, M. R. 2018. A Critical Time for Mercury Science to Inform Global Policy. *Environmental Science & Technology, 52,* 9556–9561.

Chen, J. 2015. Coupled Human and Natural Systems. *BioScience, 65,* 539–540.

Cheng, H., & Hu, Y. 2011. Mercury in Municipal Solid Waste in China and Its Control: A Review. *Environmental Science & Technology, 46,* 593–605.

Childs, J. 2014. A New Means of Governing Artisanal and Small-Scale Mining? Fairtrade Gold and Development in Tanzania. *Resources Policy, 40,* 128–136.

China Council for International Cooperation on Environment and Development. 2011. Special Policy Study on Mercury Management in China. CCIED 2011 Annual General Meeting. https://web.archive.org/web/20190203171423/http://www.cciced .net/cciceden/POLICY/rr/prr/2011/201205/P020160810466199200238.pdf

Chouinard, R., & Veiga, M. 2008. *Results of the Awareness Campaign and Technology Demonstration for Artisanal Gold Miners: Summary Report.* Vienna: UNIDO.

Clark, W. C. 2007. Sustainability Science: A Room of Its Own. *Proceedings of the National Academy of Sciences, 104,* 1737.

Clark, W. C. 2015. London: A Multi-Century Struggle for Sustainable Development in an Urban Environment. John F. Kennedy School of Government, Harvard University. Cambridge, MA. HKS Faculty Research Working Paper Series. RWP15–047. http://nrs.harvard.edu/urn-3:HUL.InstRepos:22356529

Clark, W. C., van Kerkhoff, L., Lebel, L., & Gallopin, G. C. 2016. Crafting Usable Knowledge for Sustainable Development. *Proceedings of the National Academy of Sciences, 113,* 4570–4578.

Clarkson, T. W. 2002. The Three Modern Faces of Mercury. *Environmental Health Perspectives, 110,* 11.

Clarkson, T. W., & Magos, L. 2006. The Toxicology of Mercury and Its Chemical Compounds. *Critical Reviews in Toxicology, 36,* 609–662.

Clarkson, T. W., Magos, L., & Myers, G. J. 2003. The Toxicology of Mercury—Current Exposures and Clinical Manifestations. *New England Journal of Medicine, 349,* 1731–1737.

Clarkson, T. W., Vyas, J. B., & Ballatori, N. 2007. Mechanisms of Mercury Disposition in the Body. *American Journal of Industrial Medicine, 50,* 757–764.

Claussen, M., Mysak, L., Weaver, A., Crucifix, M., Fichefet, T., Loutre, M.-F., ... Berger, A. 2002. Earth System Models of Intermediate Complexity: Closing the Gap in the Spectrum of Climate System Models. *Climate Dynamics, 18,* 579–586.

Cleveland, C. J., & Morris, C. 2014. Section 37—Lighting. *Handbook of Energy, Volume II: Chronologies, Top Ten Lists, and Word Clouds,* 649–668. Amsterdam: Elsevier.

Clifford, M. 2014. Future Strategies for Tackling Mercury Pollution in the Artisanal Gold Mining Sector: Making the Minamata Convention Work. *Futures,* 62, 106–112.

Colding, J., & Barthel, S. 2019. Exploring the Social-Ecological Systems Discourse 20 Years Later. *Ecology and Society,* 24, 2.

Commission for Environmental Cooperation. 1997. *North American Regional Action Plan on Mercury.* North American Working Group on the Sound Management of Chemicals, North American Task Force on Mercury. https://web.archive.org/web /20191224165224/http://www3.cec.org/islandora/en/item/11585-north-american -regional-action-plan-mercury-phase-i-en.pdf

Commission for Environmental Cooperation. 2000. *North American Regional Action Plan on Mercury: Phase II.* North American Implementation Task Force on Mercury. https://web.archive.org/web/20191224165320/http://www3.cec.org/islandora/en /item/3458-north-american-regional-action-plan-mercury-en.pdf

Cooke, C. A., Balcom, P. H., Biester, H., & Wolfe, A. P. 2009. Over Three Millennia of Mercury Pollution in the Peruvian Andes. *Proceedings of the National Academy of Sciences,* 106, 8830–8834.

Corbitt, E. S., Jacob, D. J., Holmes, C. D., Streets, D. G., & Sunderland, E. M. 2011. Global Source–Receptor Relationships for Mercury Deposition under Present-Day and 2050 Emissions Scenarios. *Environmental Science & Technology,* 45, 10477–10484.

Coughlan, T. 2018. Our Inconvenient Truth: NZ Will Keep Burning Coal. *Newsroom,* February 15. https://web.archive.org/web/20190725123944/https://www.newsroom .co.nz/2018/02/14/88961/our-inconvenient-truth-nz-will-keep-burning-coal

Coulson, M. 2012. *The History of Mining: The Events, Technology and People Involved in the Industry That Forged the Modern World.* Petersfield, UK: Harriman House Limited.

Cousins, I. T., Herzke, D., Goldenman, G., Lohmann, R., Miller, M., Ng, C., … Vierke, L. 2019. The Concept of Essential Use for Determining When Uses of PFASs Can Be Phased Out. *Environmental Science: Processes & Impacts,* 21, 1803–1815.

Crook, J., & Mousavi, A. 2016. The Chlor-Alkali Process: A Review of History and Pollution. *Environmental Forensics,* 17, 211–217.

Crosland, M. P. 2004. *Historical Studies in the Language of Chemistry.* Mineola, NY: Dover Publications.

Crutzen, P. J. 2006. The "Anthropocene." *Earth System Science in the Anthropocene.* Switzerland: Springer.

Crutzen, P. J., & Stoermer, E. F. 2000. The Anthropocene. *Global Change Newsletter,* 41, 17–18.

Cutcher-Gershenfeld, J., Field, F., Hall, R., Kirchain, R., Marks, D., Oye, K., & Sussman, J. 2004. Sustainability as an Organizing Design Principle for Large-Scale

Engineering Systems. Massachusetts Institute of Technology. Engineering Systems Division Working Paper Series. Engineering Systems Monograph.

Daily, G. C. 1997. *Nature's Services*. Washington, DC: Island Press.

Dally, A. 1997. The Rise and Fall of Pink Disease. *Social History of Medicine*, 10, 291–304.

Daly, H. E. 1990. Toward Some Operational Principles of Sustainable Development. *Ecological Economics*, 2, 1–6.

Davies, G. R. 2014. A Toxic-Free Future: Is There a Role for Alternatives to Mercury in Small-Scale Gold Mining? *Futures*, 62, 113–119.

Davis, D. L. 2002. A Look Back at the London Smog of 1952 and the Half Century Since. *Environmental Health Perspectives*, 110, A734–A735.

de Vries, B. J. 2012. *Sustainability Science*. Cambridge: Cambridge University Press.

de Weck, O. L., Roos, D., & Magee, C. L. 2011. *Engineering Systems: Meeting Human Needs in a Complex Technological World*. Cambridge, MA: MIT Press.

de Weck, O. L., Ross, A. M., & Rhodes, D. H. 2012. Investigating Relationships and Semantic Sets Amongst System Lifecycle Properties (Ilities). *Third International Engineering Systems Symposium CESUN 2012*. Delft University of Technology, June 18–20, 2012.

Depew, D. C., Basu, N., Burgess, N. M., Campbell, L. M., Devlin, E. W., Drevnick, P. E., … Wiener, J. G. 2012. Toxicity of Dietary Methylmercury to Fish: Derivation of Ecologically Meaningful Threshold Concentrations. *Environmental Toxicology and Chemistry*, 31, 1536–1547.

Derban, L. K. 1974. Outbreak of Food Poisoning Due to Alkyl-Mercury Fungicide. *Archives of Environmental Health: An International Journal*, 28, 49–52.

Deser, C., Phillips, A., Bourdette, V., & Teng, H. 2012. Uncertainty in Climate Change Projections: The Role of Internal Variability. *Climate Dynamics*, 38, 527–546.

Dev, A. 2015. The Unending Fallout of Unilever's Thermometer Factory in Kodaikanal. *Caravan Magazine*, August 18. https://web.archive.org/web/20191224163633 /https://caravanmagazine.in/vantage/unending-fallout-unilever-thermometer -factory-kodaikanal

Diamond, M. L., de Wit, C. A., Molander, S., Scheringer, M., Backhaus, T., Lohmann, R., … Zetzsch, C. 2015. Exploring the Planetary Boundary for Chemical Pollution. *Environment International*, 78, 8–15.

Díaz, S., Demissew, S., Carabias, J., Joly, C., Lonsdale, M., Ash, N., … Zlatinova, D. 2015. The IPBES Conceptual Framework: Connecting Nature and People. *Current Opinion in Environmental Sustainability*, 14, 1–16.

Ditchburn, R. W. 1980. Newton's Illness of 1692–3. *Notes and Records of the Royal Society of London,* 35, 1–16.

Dockery, D. W., Pope, C. A., Xu, X., Spengler, J. D., Ware, J. H., Fay, M. E., … Speizer, F. E. 1993. An Association between Air Pollution and Mortality in Six US Cities. *New England Journal of Medicine,* 329, 1753–1759.

Douglas, M. 1966. *Purity and Danger: An Analysis of Concepts of Pollution and Taboo.* New York: Routledge.

Dowlatabadi, H. 1995. Integrated Assessment Models of Climate Change: An Incomplete Overview. *Energy Policy,* 23, 289–296.

Driscoll, C., Sunderland, E., Lambert, K. F., Blum, J., Chen, C., Evers, D., … & Selin, N. E. 2018. Mercury Matters 2018: A Science Brief for Journalists and Policymakers, December 13. https://web.archive.org/web/20191229144909/https://eng-cs.syr.edu /news-events/news/mercury/

Dryzek, J. S. 2013. *The Politics of the Earth: Environmental Discourses.* Oxford: Oxford University Press.

Dubs, H. H. 1947. The Beginnings of Alchemy. *Isis,* 38, 62–86.

Duracell. 2019. Frequently Asked Questions. Retrieved February 3, 2019, from https://web.archive.org/web/20190203170457/https://www.duracell.com/en-us /help/faq/do-your-batteries-contain-mercury/

Eagles-Smith, C. A., Silbergeld, E. K., Basu, N., Bustamante, P., Diaz-Barriga, F., Hopkins, W. A., … Nyland, J. F. 2018. Modulators of Mercury Risk to Wildlife and Humans in the Context of Rapid Global Change. *Ambio,* 47, 170–197.

Earth Negotiations Bulletin. 2010. Mercury INC 1 Highlights: Wednesday, 9 June 2010. IISD Reporting Services. *Earth Negotiations Bulletin,* 28.

Earth Negotiations Bulletin. 2012. Mercury INC 4 Highlights: Friday, 29 June 2012. IISD Reporting Services. *Earth Negotiations Bulletin,* 28.

Earth Negotiations Bulletin 2013a. Mercury INC 5 Highlights: Monday, 14 January 2013. IISD Reporting Services. *Earth Negotiations Bulletin,* 28.

Earth Negotiations Bulletin 2013b. Mercury INC 5 Highlights: Tuesday, 15 January 2013. IISD Reporting Services. *Earth Negotiations Bulletin,* 28.

Eckelman, M. J., Anastas, P. T., & Zimmerman, J. B. 2008. Spatial Assessment of Net Mercury Emissions from the Use of Fluorescent Bulbs. *Environmental Science & Technology,* 42, 8564–8570.

Eckert, M., Fleischmann, G., Jira, R., Bolt, H. M., & Golka, K. 2006. Acetaldehyde. *Ullmann's Encyclopedia of Industrial Chemistry.* Weinheim, Germany: Wiley-VCH.

Edenhofer, O., Steckel, J. C., Jakob, M., & Bertram, C. 2018. Reports of Coal's Terminal Decline May Be Exaggerated. *Environmental Research Letters,* 13, 024019.

Egan, M. 2013. Communicating Knowledge: The Swedish Mercury Group and Vernacular Science, 1965–1972. *In:* Jørgensen, D., Jørgensen, F. A., & Pritchard S. B. (eds.), *New Natures: Joining Environmental History with Science and Technology Studies,* 103–117. Pittsburgh, PA: University of Pittsburgh Press.

Eisler, R. 2003. Health Risks of Gold Miners: A Synoptic Review. *Environmental Geochemistry and Health,* 25, 325–345.

Elmes, A., Yarlequé Ipanaqué, J. G., Rogan, J., Cuba, N., & Bebbington, A. 2014. Mapping Licit and Illicit Mining Activity in the Madre De Dios Region of Peru. *Remote Sensing Letters,* 5, 882–891.

EMEA (The European Agency for the Evaluation of Medicinal Products). 1999. EMEA Public Statement on Thiomersal Containing Medicinal Products. Human Medicines Evaluation Unit. London. EMEA/20962/99. https://web.archive.org/web/20190 203161138/https://www.ema.europa.eu/documents/scientific-guideline/emea -public-statement-thiomersal-containing-medicinal-products_en.pdf

Energy Policy Act. 2005. US Public Law 109–58, 119 Stat. 625, August 8.

Engstrom, D. R., Fitzgerald, W. F., Cooke, C. A., Lamborg, C. H., Drevnick, P. E., Swain, E. B., … Balcom, P. H. 2014. Atmospheric Hg Emissions from Preindustrial Gold and Silver Extraction in the Americas: A Reevaluation from Lake-Sediment Archives. *Environmental Science & Technology,* 48, 6533–6543.

Environmental Justice Australia. 2017. *Toxic and Terminal: How the Regulation of Coal-Fired Power Stations Fails Australian Communities.* Environmental Justice Australia, Carlton, Australia. https://web.archive.org/web/20191224165417/https://www .envirojustice.org.au/powerstations/

Eppinger, S., & Browning, T. 2012. *Design Structure Matrix Methods and Applications.* Cambridge, MA: MIT Press.

Epstein, G., Pittman, J., Alexander, S. M., Berdej, S., Dyck, T., Kreitmair, U., … Armitage, D. 2015. Institutional Fit and the Sustainability of Social–Ecological Systems. *Current Opinion in Environmental Sustainability,* 14, 34–40.

Epstein, K. 2019. "He Is Wrong": Robert F. Kennedy Jr.'s Family Calls Him Out for Anti-Vaccine Conspiracy Theories. *Washington Post,* May 19.

Eriksen, H. H., & Perrez, F. X. 2014. The Minamata Convention: A Comprehensive Response to a Global Problem. *Review of European, Comparative & International Environmental Law,* 23, 195–210.

Erkman, S. 1997. Industrial Ecology: An Historical View. *Journal of Cleaner Production,* 5, 1–10.

Eto, K. 2000. Minamata Disease. *Neuropathology,* 20, 14–19.

European Chemicals Agency. 2011. Background Document to the Opinions on the Annex XV Dossier Proposing Restrictions on Five Phenylmercury Compounds. ECHA/RAC/RES-O-0000001362-83-02/S1.

European Commission. 2012. Commission Regulation (EU) No 848/2012 of 19 September 2012 Amending Annex XVII to Regulation (EC) No 1907/2006 of the European Parliament and of the Council on the Registration, Evaluation, Authorisation and Restriction of Chemicals (REACH) as Regards Phenylmercury Compounds. *Official Journal of the European Union,* L253/5.

European Commission. 2014. Best Available Techniques (BAT) Reference Document for the Production of Chlor-Alkali. Joint Research Centre, Institute for Prospective Technological Studies, Sustainable Production and Consumption Unit, European IPPC Bureau. Industrial Emissions Directive 2010/75/EU (Integrated Pollution Prevention and Control).

European Commission. 2017a. EU Protects Citizens from Toxic Mercury, Paves the Way for Global Action. Press Release. May 17. Retrieved January 18, 2020, from https://web.archive.org/web/20200118142219/https://ec.europa.eu/commission /presscorner/detail/en/IP_17_1345

European Commission. 2017b. Questions and Answers: EU Mercury Policy and the Ratification of the Minamata Convention. European Commission. Brussels. Fact Sheet. May 18.

European Commission. 2018. EU Rules on Mercury in Action. https://web.archive .org/web/20200118141744/https://ec.europa.eu/environment/chemicals/mercury /pdf/ENV-17-011-IndustrialEmissionsFactsheet-MERCURY-E-web.pdf

European Environment Agency. 2018. *Mercury in Europe's Environment: A Priority for European and Global Action.* EEA. Copenhagen. EEA Report 11/2018. https://www.eea .europa.eu/publications/mercury-in-europe-s-environment

European Union. 2006. Commission Regulation (EC) No 1881/2006 of 19 December 2006 Setting Maximum Levels for Certain Contaminants in Foodstuffs. *Official Journal of the European Union,* L364/5.

European Union. 2017. Regulation (EU) 2017/852 of the European Parliament and of the Council of 17 May 2017 on Mercury, and Repealing Regulation (EC) No 1102/2008. *Official Journal of the European Union,* L137/1.

Evers, D. C., Keane, S. E., Basu, N., & Buck, D. 2016. Evaluating the Effectiveness of the Minamata Convention on Mercury: Principles and Recommendations for Next Steps. *Science of the Total Environment,* 569–570, 888–903.

Evers, D. C., Savoy, L. J., DeSorbo, C. R., Yates, D. E., Hanson, W., Taylor, K.... Major, A. 2008. Adverse Effects from Environmental Mercury Loads on Breeding Common Loons. *Ecotoxicology,* 17, 69–81.

Ezenobi, N. O., & Chinaka, C. N. 2018. Microbiological and Physicochemical Assessment of Some Brands of Gentamicin Eye Drops Marketed in Registered Retail Pharmacies in Port Harcourt, Nigeria. *Journal of Pharmacy and Bioresources,* 15, 27–36.

FAO/WHO (Food and Agriculture Organization and World Health Organization). 2011. Report of the Joint FAO/WHO Expert Consultation on the Risks and Benefits of Fish Consumption. FAO Fisheries and Aquaculture Report No. 978, World Health Organization, Geneva. https://apps.who.int/iris/bitstream/handle/10665 /44666/9789241564311_eng.pdf;jsessionid=8A4082DDEA450964953875E81F252 D32?sequence=1

Farwell, B. 1999. *Over There: The United States in the Great War, 1917–1918.* New York: W. W. Norton & Company.

Feil, R., & Fraga, M. F. 2012. Epigenetics and the Environment: Emerging Patterns and Implications. *Nature Reviews Genetics,* 13, 97–109.

Feng, X. 2005. Mercury Pollution in China: An Overview. *Dynamics of Mercury Pollution on Regional and Global Scales.* Switzerland: Springer.

Fenger, J. 2009. Air Pollution in the Last 50 Years—from Local to Global. *Atmospheric Environment,* 43, 13–22.

Feola, G. 2015. Societal Transformation in Response to Global Environmental Change: A Review of Emerging Concepts. *Ambio,* 44, 376–90.

Fernández-Llamazares, Á., Garteizgogeascoa, M., Basu, N., Brondizio, E., Cabeza, M., Martínez-Alier, J., … Reyes-García, V. 2020. A State-of-the-Art Review of Indigenous Peoples and Environmental Pollution. *Integrated Environmental Assessment and Management,* 16, 324–341.

Ferrão, P., & Fernández, J. E. 2013. *Sustainable Urban Metabolism.* Cambridge, MA: MIT Press.

Fessenden, M. 2015. How to Reconstruct Lewis and Clark's Journey: Follow the Mercury-Laden Latrine Pits. *Smithsonian,* September 8. https://web.archive.org/web /20191224165530/https://www.smithsonianmag.com/smart-news/how-reconstruct -lewis-and-clark-journey-follow-mercury-laden-latrine-pits-180956518/

Fielding, R. 2010. Environmental Change as a Threat to the Pilot Whale Hunt in the Faroe Islands. *Polar Research,* 29, 430–438.

Fielding, R., Davis, J. E., & Singleton, B. 2015. Mutual Aid, Environmental Policy, and the Regulation of Faroese Pilot Whaling. *Human Geography,* 8, 37–48.

Fiore, A. M., Dentener, F., Wild, O., Cuvelier, C., Schultz, M., Hess, P., … Horowitz, L. 2009. Multimodel Estimates of Intercontinental Source-Receptor Relationships for Ozone Pollution. *Journal of Geophysical Research: Atmospheres,* 114, D04301.

Fischer, J., Dyball, R., Fazey, I., Gross, C., Dovers, S., Ehrlich, P. R., ... Borden, R. J. 2012. Human Behavior and Sustainability. *Frontiers in Ecology and the Environment*, 10, 153–160.

Fischer-Kowalski, M., & Haberl, H. 2007. *Socioecological Transitions and Global Change: Trajectories of Social Metabolism and Land Use*. Cheltenahm, UK: Edward Elgar Publishing.

Fischer-Kowalski, M., Krausmann, F., Giljum, S., Lutter, S., Mayer, A., Bringezu, S.,... Weisz, H. 2011. Methodology and Indicators of Economy-Wide Material Flow Accounting: State of the Art and Reliability across Sources. *Journal of Industrial Ecology*, 15, 855–876.

Fisher, E., Mwaipopo, R., Mutagwaba, W., Nyange, D., & Yaron, G. 2009. "The Ladder That Sends Us to Wealth": Artisanal Mining and Poverty Reduction in Tanzania. *Resources Policy*, 34, 32–38.

Fisher, J., Arora, P., & Rhee, S. 2018. Conserving Tropical Forests: Can Sustainable Livelihoods Outperform Artisanal or Informal Mining? *Sustainability*, 10, 2586.

Fitzgerald, W. F., Engstrom, D. R., Mason, R. P., & Nater, E. A. 1998. The Case for Atmospheric Mercury Contamination in Remote Areas. *Environmental Science & Technology*, 32, 1–7.

Flynn, D. O., & Giráldez, A. 1995. Born with a "Silver Spoon": The Origin of World Trade in 1571. *Journal of World History*, 6, 201–221.

Fold, N., Jønsson, J. B., & Yankson, P. 2014. Buying into Formalization? State Institutions and Interlocked Markets in African Small-Scale Gold Mining. *Futures*, 62, 128–139.

Folke, C. 2016. Resilience (republished). *Ecology and Society*, 21, 44.

Folke, C., Pritchard Jr., L., Berkes, F., Colding, J., & Svedin, U. 2007. The Problem of Fit between Ecosystems and Institutions: Ten Years Later. *Ecology and Society*, 12, 30.

Fouquet, R. 2011. Long Run Trends in Energy-Related External Costs. *Ecological Economics*, 70, 2380–2389.

Fragnière, A. 2016. Climate Change and Individual Duties. *Wiley Interdisciplinary Reviews: Climate Change*, 7, 798–814.

Francis (pope). 2015. *Laudato Si: On Care for Our Common Home*. Vatican Press. https://web.archive.org/web/20200229191755/http://w2.vatican.va/content/dam/francesco/pdf/encyclicals/documents/papa-francesco_20150524_enciclica-laudato-si_en.pdf

Freeman, G. A. 1940. Trends in High-Intensity Mercury Lamps. *Electrical Engineering*, 59, 444–447.

French Agency for Food, Environmental, and Occupational Health and Safety. 2016. Vary Your Fish Consumption According to Your Tastes and Needs. Retrieved September

28, 2018, from https://web.archive.org/web/20190203160318/https://www.anses.fr/en/content/vary-your-fish-consumption-according-your-tastes-and-needs

Friedl, G., & Wüest, A. 2002. Disrupting Biogeochemical Cycles—Consequences of Damming. *Aquatic Science,* 64, 55–65.

Friedman, L., & Davenport, C. 2020. EPA Weakens Controls on Mercury. *New York Times*, April 16.

Fritz, M. M., Maxson, P. A., & Baumgartner, R. J. 2016. The Mercury Supply Chain, Stakeholders and Their Responsibilities in the Quest for Mercury-Free Gold. *Resources Policy,* 50, 177–192.

Funabashi, H. 2006. Minamata Disease and Environmental Governance. *International Journal of Japanese Sociology,* 15, 7–25.

Gagnon, V. S. 2016. Ojibwe Gichigami ("Ojibwa's Great Sea"): An Intersecting History of Treaty Rights, Tribal Fish Harvesting, and Toxic Risk in Keweenaw Bay, United States. *Water History,* 8, 365–384.

Gamu, J., Le Billon, P., & Spiegel, S. 2015. Extractive Industries and Poverty: A Review of Recent Findings and Linkage Mechanisms. *The Extractive Industries and Society,* 2, 162–176.

Garnham, B. L., & Langerman, K. E. 2016. Mercury Emissions from South Africa's Coal-Fired Power Stations. *The Clean Air Journal,* 26, 14–20.

GAVI (The Vaccine Alliance). 2012. Report to the GAVI Alliance Board. June 12–13, 2012. https://web.archive.org/web/20191229153003/https://www.gavi.org/sites/default/files/board/minutes/2012/12-june/02%20-%20CEO%20report.pdf

Geels, F. W. 2002. Technological Transitions as Evolutionary Reconfiguration Processes: A Multi-Level Perspective and a Case-Study. *Research Policy,* 31, 1257–1274.

Geels, F. W. 2004. From Sectoral Systems of Innovation to Socio-Technical Systems: Insights About Dynamics and Change from Sociology and Institutional Theory. *Research Policy,* 33, 897–920.

Geels, F. W. 2011. The Multi-Level Perspective on Sustainability Transitions: Responses to Seven Criticisms. *Environmental Innovation and Societal Transitions,* 1, 24–40.

Geissdoerfer, M., Savaget, P., Bocken, N. M., & Hultink, E. J. 2017. The Circular Economy—a New Sustainability Paradigm? *Journal of Cleaner Production,* 143, 757–768.

Genchi, G., Sinicropi, M., Carocci, A., Lauria, G., & Catalano, A. 2017. Mercury Exposure and Heart Diseases. *International Journal of Environmental Research and Public Health,* 14, 74.

George, T. S. 2001. *Minamata: Pollution and the Struggle for Democracy in Postwar Japan.* Cambridge, MA: Harvard University Asia Center.

George Washington Foundation. 2017. George Washington's Troublesome Teeth. Retrieved December 23, 2019, from https://web.archive.org/web/20191223163438/https://livesandlegaciesblog.org/2017/05/18/george-washingtons-troublesome-teeth/

Gettens, R. J., Feller, R. L., & Chase, W. T. 1972. Vermilion and Cinnabar. *Studies in Conservation,* 17, 45–69.

Ghisellini, P., Cialani, C., & Ulgiati, S. 2016. A Review on Circular Economy: The Expected Transition to a Balanced Interplay of Environmental and Economic Systems. *Journal of Cleaner Production,* 114, 11–32.

Giang, A., & Selin, N. E. 2016. Benefits of Mercury Controls for the United States. *Proceedings of the National Academy of Sciences,* 113, 286–291.

Giang, A., Song, S., Muntean, M., Janssens-Maenhout, G., Harvey, A., Berg, E., & Selin, N. E. 2018. Understanding Factors Influencing the Detection of Mercury Policies in Modelled Laurentian Great Lakes Wet Deposition. *Environmental Science: Processes & Impacts,* 20, 1373–1389.

Giang, A., Stokes, L. C., Streets, D. G., Corbitt, E. S., & Selin, N. E. 2015. Impacts of the Minamata Convention on Mercury Emissions and Global Deposition from Coal-Fired Power Generation in Asia. *Environmental Science & Technology,* 49, 5326–5335.

Gibb, H., & O'Leary, K. G. 2014. Mercury Exposure and Health Impacts among Individuals in the Artisanal and Small-Scale Gold Mining Community: A Comprehensive Review. *Environmental Health Perspectives,* 122, 667–672.

Gibbs, D. 2006. Prospects for an Environmental Economic Geography: Linking Ecological Modernization and Regulationist Approaches. *Economic Geography,* 82, 193–215.

Giese, A. C. 1940. Mercury Poisoning. *Science,* 91, 476–477.

Gilmour, C. C., Henry, E. A., & Mitchell, R. 1992. Sulfate Stimulation of Mercury Methylation in Freshwater Sediments. *Environmental Science & Technology,* 26, 2281–2287.

Gleick, P. H. 2018. Transitions to Freshwater Sustainability. *Proceedings of the National Academy of Sciences,* 115, 8863–8871.

Global Witnesss. 2016. *River of Gold.* Global Witness, London. https://www.globalwitness.org/en/campaigns/conflict-minerals/river-of-gold-drc/

Goldstein, R. forthcoming. Mother of God, Son of Jupiter: Mercury Rising and (Re)Producing Gendered Environmental Racisms in a Quickly Heating Planet. *American Anthropologist.*

Goldwater, L. J. 1972. *Mercury: A History of Quicksilver.* Baltimore: York Press.

Goodman, C. 1938. Mercury Poisoning: A Review of Present Knowledge. *Review of Scientific Instruments,* 9, 233–236.

Government of Canada. 2014. Canadian Environmental Protection Act: Products Containing Mercury Regulations. November 19. *Canada Gazette*, 148. https://web.archive.org/web/20190915231338/http://gazette.gc.ca/rp-pr/p2/2014/2014-11-19/html/sor-dors254-eng.html

Government of Canada. n.d. Canadian Measures to Implement the Minamata Convention on Mercury. https://web.archive.org/web/20190725120052/http://www.mercuryconvention.org/Portals/11/documents/Notifications/Canada_Hg_Summary ImplementationMeasures_Art30_4_2.pdf

Government of Japan. n.d. Information on Measures to Implement the Minamata Convention on Mercury. https://web.archive.org/web/20190725124539/http://www.mercuryconvention.org/Portals/11/documents/submissions/Japan%20declaration _Art%2030%20para%204.pdf

Grandjean, P. 2016. Paracelsus Revisited: The Dose Concept in a Complex World. *Basic & Clinical Pharmacology & Toxicology*, 119, 126–132.

Grandjean, P., & Budtz-Jørgensen, E. 2007. Total Imprecision of Exposure Biomarkers: Implications for Calculating Exposure Limits. *American Journal of Industrial Medicine*, 50, 712–719.

Grandjean, P., Weihe, P., White, R. F., Debes, F., Araki, S., Yokoyama, K., … Jørgensen, P. J. 1997. Cognitive Deficit in 7-Year-Old Children with Prenatal Exposure to Methylmercury. *Neurotoxicology and Teratology*, 19, 417–428.

Granet, E. 1940. Pruritus and the Etiologic Factors and Treatment in 100 Cases. *New England Journal of Medicine*, 223, 1015–1020.

Greer, L., Bender, M., Maxson, P., & Lennett, D. 2006. Curtailing Mercury's Global Reach. *In:* Worldwatch Institute, *State of the World 2006*. London: Norton.

Grundmann, R. 2016. Climate Change as a Wicked Social Problem. *Nature Geoscience*, 9, 562–563.

Gustin, M. S., Lindberg, S. E., Austin, K., Coolbaugh, M., Vette, A., & Zhang, H. 2000. Assessing the Contribution of Natural Sources to Regional Atmospheric Mercury Budgets. *Science of the Total Environment*, 259, 61–71.

Ha, E., Basu, N., Bose-O'Reilly, S., Dórea, J. G., McSorley, E., Sakamoto, M., & Chan, H. M. 2017. Current Progress on Understanding the Impact of Mercury on Human Health. *Environmental Research*, 152, 419–433.

Hadsund, P. 1993. The Tin-Mercury Mirror: Its Manufacturing Technique and Deterioration Processes. *Studies in Conservation*, 38, 3–16.

Hagemann, S., Oppermann, U., & Brasser, T. 2014. *Behaviour of Mercury and Mercury Compounds at the Underground Disposal in Salt Formations and Their Potential Mobilisation by Saline Solutions*. German Federal Environment Agency (Umweltbundesamt),

Dessau-Roßlau, Germany. January 2014. https://web.archive.org/web/201912291456 14/https://www.umweltbundesamt.de/sites/default/files/medien/378/publikationen /texte_07_2014_behaviour_of_mercury_and_mercury_compounds_at_the_under ground_disposal_in_salt_formations.pdf

Hall, P. A., & Lamont, M. 2013. *Social Resilience in the Neoliberal Era*. Cambridge: Cambridge University Press.

Halsey, N. A., & Goldman, L. 2001. Balancing Risks and Benefits: Primum Non Nocere Is Too Simplistic. *Pediatrics*, 108, 466–467.

Hamilton, A. 1943. *Exploring the Dangerous Trades: The Autobiography of Alice Hamilton, M.D.* Boston: Little, Brown.

Haq, I. U. 1963. Agrosan Poisoning in Man. *British Medical Journal*, 1, 1579–1582.

Harada, M. 1995. Minamata Disease: Methylmercury Poisoning in Japan Caused by Environmental Pollution. *Critical Reviews in Toxicology*, 25, 1–24.

Harada, M., Fujino, T., Oorui, T., Nakachi, S., Nou, T., Kizaki, T., … Ohno, H. 2005. Followup Study of Mercury Pollution in Indigenous Tribe Reservations in the Province of Ontario, Canada, 1975–2002. *Bulletin of Environmental Contamination and Toxicology*, 74, 689–697.

Harré, R., Brockmeier, J., & Mühlhäusler, P. 1999. *Greenspeak: A Study of Environmental Discourse*. Thousand Oaks, CA: Sage Publications.

Harremoës, P., Gee, D., MacGarvin, M., Stirling, A., Keys, J., Wynne, B., & Vaz, S. G. 2013. *The Precautionary Principle in the 20th Century: Late Lessons from Early Warnings*. New York: Routledge.

Hayes, A. W., & Kruger, C. L. 2014. *Hayes' Principles and Methods of Toxicology*. Boca Raton, FL: CRC Press.

Head, B. W., & Alford, J. 2015. Wicked Problems. *Administration & Society*, 47, 711–739.

Health Care Without Harm. 2018. Laws and Resolutions. Retrieved September 15, 2019, from https://web.archive.org/web/20190915230644/https://noharm-uscanada .org/issues/us-canada/laws-and-resolutions

Hightower, J. M. 2009. *Diagnosis: Mercury: Money, Politics, and Poison*. Washington, DC: Island Press.

Hilson, G. 2002. Harvesting Mineral Riches: 1000 years of Gold Mining in Ghana. *Resources Policy* 28, 13–26.

Hilson, G. 2012. Family Hardship and Cultural Values: Child Labor in Malian Small-Scale Gold Mining Communities. *World Development*, 40, 1663–1674.

Hilson, G. 2016. Farming, Small-Scale Mining and Rural Livelihoods in Sub-Saharan Africa: A Critical Overview. *The Extractive Industries and Society*, 3, 547–563.

Hilson, G., Amankwah, R., & Ofori-Sarpong, G. 2013. Going for Gold: Transitional Livelihoods in Northern Ghana. *The Journal of Modern African Studies*, 51, 109–137.

Hilson, G., & Banchirigah, S. M. 2009. Are Alternative Livelihood Projects Alleviating Poverty in Mining Communities? Experiences from Ghana. *Journal of Development Studies*, 45, 172–196.

Hilson, G., & Gatsinzi, A. 2014. A Rocky Road Ahead? Critical Reflections on the Futures of Small-Scale Mining in Sub-Saharan Africa. *Futures*, 62, 1–9.

Hilson, G., Goumandakoye, H., & Diallo, P. 2019. Formalizing Artisanal Mining "Spaces" in Rural Sub-Saharan Africa: The Case of Niger. *Land Use Policy*, 80, 259–268.

Hilson, G., Hilson, A., & Adu-Darko, E. 2014. Chinese Participation in Ghana's Informal Gold Mining Economy: Drivers, Implications and Clarifications. *Journal of Rural Studies*, 34, 292–303.

Hilson, G., Hilson, A., & McQuilken, J. 2016. Ethical Minerals: Fairer Trade for Whom? *Resources Policy*, 49, 232–247.

Hilson, G., & McQuilken, J. 2014. Four Decades of Support for Artisanal and Small-Scale Mining in Sub-Saharan Africa: A Critical Review. *The Extractive Industries and Society*, 1, 104–118.

Hilson, G., & Osei, L. 2014. Tackling Youth Unemployment in Sub-Saharan Africa: Is There a Role for Artisanal and Small-Scale Mining? *Futures*, 62, 83–94.

Hilson, G., & Potter, C. 2005. Structural Adjustment and Subsistence Industry: Artisanal Gold Mining in Ghana. *Development and Change*, 36, 103–131.

Hilson, G. M. 2002. The Future of Small-Scale Mining: Environmental and Socioeconomic Perspectives. *Futures*, 34, 863–872.

Hoffman, M. J. 2011. *Climate Governance at the Crossroads: Experimenting with a Global Response after Kyoto*. Oxford: Oxford University Press.

Holland, J. H. 2006. Studying Complex Adaptive Systems. *Journal of Systems Science and Complexity*, 19, 1–8.

Holling, C. S. 1973. Resilience and Stability of Ecological Systems. *Annual Review of Ecology and Systematics*, 4, 1–23.

Horowitz, H. M., Jacob, D. J., Amos, H. M., Streets, D. G., & Sunderland, E. M. 2014. Historical Mercury Releases from Commercial Products: Global Environmental Implications. *Environmental Science & Technology*, 48, 10242–10250.

Horowitz, H. M., Jacob, D. J., Zhang, Y., Dibble, T. S., Slemr, F., Amos, H. M., ... Sunderland, E. M. 2017. A New Mechanism for Atmospheric Mercury Redox Chemistry: Implications for the Global Mercury Budget. *Atmospheric Chemistry and Physics*, 17, 6353–6371.

Hou, M., Chen, L., Guo, Z., Dong, X., Wang, Y., & Xia, Y. 2018. A Clean and Membrane-Free Chlor-Alkali Process with Decoupled Cl_2 and H_2/NaOH Production. *Nature Communications, 9,* 438.

Hruschka, F. 2011. *SDC Experiences with Formalization and Responsible Environmental Practices in Artisanal and Small-Scale Gold Mining in Latin America and Asia (Mongolia).* Swiss Federal Department of Foreign Affairs, Bern, Switzerland. https://web.archive.org /web/20191229145705/https://www.files.ethz.ch/isn/154963/resource_en_216063.pdf

Hsu-Kim, H., Eckley, C. S., Achá, D., Feng, X., Gilmour, C. C., Jonsson, S., & Mitchell, C. P. 2018. Challenges and Opportunities for Managing Aquatic Mercury Pollution in Altered Landscapes. *Ambio, 47,* 141–169.

Hui, M., Wu, Q., Wang, S., Liang, S., Zhang, L., Wang, F., ... Lin, Z. 2016. Mercury Flows in China and Global Drivers. *Environmental Science & Technology, 51,* 222–231.

Hunter, D., Bomford, R. R., & Russell, D. S. 1940. Poisoning by Methyl Mercury Compounds. *QJM: An International Journal of Medicine, 9,* 193–226.

Hutcheson, M. S., Smith, C. M., Rose, J., Batdorf, C., Pancorbo, O., West, C. R., ... Francis, C. 2014. Temporal and Spatial Trends in Freshwater Fish Tissue Mercury Concentrations Associated with Mercury Emissions Reductions. *Environmental Science & Technology, 48,* 2193–2202.

Hylander, L. D. 2001. Global Mercury Pollution and Its Expected Decrease after a Mercury Trade Ban. *Water, Air, and Soil Pollution, 125,* 331–344.

Hylander, L. D., & Meili, M. 2003. 500 Years of Mercury Production: Global Annual Inventory by Region until 2000 and Associated Emissions. *Science of The Total Environment, 304,* 13–27.

Hylander, L. D., & Meili, M. 2005. The Rise and Fall of Mercury: Converting a Resource to Refuse after 500 Years of Mining and Pollution. *Critical Reviews in Environmental Science and Technology, 35,* 1–36.

IGF (Intergovernmental Forum on Mining, Minerals, Metals, and Sustainable Development). 2017. *Global Trends in Artisanal and Small-Scale Mining (ASM): A Review of Key Numbers and Issues.* IISD, Winnipeg. https://web.archive.org/web /20190624110258/https://www.iisd.org/sites/default/files/publications/igf-asm -global-trends.pdf

International Energy Agency. 2018. Coal Information 2018: Overview. https:// webstore.iea.org/coal-information-2018-overview

International Labour Office. 2006. *Minors out of Mining! Partnership for Global Action against Child Labour in Small-Scale Mining.* International Labour Organization, Geneva. https://web.archive.org/web/20200229192948/https://www.ilo.org /ipec/Informationresources/WCMS_IPEC_PUB_2519/lang--en/index.htm

International Peace Information Service. 2019. *Mapping Artisanal Mining Areas and Mineral Supply Chains in Eastern DR Congo: Impact of Armed Interference & Responsible Sourcing.* International Peace Information Service, Antwerp. Retrieved December 29, 2019, from https://web.archive.org/web/20190730161750/http://ipisresearch.be/wp -content/uploads/2019/04/1904-IOM-mapping-eastern-DRC.pdf

Irukayama, K., Kai, F., Fujiki, M., & Kondo, T. 1962. Studies on the Origin of the Causative Agent of Minamata Disease. III. Industrial Wastes Containing Mercury Compounds from Minamata Factory. *Kumamoto Medical Journal*, 15, 57–68.

Itzkoff, D. 2008. Piven Leaves Show Amid Concerns for His Health. *New York Times*, December 18.

Iverfeldt, Å. 1991. Occurrence and Turnover of Atmospheric Mercury over the Nordic Countries. *Water, Air, and Soil Pollution*, 56, 251–265.

Jacob, D. J., Prather, M. J., Wofsy, S. C., & McElroy, M. B. 1987. Atmospheric Distribution of ^{85}Kr Simulated with a General Circulation Model. *Journal of Geophysical Research: Atmospheres*, 92, 6614–6626.

Jacobs, E. T., Burgess, J. L., & Abbott, M. B. 2018. The Donora Smog Revisited: 70 Years after the Event That Inspired the Clean Air Act. *American Journal of Public Health*, 108, S85–S88.

Jaffe, D. A., Lyman, S., Amos, H. M., Gustin, M. S., Huang, J., Selin, N. E., ... Talbot, R. 2014. Progress on Understanding Atmospheric Mercury Hampered by Uncertain Measurements. *Environmental Science & Technology*, 48, 7204–7206.

Jago, R. 2018. The Warrior Society Rises: How a Mercury Spill in Canada Inspired a Movement. *The Guardian*, October 16.

Jalili, M., & Abbasi, A. 1961. Poisoning by Ethyl Mercury Toluene Sulphonanilide. *Occupational and Environmental Medicine*, 18, 303–308.

Janssen, M., Bodin, Ö., Anderies, J., Elmqvist, T., Ernstson, H., McAllister, R. R., ... Ryan, P. 2006. Toward a Network Perspective of the Study of Resilience in Social-Ecological Systems. *Ecology and Society*, 11, 15.

Jenkins, K. 2014. Women, Mining and Development: An Emerging Research Agenda. *The Extractive Industries and Society*, 1, 329–339.

Jiskra, M., Sonke, J. E., Obrist, D., Bieser, J., Ebinghaus, R., Myhre, C. L., ... Worthy, D. 2018. A Vegetation Control on Seasonal Variations in Global Atmospheric Mercury Concentrations. *Nature Geoscience*, 11, 244–250.

Joensuu, O. I. 1971. Fossil Fuels as a Source of Mercury Pollution. *Science*, 172, 1027–1028.

Johnson, C. W. 2006. What Are Emergent Properties and How Do They Affect the Engineering of Complex Systems? *Reliability Engineering and System Safety*, 91, 1475–1481.

Johnson, L., & Wolbarsht, M. L. 1979. Mercury Poisoning: A Probable Cause of Isaac Newton's Physical and Mental Ills. *Notes and Records of the Royal Society of London*, 34, 1–9.

Johnson, T. 2018. Breaking Down the Forever Chemicals—What Are PFAS? Clean Water Action. Retrieved July 22, 2019, from https://web.archive.org/web /20200229193302/https://www.cleanwateraction.org/2018/08/02/breaking-down -forever-chemicals--what-are-pfas

Joint FAO-WHO Expert Committee on Food Additives 1972. *Evaluation of Certain Food Additives: And of the Contaminants Mercury, Lead and Cadmium*. World Health Organization, Geneva. https://web.archive.org/web/20200229193609/https://apps .who.int/iris/bitstream/handle/10665/40985/WHO_TRS_505.pdf?sequence=1 &isAllowed=y

Jonsson, S., Andersson, A., Nilsson, M. B., Skyllberg, U., Lundberg, E., Schaefer, J. K., ... Björn, E. 2017. Terrestrial Discharges Mediate Trophic Shifts and Enhance Methylmercury Accumulation in Estuarine Biota. *Science Advances*, 3, e1601239.

Kamau, M., Chasek, P., & O'Connor, D. 2018. *Transforming Multilateral Diplomacy: The Inside Story of the Sustainable Development Goals*. New York: Routledge.

Kanie, N., & Biermann, F. (eds.) 2017. *Governing through Goals: Sustainable Development Goals as Governance Innovation*. Cambridge, MA: MIT Press.

Kates, R. W. 2011. What Kind of a Science Is Sustainability Science? *Proceedings of the National Academy of Sciences*, 108, 19449–19450.

Kates, R. W., Clark, W. C., Corell, R., Hall, J. M., Jaeger, C. C., Lowe, I., ... Dickson, N. M. 2001. Sustainability Science. *Science*, 292, 641–642.

Kaur, I. P., Lal, S., Rana, C., Kakkar, S., & Singh, H. 2009. Ocular Preservatives: Associated Risks and Newer Options. *Cutaneous and Ocular Toxicology*, 28, 93–103.

Keeler, G. J., Landis, M. S., Norris, G. A., Christianson, E. M., & Dvonch, J. T. 2006. Sources of Mercury Wet Deposition in Eastern Ohio, USA. *Environmental Science & Technology*, 40, 5874–5881.

Keohane, R. O. 1989. *International Institutions and State Power: Essays in International Relations Theory*. Boulder: CO: Westview Press.

Keohane, R. O., & Victor, D. G. 2011. The Regime Complex for Climate Change. *Perspectives on Politics*, 9, 7–23.

Kessler, R. 2013. The Minamata Convention on Mercury: A First Step toward Protecting Future Generations. *Environmental Health Perspectives*, 121, A304–A309.

Kirby, D. 2006. *Evidence of Harm: Mercury in Vaccines and the Autism Epidemic: A Medical Controversy*. London: Macmillan.

Kirkemo, H., Newman, W. L., & Ashley, R. P. 2001. Gold. U.S. Geological Survey. Denver, CO. https://web.archive.org/web/20191229150042/https://pubs.usgs.gov/gip /prospect1/goldgip.html

Kiser, L., & Ostrom, E. 1982. The Three Worlds of Action: A Metatheoretical Synthesis of Institutional Approaches. *In*: Ostrom, E. (ed.), *Strategies of Political Inquiry*, 179–222. Beverly Hills, CA: Sage.

Klee, R., & Graedel, T. 2004. Elemental Cycles: A Status Report on Human or Natural Dominance. *Annual Review of Environment and Resources* 29, 69–107.

Kocman, D., Horvat, M., Pirrone, N., & Cinnirella, S. 2013. Contribution of Contaminated Sites to the Global Mercury Budget. *Environmental Research, 125*, 160–170.

Kocman, D., Wilson, S. J, Amos, H. M, Telmer, K. H., Steenhuisen, F., Sunderland, E. M., … Horvat, M. 2017. Toward an Assessment of the Global Inventory of Present-Day Mercury Releases to Freshwater Environments. *International Journal of Environmental Research and Public Health, 14*, 138.

Köhler, J., Geels, F., Kern, F., Onsongo, E., & Wieczorek, A. 2017. *A Research Agenda for the Sustainability Transitions Research Network, STRN Working Group*. STRN. https:// web.archive.org/web/20191229143126/https://pure.tue.nl/ws/portalfiles/portal /101288346/STRN_Research_Agenda_2017.pdf

Komiyama, H., & Takeuchi, K. 2006. Sustainability Science: Building a New Discipline. *Sustainability Science, 1*, 1–6.

König, A., & Ravetz, J. 2017. *Sustainability Science: Key Issues*. New York: Routledge.

Korns, R. F. 1972. The Frustrations of Bettye Russow. *Nutrition Today, 7*, 21–23.

Krabbenhoft, D. P., & Sunderland, E. M. 2013. Global Change and Mercury. *Science,* 341, 1457–1458.

Krishnakumar, B., Niksa, S., Sloss, L., Jozewicz, W., & Futsaeter, G. 2012. Process Optimization Guidance (POG and IPOG) for Mercury Emissions Control. *Energy & Fuels, 26*, 4624–4634.

Kuenkel, P. 2019. *Stewarding Sustainability Transformations: An Emerging Theory and Practice of SDG Implementation*. New York: Springer.

Kumar, A., Wu, S., Huang, Y., Liao, H., & Kaplan, J. O. 2018. Mercury from Wildfires: Global Emission Inventories and Sensitivity to 2000–2050 Global Change. *Atmospheric Environment, 173*, 6–15.

Kwon, S., & Selin, N. 2016. Uncertainties in Atmospheric Mercury Modeling for Policy Evaluation. *Current Pollution Reports, 2*, 103–114.

Kwon, S., Selin, N., Giang, A., Karplus, V., & Zhang, D. 2018. Present and Future Mercury Concentrations in Chinese Rice: Insights from Modeling. *Global Biogeochemical Cycles, 32,* 437–462.

Laden, F., Schwartz, J., Speizer, F. E., & Dockery, D. W. 2006. Reduction in Fine Particulate Air Pollution and Mortality: Extended Follow-up of the Harvard Six Cities Study. *American Journal of Respiratory and Critical Care Medicine, 173,* 667–672.

Lange, G.-M., Wodon, Q., & Carey, K. 2018. *The Changing Wealth of Nations 2018: Building a Sustainable Future.* Washington, DC: The World Bank. https://web.archive .org/web/20191229143304/https://openknowledge.worldbank.org/bitstream /handle/10986/29001/9781464810466.pdf?sequence=4&isAllowed=y

Langston, J. D., Lubis, M. I., Sayer, J. A., Margules, C., Boedhihartono, A. K., & Dirks, P. H. 2015. Comparative Development Benefits from Small and Large Scale Mines in North Sulawesi, Indonesia. *The Extractive Industries and Society, 2,* 434–444.

Lee, B. 1991. Highlights of the Clean Air Act Amendments of 1990. *Journal of the Air & Waste Management Association, 41,* 16–19.

Lefohn, A. S., Husar, J. D., & Husar, R. B. 1999. Estimating Historical Anthropogenic Global Sulfur Emission Patterns for the Period 1850–1990. *Atmospheric Environment, 33,* 3435–3444.

Lenton, T. M., Held, H., Kriegler, E., Hall, J. W., Lucht, W., Rahmstorf, S., & Schellnhuber, H. J. 2008. Tipping Elements in the Earth's Climate System. *Proceedings of the National Academy of Sciences, 105,* 1786–93.

Levin, K., Cashore, B., Bernstein, S., & Auld, G. 2012. Overcoming the Tragedy of Super Wicked Problems: Constraining Our Future Selves to Ameliorate Global Climate Change. *Policy Science, 45,* 123–152.

Levin, S., Xepapadeas, T., Crépin, A.-S., Norberg, J., De Zeeuw, A., Folke, C., ... Daily, G. 2013. Social-Ecological Systems as Complex Adaptive Systems: Modeling and Policy Implications. *Environment and Development Economics, 18,* 111–132.

Lewis, S. L., & Maslin, M. A. 2015. Defining the Anthropocene. *Nature, 519,* 171–180.

Li, P., Feng, X., Qiu, G., Shang, L., & Li, Z. 2009. Mercury Pollution in Asia: A Review of the Contaminated Sites. *Journal of Hazardous Materials, 168,* 591–601.

Lin, C.-J., Shetty, S. K., Pan, L., Pongprueksa, P., Jang, C., & Chu, H.-W. 2012. Source Attribution for Mercury Deposition in the Contiguous United States: Regional Difference and Seasonal Variation. *Journal of the Air & Waste Management Association, 62,* 52–63.

Lin, Y., Wang, S., Steindal, E. H., Wang, Z., Braaten, H. F. V., Wu, Q., & Larssen, T. 2017. A Holistic Perspective Is Needed to Ensure Success of Minamata Convention on Mercury. *Environmental Science & Technology, 51,* 1070–1071.

Lindberg, S., Bullock, R., Ebinghaus, R., Engstrom, D. R., Feng, X., Fitzgerald, W. F., ... Seigneur, C. 2007. A Synthesis of Progress and Uncertainties in Attributing the Sources of Mercury in Deposition. *Ambio*, 36, 19–33.

Linnér, B.-O., & Selin, H. 2005. The Road to Rio: Early Efforts on Environment and Development. *In:* Churie Kallhauge, A., Sjöstedt, G., & Corell, E. (eds.), *Global Challenges: Furthering the Multilateral Process for Sustainable Development*, 58–73. London: Greenleaf Publishing.

Linnér, B.-O., & Wibeck, V. 2019. *Sustainability Transformations: Agents and Drivers across Societies*. Cambridge: Cambridge University Press.

Liu, J., Dietz, T., Carpenter, S. R., Folke, C., Alberti, M., Redman, C. L., ... Provencher, W. 2007. Coupled Human and Natural Systems. *Ambio*, 36, 639–649.

Liu, J., Mooney, H., Hull, V., Davis, S. J., Gaskell, J., Hertel, T., ... Li, S. 2015. Systems Integration for Global Sustainability. *Science*, 347, 1258832.

Liu, J., Shi, J.-Z., Yu, L.-M., Goyer, R. A., & Waalkes, M. P. 2008. Mercury in Traditional Medicines: Is Cinnabar Toxicologically Similar to Common Mercurials? *Experimental Biology and Medicine*, 233, 810–817.

Löfroth, G., & Duffy, M. E. 1969. Birds Give Warning. *Environment*, 11, 10–17.

Loorbach, D., Frantzeskaki, N., & Avelino, F. 2017. Sustainability Transitions Research: Transforming Science and Practice for Societal Change. *Annual Review of Environment and Resources*, 42, 599–626.

Löwy, I. 2011. "Sexual Chemistry" before the Pill: Science, Industry, and Chemical Contraceptives, 1920–1960. *British Journal for the History of Science*, 44, 245–274.

MacKenzie, E. 1945. London's Water Supply: Safeguarding Its Purity in Peace and War. *Nature*, 155, 162–164.

Mackey, T. K., Contreras, J. T., & Liang, B. A. 2014. The Minamata Convention on Mercury: Attempting to Address the Global Controversy of Dental Amalgam Use and Mercury Waste Disposal. *Science of the Total Environment*, 472, 125–129.

MacLeod, M., Breitholtz, M., Cousins, I. T., de Wit, C. A., Persson, L. M., Rudén, C., & McLachlan, M. S. 2014. Identifying Chemicals that Are Planetary Boundary Threats. *Environmental Science & Technology*, 48, 11057–11063.

Magner, L. N., & Kim, O. J. 2017. *A History of Medicine*. Boca Raton, FL: CRC Press.

Mahaffey, K. R., Clickner, R. P., & Bodurow, C. C. 2003. Blood Organic Mercury and Dietary Mercury Intake: National Health and Nutrition Examination Survey, 1999 and 2000. *Environmental Health Perspectives*, 112, 562–570.

Mahaffey, K. R., Sunderland, E. M., Chan, H. M., Choi, A. L., Grandjean, P., Mariën, K., ... Weihe, P. 2011. Balancing the Benefits of N-3 Polyunsaturated Fatty Acids and the Risks of Methylmercury Exposure from Fish Consumption. *Nutrition Reviews*, 69, 493–508.

Mahbub, K. R., Bahar, M. M., Labbate, M., Krishnan, K., Andrews, S., Naidu, R., & Megharaj, M. 2017. Bioremediation of Mercury: Not Properly Exploited in Contaminated Soils! *Applied Microbiology and Biotechnology*, 101, 963–976.

Mahdihassan, S. 1985. Cinnabar-Gold as the Best Alchemical Drug of Longevity, Called Makaradhwaja in India. *The American Journal of Chinese Medicine*, 13, 93–108.

Malhi, Y. 2017. The Concept of the Anthropocene. *Annual Review of Environment and Resources*, 42, 77–104.

Mallas, J., & Benedicto, N. 1986. Mercury and Goldmining in the Brazilian Amazon. *Ambio*, 248–249.

Malm, A., & Hornborg, A. 2014. The Geology of Mankind? A Critique of the Anthropocene Narrative. *The Anthropocene Review*, 1, 62–69.

Markard, J., Raven, R., & Truffer, B. 2012. Sustainability Transitions: An Emerging Field of Research and Its Prospects. *Research Policy*, 41, 955–967.

Marshall, B. G., & Veiga, M. M. 2017. Formalization of Artisanal Miners: Stop the Train, We Need to Get Off! *The Extractive Industries and Society*, 4, 300–303.

Martin, L. G., Labuschagne, C., Brunke, E.-G., Weigelt, A., Ebinghaus, R., & Slemr, F. 2017. Trend of Atmospheric Mercury Concentrations at Cape Point for 1995–2004 and since 2007. *Atmospheric Chemistry and Physics*, 17, 2393–2399.

Martín-Gil, J., Martín-Gil, F., Delibes-de-Castro, G., Zapatero-Magdaleno, P., & Sarabia-Herrero, F. 1995. The First Known Use of Vermillion. *Experientia*, 51, 759–761.

Masur, L. C. 2011. A Review of the Use of Mercury in Historic and Current Ritualistic and Spiritual Practices. *Alternative Medicine Review*, 16, 314–320.

Matson, P., Clark, W. C., & Andersson, K. 2016. *Pursuing Sustainability: A Guide to the Science and Practice*. Princeton, NJ: Princeton University Press.

Mayer, B., Brown, P., & Linder, M. 2002. Moving Further Upstream: From Toxics Reduction to the Precautionary Principle. *Public Health Reports*, 117, 574–586.

Mayer, J., Hopf, S., van Dijen, F., & Baldini, A. 2014. Measurement of Low Mercury Concentrations in Flue Gases of Combustion Plants. *VGB Powertech Journal*, 3, 64–68.

Mazur, A. 2004. *True Warnings and False Alarms: Evaluating Fears About the Health Risks of Technology, 1948–1971*. Washington, DC: Resources for the Future.

Mazur, T. R., Klappauf, B., & Raizen, M. G. 2014. Demonstration of Magnetically Activated and Guided Isotope Separation. *Nature Physics*, 10, 601–605.

McGlade, C., & Ekins, P. 2015. The Geographical Distribution of Fossil Fuels Unused When Limiting Global Warming to 2°C. *Nature*, 517, 187–190.

Meadowcroft, J. 2011. Engaging with the Politics of Sustainability Transitions. *Environmental Innovation and Societal Transitions*, 1, 70–75.

Meadowcroft, J. 2012. Greeing the State? *In:* Steinberg, P. F., & VanDeveer, S. D. (eds.), *Comparative Environmental Politics*, 63–88. Cambridge, MA: MIT Press.

Meadows, D. 1999. *Leverage Points: Places to Intervene in a System*. The Sustainability Institute, Hartland, VT. https://web.archive.org/web/20190203194357/http://www.donellameadows.org/wp-content/userfiles/Leverage_Points.pdf

Meier, B. 1990. Ban Is Sought on Mercury in Latex Paint. *New York Times*, April 6.

Mencimer, S. 2008. Why Mercury Tuna Is Still Legal. *Mother Jones*, September-October. https://web.archive.org/web/20191229153912/https://www.motherjones.com/politics/2008/09/why-mercury-tuna-still-legal/

Meng, M., Li, B., Shao, J.-J., Wang, T., He, B., Shi, J.-B., … Jiang, G.-B. 2014. Accumulation of Total Mercury and Methylmercury in Rice Plants Collected from Different Mining Areas in China. *Environmental Pollution*, 184, 179–186.

Mergler, D., Anderson, H. A., Chan, L. H. M., Mahaffey, K. R., Murray, M., Sakamoto, M., & Stern, A. H. 2007. Methylmercury Exposure and Health Effects in Humans: A Worldwide Concern. *Ambio*, 36, 3–11.

Meyer, H. M., & Mitchell, A. W. 1941. Mercury. *In:* Shore, F. M. (ed.), *Minerals Yearbook 1941*, 685–702. Washington, DC: US Government Printing Office.

Meyer, H. M., & Mitchell, A. W. 1945. Mercury. *In:* Pehrson, E. W. & Needham, C. E. (eds.), *Minerals Yearbook 1943*, 713–730. Washington, DC: US Government Printing Office.

Middleton, W. E. K. 1966. *A History of the Thermometer and Its Use in Meteorology*. Baltimore: Johns Hopkins University Press.

Middleton, W. E. K. 1963. The Place of Torricelli in the History of the Barometer. *Isis*, 54, 11–28.

Miles, E. L., Andresen, S., Carlin, E. M., Skjærseth, J. B., Underdal, A., & Wettestad, J. 2002. *Environmental Regime Effectiveness: Confronting Theory with Evidence*. Cambridge, MA: MIT Press.

Minamata-Juku Committee. 2006. Minamata's Pledge. 50th Anniversary of the Official Recognition of Minamata Disease Commemoration Project, October 21. https://web.archive.org/web/20200222154308/https://minamata195651.jp/yakusoku_en.html

Ministry of the Environment Japan. 2013. Lessons from Minamata Disease and Mercury Management in Japan. Environmental Health and Safety Division, Environmental Health Department. Tokyo. https://web.archive.org/web/20190203171020/https://www.env.go.jp/chemi/tmms/pr-m/mat01/en_full.pdf

Minnesota Department of Health. 2019. Skin Lightening Products Found to Contain Mercury. Retrieved February 3, 2019, from https://web.archive.org/web/20190203170616/http://www.health.state.mn.us/topics/skin/

Mitchell, R. B. 2006. Problem Structure, Institutional Design, and the Relative Effectiveness of International Environmental Agreements. *Global Environmental Politics,* 6, 72–89.

Mnookin, S. 2017. How Robert F. Kennedy, Jr., Distorted Vaccine Science. *STAT News,* January 10. https://www.scientificamerican.com/article/how-robert-f-kennedy-jr-distorted-vaccine-science1/

Mol, A. P. 2003. *Globalization and Environmental Reform: The Ecological Modernization of the Global Economy.* Cambridge, MA: MIT Press.

Moore, J. W. 2003. The Modern World-System as Environmental History? Ecology and the Rise of Capitalism. *Theory and Society,* 32, 307–377.

Mosa, A., & Duffin, J. 2017. The Interwoven History of Mercury Poisoning in Ontario and Japan. *CMAJ: Canadian Medical Association Journal,* 189, E213–E215.

Mukherjee, A. B., Zevenhoven, R., Bhattacharya, P., Sajwan, K. S., & Kikuchi, R. 2008. Mercury Flow Via Coal and Coal Utilization By-Products: A Global Perspective. *Resources, Conservation and Recycling,* 52, 571–591.

Müller, E., Hilty, L. M., Widmer, R., Schluep, M., & Faulstich, M. 2014. Modeling Metal Stocks and Flows: A Review of Dynamic Material Flow Analysis Methods. *Environmental Science & Technology,* 48, 2102–2113.

Mulvaney, K. M., Selin, N. E., Giang, A., Muntean, M., Li, C.-T., Zhang, D., … Karplus, V. J. 2020. Mercury Benefits of Climate Policy in China: Addressing the Paris Agreement and the Minamata Convention Simultaneously. *Environmental Science & Technology,* 54, 1326–1335.

Muntean, M., Janssens-Maenhout, G., Song, S., Giang, A., Selin, N. E., Zhong, H., … Crippa, M. 2018. Evaluating EDGARv4.tox2 Speciated Mercury Emissions Ex-Post Scenarios and Their Impacts on Modelled Global and Regional Wet Deposition Patterns. *Atmospheric Environment,* 184, 56–68.

Myers, D. K. 1951. History of the Mercury Flask. *Journal of Chemical Education,* 28, 127.

Najam, A., & Selin, H. 2011. Institutions for a Green Economy. *Review of Policy Research,* 28, 451–457.

National Academy of Sciences, National Academy of Engineering & Institute of Medicine. 2005. *Facilitating Interdisciplinary Research*. Washington, DC: The National Academies Press.

National Emission Standards for Hazardous Air Pollutants. 1973. Asbestos, Beryllium, and Mercury. 38 *Federal Register* 8819 (April 6).

Natural Resources Defense Council. n.d. The Facts about Light Bulbs and Mercury. https://web.archive.org/web/20190915231748/https://www.nrdc.org/sites/default /files/lightbulbmercury.pdf

Naylor, D. 2002. Mercuric Oxide Batteries. *In:* Linden, D., & Reddy, T. B. (eds.), *Handbook of Batteries, Third Edition*, 11.1–11.5. New York: McGraw-Hill.

Nemet, G. F., Holloway, T., & Meier, P. 2010. Implications of Incorporating Air-Quality Co-Benefits into Climate Change Policymaking. *Environmental Research Letters*, 5, 014007.

Neumayer, E. 2003. *Weak versus Strong Sustainability: Exploring the Limits of Two Opposing Paradigms*. Cheltenham, UK: Edward Elgar Publishing.

New England Governors and Eastern Canadian Premiers. 2011. Resolution 35-6: Resolution Regarding Efforts to Reduce Mercury Pollution. Halifax, Nova Scotia. July 10–12, 2011. https://web.archive.org/web/20190202181758/https://www.cap-cpma .ca/images/CAP/35-6%20Mercury1.pdf

Newby, C. A., Riley, D. M., & Leal-Almeraz, T. O. 2006. Mercury Use and Exposure among Santeria Practitioners: Religious Versus Folk Practice in Northern New Jersey, USA. *Ethnicity and Health*, 11, 287–306.

Nordhaus, T., Shellenberger, M., & Blomqvist, L. 2012. *The Planetary Boundaries Hypothesis: A Review of the Evidence*. Breakthrough Institute, Oakland, CA. https://web .archive.org/web/20191229143425/https://s3.us-east-2.amazonaws.com/uploads .thebreakthrough.org/legacy/blog/Planetary%20Boundaries%20web.pdf

Northeast Waste Management Officials Association. 2010. IMERC Fact Sheet: Mercury Use in Batteries. Retrieved February 3, 2019, from https://web.archive.org/web /20190203170710/http://www.newmoa.org/prevention/mercury/imerc/factsheets /batteries.cfm

Novick, S. 1969. A New Pollution Problem. *Environment*, 11, 2–9.

Nriagu, J. O. 1979. Production and Uses of Mercury. *In:* Nriagu, J. O. (ed.), *Biogeochemistry of Mercury in the Environment*, 23–40. Amsterdam: Elsevier/North-Holland Biomedical Press.

Nriagu, J. O. 1993. Legacy of Mercury Pollution. *Nature*, 363, 589–589.

Nriagu, J. O. 1994. Mercury Pollution from the Past Mining of Gold and Silver in the Americas. *Science of the Total Environment*, 149, 167–181.

Nyame, F. K., & Grant, J. A. 2012. From Carats to Karats: Explaining the Shift from Diamond to Gold Mining by Artisanal Miners in Ghana. *Journal of Cleaner Production*, 29, 163–172.

O'Brien, T. F., Bommaraju, T. V., & Hine, F. 2005. History of the Chlor-Alkali Industry. In: O'Brien, T. F., Bommaraju, T. V., & Hine, F. (eds.), *Handbook of Chlor-Alkali Technology*, 17–36. Boston: Springer.

Obrist, D., Kirk, J. L., Zhang, L., Sunderland, E. M., Jiskra, M., & Selin, N. E. 2018. A Review of Global Environmental Mercury Processes in Response to Human and Natural Perturbations: Changes of Emissions, Climate, and Land Use. *Ambio*, 47, 116–140.

Odén, S. 1967. Nederbördens Försurning. *Dagens Nyheter*, October 24.

Odén, S. 1976. The Acidity Problem—an Outline of Concepts. *Water, Air, and Soil Pollution*, 6, 137–166.

Oken, E., Kleinman, K. P., Berland, W. E., Simon, S. R., Rich-Edwards, J. W., & Gillman, M. W. 2003. Decline in Fish Consumption among Pregnant Women after a National Mercury Advisory. *Obstetrics & Gynecology*, 102, 346–351.

Olivetti, E. A., & Cullen, J. M. 2018. Toward a Sustainable Materials System. *Science*, 360, 1396–1398.

Olsson, P., Galaz, V., & Boonstra, W. 2014. Sustainability Transformations: A Resilience Perspective. *Ecology and Society*, 19, 1.

Organisation for Economic Co-operation and Development. 2016. *OECD Due Diligence Guidance for Responsible Supply Chains of Minerals from Conflict-Affected and High-Risk Areas: Third Edition*. OECD Publishing, Paris.

Ostrom, E. 2005. *Understanding Institutional Diversity*. Princeton, NJ: Princeton University Press.

Ostrom, E. 2007. A Diagnostic Approach for Going Beyond Panaceas. *Proceedings of the National Academy of Sciences*, 104, 15181–15187.

Ostrom, E. 2009. A General Framework for Analyzing Sustainability of Social-Ecological Systems. *Science*, 325, 419–422.

Ostrom, E. 2010. Polycentric Systems for Coping with Collective Action and Global Environmental Change. *Global Environmental Change*, 20, 550–557.

Ostrom, E. 2011. Background on the Institutional Analysis and Development Framework. *Policy Studies Journal*, 39, 7–27.

Othmer, D., Kon, K., & Igarashi, T. 1956. Acetaldehyde by the Chisso Process. *Industrial & Engineering Chemistry*, 48, 1258–1262.

Outridge, P. M., Mason, R. P., Wang, F., Guerrero, S., & Heimbürger-Boavida, L. 2018. Updated Global and Oceanic Mercury Budgets for the United Nations Global Mercury Assessment 2018. *Environmental Science & Technology,* 52, 11466–11477.

Overton, P. 2016. Mercury Findings Prompt State to Widen Lobster Fishing Ban in Penobscot River Estuary. *Portland Press Herald,* June 22. https://web.archive.org/web /20190203160706/https://www.pressherald.com/2016/06/22/state-widens-lobster -fishing-ban-in-the-penobscot-river-estuary/

Panasonic. 2018. Which Products Have No Mercury Added? Retrieved February 3, 2019, from https://web.archive.org/web/20190203171134/https://www.panasonic -batteries.com/en/faq/which-products-have-no-mercury-added

Parascandola, J. 2009. From Mercury to Miracle Drugs: Syphilis Therapy over the Centuries. *Pharmacy in History,* 51, 14–23.

Parker, K. R. (ed.) 1997. *Applied Electrostatic Precipitation.* London: Chapman and Hall.

Parris, T. M., & Kates, R. W. 2003. Characterizing and Measuring Sustainable Development. *Annual Review of Environment and Resources,* 28, 559–586.

Patel, K., Rogan, J., Cuba, N., & Bebbington, A. 2016. Evaluating Conflict Surrounding Mineral Extraction in Ghana: Assessing the Spatial Interactions of Large and Small-Scale Mining. *The Extractive Industries and Society,* 3, 450–463.

Patterson, J., Schulz, K., Vervoort, J., van der Hel, S., Widerberg, O., Adler, C., ... Barau, A. 2017. Exploring the Governance and Politics of Transformations Towards Sustainability. *Environmental Innovation and Societal Transitions,* 24, 1–16.

Pelling, M. 2010. *Adaptation to Climate Change: From Resilience to Transformation.* New York: Routledge.

Perkin, F. M. 1911. Mercury Vapour Lamps and Action of Ultra Violet Rays. *Transactions of the Faraday Society,* 6, 199–204.

Perks, R. 2013. The Inclusion of Artisanal and Small-Scale Mining in National Legislation: Case Studies from Sub-Saharan Africa. *Rocky Mountain Mineral Law Institute,* 19.

Pliny the Elder. n.d. *The Natural History.* London: Taylor and Francis.

Polasky, S., Bryant, B., Hawthorne, P., Johnson, J., Keeler, B., & Pennington, D. 2015. Inclusive Wealth as a Metric of Sustainable Development. *Annual Review of Environment and Resources,* 40, 445–466.

Pollack, A. 1984. Battery Pollution Worries Japanese. *New York Times,* June 25.

Pradhan, P., Costa, L., Rybski, D., Lucht, W., & Kropp, J. P. 2017. A Systematic Study of Sustainable Development Goal (SDG) Interactions. *Earth's Future,* 5, 1169–1179.

Prakash, M., & Johnny, J. C. 2015. Things You Don't Learn in Medical School: Caduceus. *Journal of Pharmacy & Bioallied Sciences*, 7, S49–S50.

Prestbo, E. M., & Gay, D. A. 2009. Wet Deposition of Mercury in the US and Canada, 1996–2005: Results and Analysis of the NADP Mercury Deposition Network (MDN). *Atmospheric Environment*, 43, 4223–4233.

Princen, T. 2005. *The Logic of Sufficiency*. Cambridge, MA: MIT Press.

Rabinowitch, I. 1934. Mercurial Poisoning. *Canadian Medical Association Journal*, 30, 386–393.

Rafaj, P., Cofala, J., Kuenen, J., Wyrwa, A., & Zyśk, J. 2014. Benefits of European Climate Policies for Mercury Air Pollution. *Atmosphere*, 5, 45–59.

Rasmussen, P. 1998. Long-Range Atmospheric Transport of Trace Metals: The Need for Geoscience Perspectives. *Environmental Geology*, 33, 96–108.

Rasmussen, S. C. 2012. *How Glass Changed the World: The History and Chemistry of Glass from Antiquity to the 13th Century*. New York: Springer Science & Business Media.

Redacción Gestión. 2019. Minería Ilegal: Bienes Incautados En La Pampa Ascienden a S/ 54 Millones. *Gestión*, March 6. https://web.archive.org/web/20190725090928 /https://gestion.pe/economia/mineria-ilegal-bienes-incautados-pampa-ascienden-s -54-millones-260549

Rees, D., Murray, J., Nelson, G., & Sonnenberg, P. 2010. Oscillating Migration and the Epidemics of Silicosis, Tuberculosis, and HIV Infection in South African Gold Miners. *American Journal of Industrial Medicine*, 53, 398–404.

Reid, W. V., Chen, D., Goldfarb, L., Hackmann, H., Lee, Y. T., Mokhele, K., ... Whyte, A. 2010. Earth System Science for Global Sustainability: Grand Challenges. *Environment and Development*, 330, 916–917.

Repko, A. F. 2008. *Interdisciplinary Research: Process and Theory*. New York: Sage.

Republic of the Philippines. 1999. Republic Act No. 8749.

Rhoten, D., & Parker, A. 2004. Risks and Rewards of an Interdisciplinary Research Path. *Science*, 306, 2046.

Rice, K. M., Walker Jr, E. M., Wu, M., Gillette, C., & Blough, E. R. 2014. Environmental Mercury and Its Toxic Effects. *Journal of Preventive Medicine and Public Health*, 47, 74–83.

Ritchie, H., & Roser, M. 2019. Energy Production & Changing Energy Sources. Retrieved July 21, 2019, from https://web.archive.org/web/20190721085733/https:// ourworldindata.org/energy-production-and-changing-energy-sources

Robins, N. A. 2011. *Mercury, Mining, and Empire: The Human and Ecological Cost of Colonial Silver Mining in the Andes*. Bloomington: Indiana University Press.

Rochabrún, M. 2018. In Peru Jungle, Francis Offers a Stirring Defense of Indigenous Peoples. *New York Times*, January 19.

Rochlin, J. 2018. Informal Gold Miners, Security, and Development in Colombia: Charting the Way Forward. *The Extractive Industries and Society*, 5, 330–339.

Rockström, J. 2009. Safe Operating Space for Humanity. *Nature*, 461, 472–476.

Rockström, J., Steffen, W., Noone, K., Persson, A., Chapin III, F. S., Lambin, E., … Walker, B. 2009. Planetary Boundaries: Exploring the Safe Operating Space for Humanity. *Ecology and Society*, 14, 32.

Roguin, A. 2005. Scipione Riva-Rocci and the Men Behind the Mercury Sphygmomanometer. *The International Journal of Clinical Practice*, 60, 73–79.

Roman, H. A., Walsh, T. L., Coull, B. A., Dewailly, É., Guallar, E., Hattis, D., … Virtanen, J. K. 2011. Evaluation of the Cardiovascular Effects of Methylmercury Exposures: Current Evidence Supports Development of a Dose–Response Function for Regulatory Benefits Analysis. *Environmental Health Perspectives*, 119, 607–614.

Roser, M. 2019. Economic Growth. Retrieved July 21, 2019, from https://web.archive .org/web/20190721090004/https://ourworldindata.org/economic-growth

Ross, A. M., Rhodes, D. H., & Hastings, D. E. 2007. Defining System Changeability: Reconciling Flexibility, Adaptability, Scalability, and Robustness for Maintaining System Lifecycle Value. *17th INCOSE International Symposium*. San Diego, CA.

Rothenberg, S. E., Windham-Myers, L., & Creswell, J. E. 2014. Rice Methylmercury Exposure and Mitigation: A Comprehensive Review. *Environmental Research*, 133, 407–423.

Rowlatt, J. 2013. Mercury: A Beautiful but Poisonous Metal. *BBC News*, November 30. https://web.archive.org/web/20190203143342/https://www.bbc.com/news/magazine -25130770

Roylance, F. 2011. Say Goodbye to Mercury Thermometers. *The Seattle Times*, February 27. https://web.archive.org/web/20190915230842/https://www.seattletimes.com /seattle-news/health/say-goodbye-to-mercury-thermometers/

Ruben, S. 1945. Alkaline Dry Cell. United States Patent Application. Patented June 1947.

Ruben, S. 1947. Balanced Alkaline Dry Cells. *Transactions of the Electrochemical Society*, 92, 183–193.

Rustam, H., & Hamdi, T. 1974. Methyl Mercury Poisoning in Iraq: A Neurological Study. *Brain*, 97, 499–510.

Sage, A. P., & Rouse, W. B. 2009. *Handbook of Systems Engineering and Management*. Hoboken, NJ: John Wiley & Sons.

Schartup, A. T., Thackray, C. P., Qureshi, A., Dassuncao, C., Gillespie, K., Hanke, A., & Sunderland, E. M. 2019. Climate Change and Overfishing Increase Neurotoxicant in Marine Predators. *Nature,* 572, 648–650.

Scheuhammer, A. M., Meyer, M. W., Sandheinrich, M. B., & Murray, M. W. 2007. Effects of Environmental Methylmercury on the Health of Wild Birds, Mammals, and Fish. *Ambio,* 36, 12–18.

Schipper, E. L. F. 2006. Conceptual History of Adaptation in the UNFCCC Process. *Review of European Community & International Environmental Law,* 15, 82–92.

Schlesinger, W. H., Klein, E. M., & Vengosh, A. 2017. Global Biogeochemical Cycle of Vanadium. *Proceedings of the National Academy of Sciences,* 114, E11092–E11100.

Schlüter, M., McAllister, R., Arlinghaus, R., Bunnefeld, N., Eisenack, K., Hoelker, F., ... Quaas, M. 2012. New Horizons for Managing the Environment: A Review of Coupled Social-Ecological Systems Modeling. *Natural Resource Modeling,* 25, 219–272.

Schmalensee, R., & Stavins, R. N. 2013. The SO_2 Allowance Trading System: The Ironic History of a Grand Policy Experiment. *Journal of Economic Perspectives,* 27, 103–122.

Schuetze, C. F. 2019. German Man Who Poisoned Colleagues' Sandwiches Gets Life in Prison. *New York Times,* March 8.

Schuster, P. F., Schaefer, K. M., Aiken, G. R., Antweiler, R. C., Dewild, J. F., Gryziec, J. D., ... Krabbenhoft, D. P. 2018. Permafrost Stores a Globally Significant Amount of Mercury. *Geophysical Research Letters,* 45, 1463–1471.

Science for Environment Policy. 2017. *Tackling Mercury Pollution in the EU and Worldwide.* In-depth Report 15. Produced for the European Commission, DG Environment by the Science Communication Unit, UWE, Bristol. https://web.archive.org/web/20191229143706/https://ec.europa.eu/environment/chemicals/mercury/pdf/tackling_mercury_pollution_EU_and_worldwide_IR15_en.pdf

Science News Staff. 1997. Mercury Poisoning Kills Lab Chemist. *Science,* June 11.

Seigneur, C., Vijayaraghavan, K., Lohman, K., Karamchandani, P., & Scott, C. 2004. Global Source Attribution for Mercury Deposition in the United States. *Environmental Science & Technology,* 38, 555–569.

Selin, H. 2010. *Global Governance of Hazardous Chemicals: Challenges of Multilevel Management.* Cambridge, MA: MIT Press.

Selin, H. 2014. Global Environmental Law and Treaty-Making on Hazardous Substances: The Minamata Convention and Mercury Abatement. *Global Environmental Politics,* 14, 1–19.

Selin, H., Keane, S. E., Wang, S., Selin, N. E., Davis, K., & Bally, D. 2018. Linking Science and Policy to Support the Implementation of the Minamata Convention on Mercury. *Ambio,* 47, 198–215.

Selin, H., & VanDeveer, S. D. 2005. Canadian-US Environmental Cooperation: Climate Change Networks and Regional Action. *The American Review of Canadian Studies*, 35, 353–378.

Selin, H., & VanDeveer, S. D. 2006. Raising Global Standards: Hazardous Substances and E-Waste Management in the European Union. *Environment: Science and Policy for Sustainable Development*, 48, 6–18.

Selin, N. E. 2009. Global Biogeochemical Cycling of Mercury: A Review. *Annual Review of Environment and Resources*, 34, 43–63.

Selin, N. E. 2014. Global Change and Mercury Cycling: Challenges for Implementing a Global Mercury Treaty. *Environmental Toxicology and Chemistry*, 33, 1202–1210.

Selin, N. E. 2018. A Proposed Global Metric to Aid Mercury Pollution Policy. *Science*, 360, 607–609.

Selin, N. E., & Friedman, C. L. 2012. The Earth as an Engineering System: Addressing Sustainability through Science, Technology and Policy. Massachusetts Institute of Technology. ESD Working Paper Series ESD-WP-2012–06. http://hdl.handle.net/1721 .1/102921

Selin, N. E., & Jacob, D. J. 2008. Seasonal and Spatial Patterns of Mercury Wet Deposition in the United States: Constraints on the Contribution from North American Anthropogenic Sources. *Atmospheric Environment*, 42, 5193–5204.

Selin, N. E., Jacob, D. J., Yantosca, R. M., Strode, S., Jaeglé, L., & Sunderland, E. M. 2008. Global 3-D Land-Ocean-Atmosphere Model for Mercury: Present-Day Versus Preindustrial Cycles and Anthropogenic Enrichment Factors for Deposition. *Global Biogeochemical Cycles*, 22, GB2011.

Selin, N. E., & Selin, H. 2006. Global Politics of Mercury Pollution: The Need for Multi-Scale Governance. *Review of European Community & International Environmental Law*, 15, 258–269.

Sen, I. S., & Peucker-Ehrenbrink, B. 2012. Anthropogenic Disturbance of Element Cycles at the Earth's Surface. *Environmental Science & Technology*, 46, 8601–8609.

Severyanov, M. D., & Anisimova, L. U. 2013. Abortion as a Means of Family Planning in Russia in the First Quarter of the Twentieth Century. *Journal of Siberian Federal University. Humanities & Social Sciences*, 6, 1066–1074.

Shaffer, C. W. 1920. The Problem of Administering Venereal Prophylaxis in Cities Adjacent to Army Camps. *The Military Surgeon*, 47, 574–579.

Shaman, J., Solomon, S., Colwell, R. R., & Field, C. B. 2013. Fostering Advances in Interdisciplinary Climate Science. *Proceedings of the National Academy of Sciences*, 110, 3653–3656.

Sheehan, M. C., Burke, T. A., Navas-Acien, A., Breysse, P. N., McGready, J., & Fox, M. A. 2014. Global Methylmercury Exposure from Seafood Consumption and Risk of Developmental Neurotoxicity: A Systematic Review. *Bulletin of the World Health Organization*, 92, 254–269F.

Sherman, L. S., Blum, J. D., Dvonch, J. T., Gratz, L. E., & Landis, M. S. 2015. The Use of Pb, Sr, and Hg Isotopes in Great Lakes Precipitation as a Tool for Pollution Source Attribution. *Science of the Total Environment*, 502, 362–374.

Sicherman, B. 1984. *Alice Hamilton: A Life in Letters*. Cambridge, MA: Harvard University Press.

Siegel, S., & Veiga, M. M. 2010. The Myth of Alternative Livelihoods: Artisanal Mining, Gold and Poverty. *International Journal of Environment and Pollution*, 41, 272–288.

Silbernagel, S. M., Carpenter, D. O., Gilbert, S. G., Gochfeld, M., Groth, E., Hightower, J. M., & Schiavone, F. M. 2011. Recognizing and Preventing Overexposure to Methylmercury from Fish and Seafood Consumption: Information for Physicians. *Journal of Toxicology*, 2011, 983072.

Sippl, K. 2015. Private and Civil Society Governors of Mercury Pollution from Artisanal and Small-Scale Gold Mining: A Network Analytic Approach. *The Extractive Industries and Society*, 2, 198–208.

Sippl, K. 2016. *Civil Society Governance Decisions: Certification Organization Response to Artisanal and Small-Scale Gold Mining*. PhD dissertation, Boston University.

Sippl, K. 2020. Southern Responses to Fair Trade Gold: Cooperation, Complaint, Competition, Supplementation. *Ecological Economics*, 169, 106377.

Sippl, K., & Selin, H. 2012. Global Policy for Local Livelihoods: Phasing out Mercury in Artisanal and Small-Scale Gold Mining. *Environment: Science and Policy for Sustainable Development*, 54, 18–29.

Slemr, F., Brunke, E. G., Ebinghaus, R., Temme, C., Munthe, J., Wängberg, I., … Berg, T. 2003. Worldwide Trend of Atmospheric Mercury since 1977. *Geophysical Research Letters*, 30, 1516.

Slemr, F., & Scheel, H. E. 1998. Trends in Atmospheric Mercury Concentrations at the Summit of the Wank Mountain, Southern Germany. *Atmospheric Environment*, 32, 845–853.

Sloss, L. 2012. *Legislation, Standards and Methods for Mercury Emissions Control*. IEA Clean Coal Centre. CCC/195. https://www.usea.org/sites/default/files/042012_Legislation%2C%20standards%20and%20methods%20for%20mercury%20emission%20control_ccc195.pdf

Sloss, L. L. 2008. *Economics of Mercury Control*. IEA Clean Coal Centre. CCC/134. https://wedocs.unep.org/bitstream/handle/20.500.11822/11681/Economics_of_Hg_control_in_Coal_power_plants.pdf

Smart, N. A. 1968. Use and Residues of Mercury Compounds in Agriculture. *In:* Gunther, F. A. (ed.), *Residue Reviews/Rückstands-Berichte*, 1–36. New York: Springer.

Smit, B., & Wandel, J. 2006. Adaptation, Adaptive Capacity and Vulnerability. *Global Environmental Change*, 16, 282–292.

Smith, C. M., & Trip, L. J. 2005. Mercury Policy and Science in Northeastern North America: The Mercury Action Plan of the New England Governors and Eastern Canadian Premiers. *Ecotoxicology*, 14, 19–35.

Smithsonian National Museum of American History. 2017. Lighting a Revolution. Retrieved September 15, 2019, from https://web.archive.org/web/20190915145714/https://americanhistory.si.edu/lighting/20thcent/invent20.htm

Snowden, F. 2006. *The Conquest of Malaria: Italy, 1900–1962*. New Haven, CT: Yale University Press.

Snyder, L. D., Miller, N. H., & Stavins, R. N. 2003. The Effects of Environmental Regulation on Technology Diffusion: The Case of Chlorine Manufacturing. *American Economic Review*, 93, 431–435.

Soerensen, A. L., Jacob, D. J., Streets, D. G., Witt, M. L., Ebinghaus, R., Mason, R. P., ... Sunderland, E. M. 2012. Multi-Decadal Decline of Mercury in the North Atlantic Atmosphere Explained by Changing Subsurface Seawater Concentrations. *Geophysical Research Letters*, 39, L21810.

Sokol, J. 2017. Something in the Water: Life after Mercury Poisoning. *Mosaic*, September 26. https://web.archive.org/web/20190901152850/https://mosaicscience.com/story/mercury-poisoning-minamata-disaster-environment/

Sousa, R. N., & Veiga, M. M. 2009. Using Performance Indicators to Evaluate an Environmental Education Program in Artisanal Gold Mining Communities in the Brazilian Amazon. *Ambio*, 38, 40–46.

Spaargaren, G., & Mol, A. P. 1992. Sociology, Environment, and Modernity: Ecological Modernization as a Theory of Social Change. *Society & Natural Resources*, 5, 323–344.

Spangenberg, J. H. 2011. Sustainability Science: A Review, an Analysis and Some Empirical Lessons. *Environmental Conservation*, 38, 275–287.

Speth, J. G. 2008. *The Bridge at the Edge of the World: Capitalism, the Environment, and Crossing from Crisis to Sustainability*. New Haven, CT: Yale University Press.

Spiegel, S., Keane, S., Metcalf, S., & Veiga, M. 2015. Implications of the Minamata Convention on Mercury for Informal Gold Mining in Sub-Saharan Africa: From

Global Policy Debates to Grassroots Implementation? *Environment, Development and Sustainability*, 17, 765–785.

Spiegel, S. J. 2009. Socioeconomic Dimensions of Mercury Pollution Abatement: Engaging Artisanal Mining Communities in Sub-Saharan Africa. *Ecological Economics*, 68, 3072–3083.

Spiegel, S. J. 2015. Shifting Formalization Policies and Recentralizing Power: The Case of Zimbabwe's Artisanal Gold Mining Sector. *Society & Natural Resources*, 28, 543–558.

Spiegel, S. J., Agrawal, S., Mikha, D., Vitamerry, K., Le Billon, P., Veiga, M., ... Paul, B. 2018. Phasing Out Mercury? Ecological Economics and Indonesia's Small-Scale Gold Mining Sector. *Ecological Economics*, 144, 1–11.

Spiegel, S. J., & Veiga, M. M. 2010. International Guidelines on Mercury Management in Small-Scale Gold Mining. *Journal of Cleaner Production*, 18, 375–385.

Sprovieri, F., Pirrone, N., Bencardino, M., D'Amore, F., Carbone, F., Cinnirella, S., ... Weigelt, A. 2016. Atmospheric Mercury Concentrations Observed at Ground-Based Monitoring Sites Globally Distributed in the Framework of the GMOS Network. *Atmospheric Chemistry and Physics*, 16, 11915–11935.

Srivastava, R. K., Hutson, N., Martin, B., Princiotta, F., & Staudt, J. 2006. Control of Mercury Emissions from Coal-Fired Electric Utility Boilers. *Environmental Science & Technology*, 40, 1385–1393.

Steffen, W., Broadgate, W., Deutsch, L., Gaffney, O., & Ludwig, C. 2015a. The Trajectory of the Anthropocene: The Great Acceleration. *The Anthropocene Review*, 2, 81–98.

Steffen, W., Crutzen, P. J., & McNeill, J. R. 2007. The Anthropocene: Are Humans Now Overwhelming the Great Forces of Nature? *Ambio*, 36, 614–621.

Steffen, W., Richardson, K., Rockström, J., Cornell, S. E., Fetzer, I., Bennett, E. M., ... Sörlin, S. 2015b. Planetary Boundaries: Guiding Human Development on a Changing Planet. *Science*, 347, 1259855.

Stegemann, J. 2019. Giftiger Als Kampfstoffe Im Ersten Weltkrieg. *Süddeutsche Zeitung*, March 7.

Stephens, L., Fuller, D., Boivin, N., Rick, T., Gauthier, N., Kay, A., ... Barton, C. M. J. S. 2019. Archaeological Assessment Reveals Earth's Early Transformation through Land Use. *Science*, 365, 897–902.

Sterman, J. D. 2011. Sustaining Sustainability: Creating a Systems Science in a Fragmented Academy and Polarized World. *In:* Weinstein, M. P., & Turner, R. E. (eds.), *Sustainability Science: The Emerging Paradigm and the Urban Environment*, 21–58. New York: Springer.

Stewart, S. 2014. "Gleaming and Deadly White": Toxic Cosmetics in the Roman World. *In:* Wexler, P. (ed.), *History of Toxicology and Environmental Health: Toxicology in Antiquity II*, 301–311. London: Elsevier.

Stock, A. 1926. Die Gefaehrlichkeit Des Quecksilberdampfes (the Dangerousness of Mercury Vapor). *Zeitschrift fuer angewandte Chemie, 29*, 461–466.

Stocklin-Weinberg, R., Veiga, M. M., & Marshall, B. G. 2019. Training Artisanal Miners: A Proposed Framework with Performance Evaluation Indicators. *Science of the Total Environment, 660*, 1533–1541.

Storrow, B. 2017. Big, Young Power Plants Are Closing. Is It a New Trend? *E&E News*, April 27.

Streets, D. G., Horowitz, H. M., Jacob, D. J., Lu, Z., Levin, L., ter Schure, A. F., & Sunderland, E. M. 2017. Total Mercury Released to the Environment by Human Activities. *Environmental Science & Technology, 51*, 5969–5977.

Streets, D. G., Lu, Z., Levin, L., Ter Schure, A. F. H., & Sunderland, E. M. 2018. Historical Releases of Mercury to Air, Land, and Water from Coal Combustion. *Science of the Total Environment, 615*, 131–140.

Streets, D. G., Zhang, Q., & Wu, Y. 2009. Projections of Global Mercury Emissions in 2050. *Environmental Science & Technology, 43*, 2983–2988.

Strynar, M., Dagnino, S., McMahen, R., Liang, S., Lindstrom, A., Andersen, E., ... Ball, C. 2015. Identification of Novel Perfluoroalkyl Ether Carboxylic Acids (PFECAs) and Sulfonic Acids (PFESAs) in Natural Waters Using Accurate Mass Time-of-Flight Mass Spectrometry (TOFMS). *Environmental Science & Technology, 49*, 11622–11630.

Subramanian, M. 2019. Anthropocene Now: Influential Panel Votes to Recognize Earth's New Epoch. *Nature,* May 21.

Sun, Y. 2017. Transnational Public-Private Partnerships as Learning Facilitators: Global Governance of Mercury. *Global Environmental Politics, 17*, 21–44.

Sunderland, E. M., Driscoll Jr, C. T., Hammitt, J. K., Grandjean, P., Evans, J. S., Blum, J. D., ... Mason, R. P. 2016. Benefits of Regulating Hazardous Air Pollutants from Coal and Oil-Fired Utilities in the United States. *Environmental Science & Technology, 50*, 2117–2120.

Sunderland, E. M., Li, M., & Bullard, K. 2018. Decadal Changes in the Edible Supply of Seafood and Methylmercury Exposure in the United States. *Environmental Health Perspectives, 126*, 017006.

Svidén, J., & Jonsson, A. 2001. Urban Metabolism of Mercury Turnover, Emissions and Stock in Stockholm 1795–1995. *Water, Air, and Soil Pollution: Focus, 1*, 179–196.

Swain, E. B., Jakus, P. M., Rice, G., Lupi, F., Maxson, P. A., Pacyna, J. M., ... Veiga, M. M. 2007. Socioeconomic Consequences of Mercury Use and Pollution. *Ambio, 36*, 45–61.

Sweden National Food Agency. 2019. Kvicksilver. Retrieved February 3, 2019, from https://web.archive.org/web/20190203161035/https://www.livsmedelsverket.se /livsmedel-och-innehall/oonskade-amnen/metaller1/kvicksilver

Swedish Chemicals Inspectorate. 2004. *Kvicksilver—Utredning om ett Generellt Nationellt Förbud.* Rapport. 2/04. https://web.archive.org/web/20190915231031/https://www.kemi .se/global/rapporter/2004/rapport-2-04.pdf

Swedish Chemicals Inspectorate. 2017. Undantag för Användning av Amalgam och Sömsvets med Kvicksilver Upphör. https://web.archive.org/web/20190203160514 /https://www.kemi.se/nyheter-fran-kemikalieinspektionen/2017/undantag-for -anvandning-av-amalgam-och-somsvets-med-kvicksilver-upphor/

Swedish Expert Group. 1971. Methyl Mercury in Fish: A Toxicologic-Epidemiologic Evaluation of Risks. Report from an Expert Group. *Nordisk Hygienisk Tidskrift,* 4, 155–362.

Sykes, L. K., Geier, D. A., King, P. G., Kern, J. K., Haley, B. E., Chaigneau, C. G., ... Geier, M. R. 2014. Thimerosal as Discrimination: Vaccine Disparity in the UN Minamata Convention on Mercury. *Indian Journal of Medical Ethics,* 11, 206–218.

Sznopek, J. L., & Goonan, T. G. 2000. *The Materials Flow of Mercury in the Economies of the United States and the World.* U.S. Geological Survey, U.S. Geological Survey Circular 1197, Denver, CO. https://pubs.usgs.gov/circ/2000/c1197/c1197.pdf

Takaoka, S., Fujino, T., Hotta, N., Ueda, K., Hanada, M., Tajiri, M., & Inoue, Y. 2014. Signs and Symptoms of Methylmercury Contamination in a First Nations Community in Northwestern Ontario, Canada. *Science of the Total Environment,* 468, 950–957.

Takeuchi, T., Frank, M., Fischer, P. V., Annett, C. S., & Okabe, M. 1977. The Outbreak of Minamata Disease (Methyl Mercury Poisoning) in Cats on Northwestern Ontario Reserves. *Environmental Research,* 13, 215–228.

Takeuchi, T., Kambara, T., Morikawa, N., Matsumoto, H., Shiraishi, Y., & Ito, H. 1959. Pathologic Observations of the Minamata Disease. *Pathology International,* 9, 769–783.

Tang, Y., Wang, S., Wu, Q., Liu, K., Wang, L., Li, S., ... Hao, J. 2018. Recent Decrease Trend of Atmospheric Mercury Concentrations in East China: The Influence of Anthropogenic Emissions. *Atmospheric Chemistry and Physics,* 18, 8279–8291.

Tavernise, S. 2012. Vaccine Rule Is Said to Hurt Health Efforts. *New York Times,* December 17.

Tegel, S. 2016. Peru Declares State of Emergency in Its Jungles Due to Rampant Mercury Poisoning. *Vice News,* May 24. https://web.archive.org/web/20190203193359 /https://news.vice.com/en_us/article/8x37eg/peru-declares-state-of-emergency-in-its -jungles-due-to-rampant-mercury-poisoning

Teleky, L., & Kober, G. M. 1916. Processes Involving Exposure to Mercurial Poisoning. *In:* Kober, G. M., & Hanson, W. C. (eds.), *Diseases of Occupational and Vocational Hygiene,* 521–534. Philadelphia: P. Blakiston's Son & Co.

Telmer, K. H., & Veiga, M. M. 2009. World Emissions of Mercury from Artisanal and Small Scale Gold Mining. *In:* N. Pirrone and R. Mason (eds.), *Mercury Fate and Transport in the Global Atmosphere*, 131–172. New York: Springer.

Tepper, L. B. 2010. Industrial Mercurialism: Agricola to the Danbury Shakes. *IA: The Journal of the Society for Industrial Archaeology*, 36, 47–63.

Tone, A. 2002. *Devices and Desires: A History of Contraceptives in America*. New York: Macmillan.

Tørseth, K., Aas, W., Breivik, K., Fjæraa, A. M., Fiebig, M., Hjellbrekke, A. G., … Yttri, K. E. 2012. Introduction to the European Monitoring and Evaluation Programme (EMEP) and Observed Atmospheric Composition Change During 1972–2009. *Atmospheric Chemistry and Physics*, 12, 5447–5481.

Trasande, L., Landrigan, P. J., & Schechter, C. 2005. Public Health and Economic Consequences of Methyl Mercury Toxicity to the Developing Brain. *Environmental Health Perspectives*, 113, 590–596.

Travnikov, O. 2005. Contribution of the Intercontinental Atmospheric Transport to Mercury Pollution in the Northern Hemisphere. *Atmospheric Environment*, 39, 7541–7548.

Trotuş, I.-T., Zimmermann, T., & Schüth, F. 2013. Catalytic Reactions of Acetylene: A Feedstock for the Chemical Industry Revisited. *Chemical Reviews*, 114, 1761–1782.

Tschakert, P. 2009. Recognizing and Nurturing Artisanal Mining as a Viable Livelihood. *Resources Policy*, 34, 24–31.

Turner, B., Clark, W., Kates, R., Richards, J., Mathews, J., & Meyer, W. 1990. *The Earth as Transformed by Human Action: Global and Regional Changes in the Biosphere over the Past 300 Years*. Cambridge: Cambridge University Press.

Turnheim, B., Berkhout, F., Geels, F., Hof, A., McMeekin, A., Nykvist, B., & van Vuuren, D. 2015. Evaluating Sustainability Transitions Pathways: Bridging Analytical Approaches to Address Governance Challenges. *Global Environmental Change*, 35, 239–253.

UNDP (United Nations Development Programme). 2016. *Mercury Management for Sustainable Development*. UNDP, New York. https://web.archive.org/web/20191229144034/https://www.undp.org/content/dam/undp/library/Environment%20and%20Energy/Chemicals%20and%20Waste%20Management/Mercury%20publication_ENGLISH.pdf

UNEP (United Nations Environment Programme). 2002. *Global Mercury Assessment*. Geneva: UNEP Chemicals Branch. http://hdl.handle.net/20.500.11822/12297

UNEP (United Nations Environment Programme). 2012. *Reducing Mercury Use in Artisanal and Small-Scale Gold Mining: A Practical Guide*. UNEP, Geneva. http://hdl.handle.net/20.500.11822/11524

UNEP (United Nations Environment Programme). 2015. *Practical Sourcebook on Mercury Waste Storage and Disposal.* UNEP. http://hdl.handle.net/20.500.11822/9839

UNEP (United Nations Environment Programme). 2016. Guidance on Best Available Techniques and Best Environmental Practices to Control Mercury Emissions from Coal-Fired Power Plants and Coal-Fired Industrial Boilers. https://web.archive.org /web/20191229144243/http://mercuryconvention.org/Portals/11/documents/BAT -BEP%20draft%20guidance/Coal_burning_power_stations_and_industrial_boilers.pdf

UNEP (United Nations Environment Programme). 2017. *Global Mercury Supply, Trade and Demand.* UNEP, Geneva. https://web.archive.org/web/20191229150757 /https://wedocs.unep.org/bitstream/handle/20.500.11822/21725/global_mercury .pdf?sequence=1&isAllowed=y

UNEP (United Nations Environment Programme). 2019. *Global Mercury Assessment 2018.* UNEP Chemicals and Health Branch, Geneva. http://hdl.handle.net/20.500 .11822/27579

UNEP & AMAP (United Nations Environment Programme and Arctic Monitoring and Assessment Programme). 2015. *Global Mercury Modelling: Update of Modelling Results in the Global Mercury Assessment 2013.* AMAP and UNEP, Oslo, Norway, and Geneva, Switzerland. http://hdl.handle.net/20.500.11822/13772

UNEP/FAO (United Nations Environment Programme and Food and Agriculture Organization). 1996. Decision Guidance Documents: Mercury Compounds. Joint FAO/UNEP Programme for the Operation of Prior Informed Consent. Operation of the Prior Informed Consent Procedure for Banned or Severely Restricted Chemicals in International Trade. Rome, Italy, and Geneva, Switzerland.

United Nations. 1972. *Small-Scale Mining in the Developing Countries.* Social Affairs Resources Transport Division, New York.

United Press International. 1976. A Ban on Most Pesticides with Mercury Is Ordered. *New York Times*, February 19.

US Agency for International Development. 2018. Illegal Gold Mining. Retrieved February 3, 2019, from https://web.archive.org/web/20190203193430/https://www .usaid.gov/peru/our-work/illegal-gold-mining

US Court of Appeals. 1980. United States of America v. Anderson Seafoods, Inc.: US Court of Appeals, Fifth Circuit.

US Department of the Treasury. 2018. Status Report of U.S. Government Gold Reserve. Retrieved February 3, 2019, from https://web.archive.org/web/20190203193521/https:// www.fiscal.treasury.gov/reports-statements/gold-report/current.html

US Environmental Protection Agency. 2000. Regulatory Finding on the Emissions of Hazardous Air Pollutants from Electric Utility Steam Generating Units. 65 *Federal Register* 79825 (December 20).

US Environmental Protection Agency. 2013. *Trends in Blood Mercury Concentrations and Fish Consumption among US Women of Childbearing Age*. NHANES 1999–2010. EPA-823-R-13-002.

US Environmental Protection Agency. 2014. EPA Strategy to Address Mercury-Containing Products. https://web.archive.org/web/20191229144339/https://www.epa .gov/sites/production/files/2015-10/documents/productsstrategy.pdf

US Environmental Protection Agency. 2015. Mercury Emissions. Report on the Environment. September 24. https://web.archive.org/web/20190203164844/https://cfpub .epa.gov/roe/indicator.cfm?i=14

US Environmental Protection Agency. 2020. Mercury in Consumer Products. Retrieved January 18, 2020, from https://web.archive.org/web/20200118125331/https:// www.epa.gov/mercury/mercury-consumer-products

US Food and Drug Administration. 1980. Vaginal Contraceptive Drug Products for Over-the-Counter Human Use; Establishment of a Monograph; Proposed Rulemaking. FDA Proposed Rule. 45 *Federal Register* 82015 (December 12).

US Food and Drug Administration. 1998. Status of Certain Additional Over-The-Counter Drug Category II and III Active Ingredients. FDA Final Rule. 63 *Federal Register* 19799 (April 22).

US Geological Survey. 2014. Mercury Statistics. US Geological Survey. Historical Statistics for Mineral and Material Commodities in the United States. Data Series 140. http://minerals.usgs.gov/minerals/pubs/historical-statistics

US Geological Survey. 2017. Mineral Commodity Summaries: Mercury. https:// minerals.usgs.gov/minerals/pubs/commodity/mercury/mcs-2017-mercu.pdf

US Geological Survey. 2019. Mineral Commodity Summaries: Mercury. https:// minerals.usgs.gov/minerals/pubs/commodity/mercury/mcs-2019-mercu.pdf

US National Archives. 2019. Marriage and Law Practice, 1764–1765. From the Autobiography of John Adams. *Founders Online*. https://web.archive.org/web/20191229151131 /https://founders.archives.gov/documents/Adams/01-03-02-0016-0011

US National Research Council. 2000. *Toxicological Effects of Methylmercury*. National Academies Press, Washington, DC.

Vaccine News Net. 2013. GAVI Praises Exclusion of Vaccine Preservative from Ban. Retrieved February 2, 2019, from https://web.archive.org/web/20190723083232 /http://www.vaccinews.net/2013/01/gavi-praises-exclusion-of-vaccine-preservative -from-ban/

Vandal, G. M., Fitzgerald, W., Boutron, C. F., & Candelone, J.-P. 1993. Variations in Mercury Deposition to Antarctica over the Past 34,000 Years. *Nature, 362*, 621–623.

Veiga, M. M., Angeloci, G., Ñiquen, W., & Seccatore, J. 2015. Reducing Mercury Pollution by Training Peruvian Artisanal Gold Miners. *Journal of Cleaner Production, 94,* 268–277.

Veiga, M. M., Masson, P., Perron, D., Laflamme, A.-C., Gagnon, R., Jimenez, G., & Marshall, B. G. 2018. An Affordable Solution for Micro-Miners in Colombia to Process Gold Ores without Mercury. *Journal of Cleaner Production, 205,* 995–1005.

Veiga, M. M., Nunes, D., Klein, B., Shandro, J. A., Velasquez, P. C., & Sousa, R. N. 2009. Mill Leaching: A Viable Substitute for Mercury Amalgamation in the Artisanal Gold Mining Sector? *Journal of Cleaner Production,* 17, 1373–1381.

Veland, S., Scoville-Simonds, M., Gram-Hanssen, I., Schorre, A. K., El Khoury, A., Nordbø, M. J., ... Bjørkan, M. 2018. Narrative Matters for Sustainability: The Transformative Role of Storytelling in Realizing 1.5°C Futures. *Current Opinion in Environmental Sustainability,* 31, 41–47.

Victor, D. G., & Raustiala, K. 2004. The Regime Complex for Plant Genetic Resources. *International Organization,* 32, 147–154.

Wajda, S. T. 2019. Ending the Danbury Shakes: A Story of Workers' Rights and Corporate Responsibility. Connecticut History Online. Retrieved January 18, 2020, from https://web.archive.org/web/20200118135214/https://connecticuthistory.org/ending-the-danbury-shakes-a-story-of-workers-rights-and-corporate-responsibility/

Waldron, H. A. 1983. Did the Mad Hatter Have Mercury Poisoning? *British Medical Journal (Clinical Research Edition),* 287, 1961.

Waldron, M. 1970. Mercury in Food: A Family Tragedy. *New York Times,* August 10.

Walke, J. 2011a. EPA's Mercury and Air Toxics Rule: Bottom Lines and Background. Natural Resources Defense Council. Retrieved July 25, 2019, from https://web.archive.org/web/20190725113910/https://www.nrdc.org/experts/john-walke/epas-mercury-and-air-toxics-rule-bottom-lines-and-background

Walke, J. 2011b. A Little Background on the EPA's New Mercury and Air Toxics Rule. Grist. Retrieved July 25, 2019, from https://web.archive.org/web/20190725114247/https://grist.org/climate-policy/2011-03-16-epas-mercury-and-air-toxics-rule-bottom-lines-and-background/

Walker, B., Holling, C. S., Carpenter, S. R., & Kinzig, A. 2004. Resilience, Adaptability and Transformability in Social–Ecological Systems. *Ecology and Society,* 9, 5.

Walker, K. R., Ricciardone, M. D., & Jensen, J. 2003. Developing an International Consensus on DDT: A Balance of Environmental Protection and Disease Control. *International Journal of Hygiene and Environmental Health,* 206, 423–435.

Wallington, T. J., Schneider, W. F., Worsnop, D. R., Nielsen, O. J., Sehested, J., Debruyn, W. J., & Shorter, J. A. 1994. The Environmental Impact of CFC Replacements HFCs and HCFCs. *Environmental Science & Technology*, 28, 320A–326A.

Wang, F., Outridge, P. M, Feng, X., Meng, B., Heimbürger-Boavida, L.-E., & Mason, R. P. 2019. How Closely Do Mercury Trends in Fish and Other Aquatic Wildlife Track Those in the Atmosphere?—Implications for Evaluating the Effectiveness of the Minamata Convention. *Science of the Total Environment*, 674, 58–70.

Wang, F., Wang, S., Zhang, L., Yang, H., Wu, Q., & Hao, J. 2014. Mercury Enrichment and Its Effects on Atmospheric Emissions in Cement Plants of China. *Atmospheric Environment*, 92, 421–428.

Wang, Z., DeWitt, J. C., Higgins, C. P., & Cousins, I. T. 2017. A Never-Ending Story of Per- and Polyfluoroalkyl Substances (PFASs)? *Environmental Science & Technology*, 51, 2508–2518.

Weiss, H., Bertine, K., Koide, M., & Goldberg, E. D. 1975. The Chemical Composition of a Greenland Glacier. *Geochimica et Cosmochimica Acta*, 39, 1–10.

Weiss, H. V., Koide, M., & Goldberg, E. D. 1971. Mercury in a Greenland Ice Sheet: Evidence of Recent Input by Man. *Science*, 174, 692–694.

Weiss-Penzias, P. S., Gay, D. A., Brigham, M. E., Parsons, M. T., Gustin, M. S., & ter Schure, A. 2016. Trends in Mercury Wet Deposition and Mercury Air Concentrations across the US and Canada. *Science of the Total Environment*, 568, 546–556.

Weiss-Penzias, P., Jaffe, D. A., Swartzendruber, P., Dennison, J. B., Chand, D., Hafner, W., & Prestbo, E. 2006. Observations of Asian Air Pollution in the Free Troposphere at Mount Bachelor Observatory During the Spring of 2004. *Journal of Geophysical Research: Atmospheres*, 111, D10304.

Wendroff, A. P. 2005. Environmental Review: Magico-Religious Mercury Use in Caribbean and Latino Communities: Pollution, Persistence, and Politics. *Environmental Practice*, 7, 87–96.

Werthmann, K. 2009. Working in a Boom-Town: Female Perspectives on Gold-Mining in Burkina Faso. *Resources Policy*, 34, 18–23.

Werthmann, K. 2017. The Drawbacks of Privatization: Artisanal Gold Mining in Burkina Faso 1986–2016. *Resources Policy*, 52, 418–426.

West, J. B. 2013. Torricelli and the Ocean of Air: The First Measurement of Barometric Pressure. *Physiology*, 28, 66–73.

Wexler, J. 2016. *When God Isn't Green: A World-Wide Journey to Places Where Religious Practice and Environmentalism Collide*. Boston: Beacon Press.

Wilburn, D. R. 2013. *Changing Patterns in the Use, Recycling, and Material Substitution of Mercury in the United States*. US Geological Survey Scientific Investigations Report

2013-5137. https://web.archive.org/web/20191229151513/https://pubs.usgs.gov/sir /2013/5137/pdf/sir2013-5137.pdf

Williston, S. H. 1968. Mercury in the Atmosphere. *Journal of Geophysical Research, 73,* 7051–7055.

Wilson, M. L., Renne, E., Roncoli, C., Agyei-Baffour, P., & Tenkorang, E. Y. 2015. Integrated Assessment of Artisanal and Small-Scale Gold Mining in Ghana—Part 3: Social Sciences and Economics. *International Journal of Environmental Research and Public Health,* 12, 8133–8156.

Wöhrnschimmel, H., Scheringer, M., Bogdal, C., Hung, H., Salamova, A., Venier, M., ... Fiedler, H. 2016. Ten Years after Entry into Force of the Stockholm Convention: What Do Air Monitoring Data Tell About Its Effectiveness? *Environmental Pollution, 217,* 149–158.

Wolfe, M. F., Schwarzbach, S., & Sulaiman, R. A. 1998. Effects of Mercury on Wildlife: A Comprehensive Review. *Environmental Toxicology and Chemistry,* 17, 146–160.

Woodger, E., & Toropov, B. 2004. *Encyclopedia of the Lewis and Clark Expedition.* New York: Facts on File, Inc.

World Bank. 1992. Strategy for African Mining. The World Bank, Mining Unit, Industry and Energy Division. Washington, DC: Africa Technical Department Series. World Bank Technical Paper Number 181.

World Coal Association. 2019. Where Is Coal Found? Retrieved February 3, 2019, from https://web.archive.org/web/20190203165021/https://www.worldcoal.org/coal/where -coal-found

World Commission on Environment and Development. 1987. *Our Common Future.* Oxford: Oxford University Press.

World Gold Council. 2019a. Gold Supply. Retrieved February 3, 2019, from https:// web.archive.org/web/20190203193645/https://www.gold.org/about-gold/gold -supply

World Gold Council. 2019b. Gold Supply and Demand Statistics. https://web .archive.org/web/20200112193739/https://www.gold.org/goldhub/data/gold-supply -and-demand-statistics

World Gold Council. 2019c. How Much Gold Has Been Mined? Retrieved February 3, 2019, from https://web.archive.org/web/20190203193807/https://www.gold.org /about-gold/gold-supply/gold-mining/how-much-gold

World Health Assembly. 2014. Public Health Impacts of Exposure to Mercury and Mercury Compounds: The Role of WHO and Ministries of Public Health in the Implementation of the Minamata Convention WHA67.11. http://apps.who.int/gb /ebwha/pdf_files/WHA67-REC1/A67_2014_REC1-en.pdf#page=1

World Health Organization. 2010. *Future Use of Materials for Dental Restoration.* WHO, Geneva. https://web.archive.org/web/20190203161205/https://www.who.int/oral_health/publications/dental_material_2011.pdf

World Health Organization. 2011a. Mercury in Skin Lightening Products. WHO, Geneva. https://web.archive.org/web/20190203171242/https://www.who.int/ipcs/assessment/public_health/mercury_flyer.pdf

World Health Organization. 2011b. Thiomersal—Questions and Answers. Retrieved July 18, 2019, from https://web.archive.org/web/20191229144516/https://www.who.int/immunization/newsroom/thiomersal_questions_and_answers/en/

Wozniak, R. J., Hirsch, A. E., Bush, C. R., Schmitz, S., & Wenzel, J. 2017. Mercury Spill Responses—Five States, 2012–2015. *Morbidity and Mortality Weekly Report (MMWR),* 66, 274–277.

Wright, D. C. 2001. *The History of China.* Westport, CT: Greenwood Press.

Wright, R., & Elcock, K. 2006. The RoHS and WEEE Directives: Environmental Challenges for the Electrical and Electronic Products Sector. *Environmental Quality Management,* 15, 9–24.

Wright, W. F., & Mackowiak, P. A. 2016. Origin, Evolution and Clinical Application of the Thermometer. *The American Journal of the Medical Sciences,* 351, 526–534.

Wu, Q., Wang, S., Zhang, L., Hui, M., Wang, F., & Hao, J. 2016. Flow Analysis of the Mercury Associated with Nonferrous Ore Concentrates: Implications on Mercury Emissions and Recovery in China. *Environmental Science & Technology,* 50, 1796–1803.

Wynes, S., & Nicholas, K. A. 2017. The Climate Mitigation Gap: Education and Government Recommendations Miss the Most Effective Individual Actions. *Environmental Research Letters,* 12, 074024.

Yarime, M. 2007. Promoting Green Innovation or Prolonging the Existing Technology. *Journal of Industrial Ecology,* 11, 117–139.

Yehle, E. 2011. EPA Weighs Threats Posed by Mercury Used in Religious Rituals. *New York Times/Greenwire,* May 18. https://archive.nytimes.com/www.nytimes.com/gwire/2011/05/18/18greenwire-epa-weighs-threats-posed-by-mercury-used-in-re-46372.html

Young, A., Taylor, F., & Merritt, H. 1930. The Distribution and Excretion of Mercury. *Archives of Dermatology and Syphilology,* 21, 539–551.

Young, H. H., White, E. C., & Swartz, E. O. 1919. A New Germicide for Use in the Genito-Urinary Tract: "Mercurochrome-220." *The Journal of the American Medical Association,* 73, 1483–1491.

Young, O. R. 2002. *The Institutional Dimensions of Environmental Change: Fit, Interplay, and Scale.* Cambridge, MA: MIT Press.

Young, O. R. 2011. Effectiveness of International Environmental Regimes: Existing Knowledge, Cutting-Edge Themes, and Research Strategies. *Proceedings of the National Academy of Sciences,* 108, 19853–19860.

Young, O. R. 2017. *Governing Complex Systems: Social Capital for the Anthropocene.* Cambridge, MA: MIT Press.

Young, O. R. 2018. Research Strategies to Assess the Effectiveness of International Environmental Regimes. *Nature Sustainability,* 1, 461–465.

Young, O. R., Webster, D., Cox, M. E., Raakjær, J., Blaxekjær, L. Ø., Einarsson, N., … Cardwell, E. 2018. Moving Beyond Panaceas in Fisheries Governance. *Proceedings of the National Academy of Sciences,* 115, 9065–9073.

Zaferani, S., Pérez-Rodríguez, M., & Biester, H. 2018. Diatom Ooze—a Large Marine Mercury Sink. *Science,* 361, 797–800.

Zaitchik, A. 2018. How Conservation Became Colonialism. *Foreign Policy,* July.

Zalasiewicz, J. 2015. Epochs: Disputed Start Dates for Anthropocene. *Nature,* 520, 436.

Zalasiewicz, J., Williams, M., Waters, C. N., Barnosky, A. D., Palmesino, J., Rönnskog, A.-S., … Ellis, E. C. 2017. Scale and Diversity of the Physical Technosphere: A Geological Perspective. *The Anthropocene Review,* 4, 9–22.

Zero Mercury Working Group. 2010. Cosmetics, Soaps and Creams. Retrieved February 3, 2019, from https://web.archive.org/web/20190203170804/http://www.zeromercury.org/index.php?option=com_content&view=article&id=138&Itemid=86

Zero Mercury Working Group. 2012. Phasing out Mercury Use in Button Cell Batteries. INC 4 Briefing Paper Series. https://web.archive.org/web/20190203171508/http://www.zeromercury.org/phocadownload/Developments_at_UNEP_level/INC4/ZMWG_Button_Cell_INC4_final.pdf

Zhang, J., Liu, N., Li, W., & Dai, B. 2011. Progress on Cleaner Production of Vinyl Chloride Monomers over Non-Mercury Catalysts. *Frontiers of Chemical Science and Engineering,* 5, 514–520.

Zhang, W., Zhen, G., Chen, L., Wang, H., Li, Y., Ye, X., … Wang, X. 2017. Economic Evaluation of Health Benefits of Mercury Emission Controls for China and the Neighboring Countries in East Asia. *Energy Policy,* 106, 579–587.

Zhang, Y., Jacob, D. J., Horowitz, H. M., Chen, L., Amos, H. M., Krabbenhoft, D. P., … Sunderland, E. M. 2016. Observed Decrease in Atmospheric Mercury Explained by Global Decline in Anthropogenic Emissions. *Proceedings of the National Academy of Sciences,* 113, 526–531.

Zhou, H., Zhou, C., Lynam, M. M., Dvonch, J. T., Barres, J. A., Hopke, P. K., ... Holsen, T. M. 2017. Atmospheric Mercury Temporal Trends in the Northeastern United States from 1992 to 2014: Are Measured Concentrations Responding to Decreasing Regional Emissions? *Environmental Science & Technology Letters,* 4, 91–97.

Zinsuur, U. S. 2018. The Effectiveness of "Operation Vanguard" against Illegal Mining: Survival of Security Personnel. *MyJoyOnline.* Retrieved December 29, 2019, from https://web.archive.org/web/20191229144613/https://www.myjoyonline.com /opinion/2018/march-23rd/the-effectiveness-of-operation-vanguard-against-illegal -mining-survival-of-security-personnel.php

Zolnikov, T. R. 2012. Limitations in Small Artisanal Gold Mining Addressed by Educational Components Paired with Alternative Mining Methods. *Science of the Total Environment,* 419, 1–6.

Index